Graduate Texts in Contemporary Physics

Series Editors:

R. Stephen Berry
Joseph L. Birman
Mark P. Silverman
H. Eugene Stanley
Mikhail Voloshin

Springer

*New York
Berlin
Heidelberg
Hong Kong
London
Milan
Paris
Tokyo*

Graduate Texts in Contemporary Physics

S.T. Ali, J.P. Antoine, and J.P. Gazeau: **Coherent States, Wavelets and Their Generalizations**

A. Auerbach: **Interacting Electrons and Quantum Magnetism**

N. Boccara: **Modeling Complex Systems**

T.S. Chow: **Mesoscopic Physics of Complex Materials**

B. Felsager: **Geometry, Particles, and Fields**

P. Di Francesco, P. Mathieu, and D. Sénéchal: **Conformal Field Theories**

A. Gonis and W.H. Butler: **Multiple Scattering in Solids**

K. Gottfried and T-M. Yan: **Quantum Mechanics: Fundamentals, 2nd Edition**

K.T. Hecht: **Quantum Mechanics**

J.H. Hinken: **Superconductor Electronics: Fundamentals and Microwave Applications**

J. Hladik: **Spinors in Physics**

Yu.M. Ivanchenko and A.A. Lisyansky: **Physics of Critical Fluctuations**

M. Kaku: **Introduction to Superstrings and M-Theory, 2nd Edition**

M. Kaku: **Strings, Conformal Fields, and M-Theory, 2nd Edition**

H.V. Klapdor (ed.): **Neutrinos**

R.L. Liboff (ed): **Kinetic Theory: Classical, Quantum, and Relativistic Descriptions, 3rd Edition**

J.W. Lynn (ed.): **High-Temperature Superconductivity**

H.J. Metcalf and P. van der Straten: **Laser Cooling and Trapping**

R.N. Mohapatra: **Unification and Supersymmetry: The Frontiers of Quark-Lepton Physics, 3rd Edition**

R.G. Newton: **Quantum Physics: A Text for Graduate Students**

H. Oberhummer: **Nuclei in the Cosmos**

G.D.J. Phillies: **Elementary Lectures in Statistical Mechanics**

R.E. Prange and S.M. Girvin (eds.): **The Quantum Hall Effect**

S.R.A. Salinas: **Introduction to Statistical Physics**

(continued after index)

Nino Boccara

Modeling Complex Systems

With 158 Illustrations

 Springer

Nino Boccara
Department of Physics
University of Illinois at Chicago
845 West Taylor Street
Chicago, IL 60607
USA
boccara@uic.edu

Series Editors
R. Stephen Berry
Department of Chemistry
University of Chicago
Chicago, IL 60637
USA

Joseph L. Birman
Department of Physics
City College of CUNY
New York, NY 10031
USA

Mark P. Silverman
Department of Physics
Trinity College
Hartford, CT 06106
USA

H. Eugene Stanley
Center for Polymer Studies
Physics Department
Boston University
Boston, MA 02215
USA

Mikhail Voloshin
Theoretical Physics Institute
Tate Laboratory of Physics
The University of Minnesota
Minneapolis, MN 55455
USA

Library of Congress Cataloging-in-Publication Data
Boccara, Nino.
    Modeling complex systems / Nino Boccara.
        p. cm. — (Graduate texts in contemporary physics)
    Includes bibliographical references and index.
    ISBN 0-387-40462-7 (alk. paper)
        1. System theory—Mathematical models.   2. System analysis—Mathematical models.
    I. Title.   II. Series.
    Q295.B59 2004
    003—dc21   2003054791

ISBN 0-387-40462-7   Printed on acid-free paper.

Printed in the United States of America.

9 8 7 6 5 4 3 2 1   SPIN 10941300

www.springer-ny.com

Springer-Verlag   New York Berlin Heidelberg
A member of BertelsmannSpringer Science+Business Media GmbH

Pyé-koko di i ka vwè lwen, maché ou ké vwè pli lwen.[1]

---

[1] Creole proverb from Guadeloupe that can be translated: The coconut palm says it sees far away, walk and you will see far beyond.

Preface

The purpose of this book is to show how models of complex systems are built up and to provide the mathematical tools indispensable for studying their dynamics. This is not, however, a book on the theory of dynamical systems illustrated with some applications; the focus is on modeling, so, in presenting the essential results of dynamical system theory, technical proofs of theorems are omitted, but references for the interested reader are indicated. While mathematical results on dynamical systems such as differential equations or recurrence equations abound, this is far from being the case for spatially extended systems such as automata networks, whose theory is still in its infancy. Many illustrative examples taken from a variety of disciplines, ranging from ecology and epidemiology to sociology and seismology, are given.

This is an introductory text directed mainly to advanced undergraduate students in most scientific disciplines, but it could also serve as a reference book for graduate students and young researchers. The material has been taught to junior students at the École de Physique et de Chimie in Paris and the University of Illinois at Chicago. It assumes that the reader has certain fundamental mathematical skills, such as calculus.

Although there is no universally accepted definition of a complex system, most researchers would describe as complex a system of connected agents that exhibits an emergent global behavior not imposed by a central controller, but resulting from the interactions between the agents. These agents may be insects, birds, people, or companies, and their number may range from a hundred to millions.

Finding the emergent global behavior of a large system of interacting agents using analytical methods is usually hopeless, and researchers therefore must rely on computer-based methods. Apart from a few exceptions, most properties of spatially extended systems have been obtained from the analysis of numerical simulations.

Although simulations of interacting multiagent systems are thought experiments, the aim is not to study accurate representations of these systems. The main purpose of a model is to broaden our understanding of general principles valid for the largest variety of systems. Models have to be as simple as possible. What makes the study of complex systems fascinating is not the study of complicated models but the complexity of unsuspected results of numerical simulations.

As a multidisciplinary discipline, the study of complex systems attracts researchers from many different horizons who publish in a great variety of scientific journals. The literature is growing extremely fast, and it would be a hopeless task to try to attain any kind of comprehensive completeness. This book only attempts to supply many diverse illustrative examples to exhibit that common modeling techniques can be used to interpret the behavior of apparently completely different systems.

After a general introduction followed by an overview of various modeling techniques used to explain a specific phenomenon, namely the observed coupled oscillations of predator and prey population densities, the book is divided into two parts. The first part describes models formulated in terms of differential equations or recurrence equations in which local interactions between the agents are replaced by uniform long-range ones and whose solutions can only give the time evolution of spatial averages. Despite the fact that such models offer rudimentary representations of multiagent systems, they are often able to give a useful qualitative picture of the system's behavior. The second part is devoted to models formulated in terms of automata networks in which the local character of the interactions between the individual agents is explicitly taken into account. Chapters of both parts include a few exercises that, as well as challenging the reader, are meant to complement the material in the text. Detailed solutions of all exercises are provided.

Nino Boccara

Contents

Preface vii

1   Introduction 1
    1.1   What is a complex system? 1
    1.2   What is a model? 4
    1.3   What is a dynamical system? 9

2   How to Build Up a Model 17
    2.1   Lotka-Volterra model 17
    2.2   More realistic predator-prey models 24
    2.3   A model with a stable limit cycle 25
    2.4   Fluctuating environments 27
    2.5   Hutchinson's time-delay model 28
    2.6   Discrete-time models 31
    2.7   Lattice models 33

Part I Mean-Field Type Models

3   Differential Equations 41
    3.1   Flows 41
    3.2   Linearization and stability 51
          3.2.1   Linear systems 51
          3.2.2   Nonlinear systems 56
    3.3   Graphical study of two-dimensional systems 66
    3.4   Structural stability 69
    3.5   Local bifurcations of vector fields 71
          3.5.1   One-dimensional vector fields 73
          3.5.2   Equivalent families of vector fields 82
          3.5.3   Hopf bifurcation 83
          3.5.4   Catastrophes 85

    3.6   Influence of diffusion 91
          3.6.1   Random walk and diffusion 91
          3.6.2   One-population dynamics with dispersal.............. 92
          3.6.3   Critical patch size 94
          3.6.4   Diffusion-induced instability 95
    Exercises 98
    Solutions 100

4   Recurrence Equations 107
    4.1   Iteration of maps 107
    4.2   Stability..110
    4.3   Poincaré maps 118
    4.4   Local bifurcations of maps 120
          4.4.1   Maps on $\mathbb{R}$ 120
          4.4.2   The Hopf bifurcation 125
    4.5   Sequences of period-doubling bifurcations................127
          4.5.1   Logistic model 128
          4.5.2   Universality 131
    Exercises 136
    Solutions 139

5   Chaos 145
    5.1   Defining chaos 147
          5.1.1   Dynamics of the logistic map f_4 149
          5.1.2   Definition of chaos...............................153
    5.2   Routes to chaos 154
    5.3   Characterizing chaos 156
          5.3.1   Stochastic properties............................156
          5.3.2   Lyapunov exponent................................158
          5.3.3   "Period three implies chaos" 159
          5.3.4   Strange attractors 161
    5.4   Chaotic discrete-time models.............................169
          5.4.1   One-population models.............................169
          5.4.2   The Hénon map.....................................169
    5.5   Chaotic continuous-time models 174
          5.5.1   The Lorenz model 174
    Exercises 176
    Solutions 177

Part II Agent-Based Models

6   Cellular Automata 191
 6.1 Cellular automaton rules 191
 6.2 Number-conserving cellular automata 194
 6.3 Approximate methods 207
 6.4 Generalized cellular automata 213
 6.5 Kinetic growth phenomena 221
 6.6 Site-exchange cellular automata 228
 6.7 Artificial societies 243
 Exercises 258
 Solutions 262

7   Networks 275
 7.1 The small-world phenomenon 275
 7.2 Graphs 276
 7.3 Random networks 279
 7.4 Small-world networks 283
  7.4.1 Watts-Strogatz model 283
  7.4.2 Newman-Watts model 286
  7.4.3 Highly connected extra vertex model 287
 7.5 Scale-free networks 288
  7.5.1 Empirical results 288
  7.5.2 A few models 292
 Exercises 299
 Solutions 301

8   Power-Law Distributions 311
 8.1 Classical examples 311
 8.2 A few notions of probability theory 317
  8.2.1 Basic definitions 317
  8.2.2 Central limit theorem 320
  8.2.3 Lognormal distribution 324
  8.2.4 Lévy distributions 325
  8.2.5 Truncated Lévy distributions 328
  8.2.6 Student's t-distribution 330
  8.2.7 A word about statistics 331
 8.3 Empirical results and tentative models 334
  8.3.1 Financial markets 335
  8.3.2 Demographic and area distribution 339
  8.3.3 Family names 340
  8.3.4 Distribution of votes 340
 8.4 Self-organized criticality 341
  8.4.1 The sandpile model 341
  8.4.2 Drossel-Schwabl forest fire model 343
  8.4.3 Punctuated equilibria and Darwinian evolution 345
  8.4.4 Real life phenomena 347

Exercises 352
Solutions 354

Notations 367

References 371

Index 389

1

Introduction

This book is about the dynamics of complex systems. Roughly speaking, a system is a collection of interacting elements making up a whole such as, for instance, a mechanical clock. While many systems may be quite complicated, they are not necessarily considered to be complex. There is no precise definition of complex systems. Most authors, however, agree on the essential properties a system has to possess to be called complex. The first section is devoted to the description of these properties.

To interpret the time evolution of a system, scientists build up models, which are simplified mathematical representations of the system. The exact purpose of a model and what its essential features should be is explained in the second section.

The mathematical models that will be discussed in this book are dynamical systems. A dynamical system is essentially a set of equations whose solution describes the evolution, as a function of time, of the state of the system. There exist different types of dynamical systems. Some of them are defined in the third section.

1.1 What is a complex system?

Outside the nest, the members of an ant colony accomplish a variety of fascinating tasks, such as foraging and nest maintenance. Gordon's work [152] on harvester ants[1] has shed considerable light on the processes by which members of an ant colony assume various roles. Outside the nest, active ant workers can perform four distinct tasks: foraging, nest maintenance, patrolling, and midden work. Foragers travel along cleared trails around the nest to collect mostly seeds and, occasionally, insect parts. Nest-maintenance workers modify the nest's chambers and tunnels and clear sand out of the nest or vegetation

---

[1] *Pogonomyrmex barbatus.* They are called harvester ants because they eat mostly seeds, which they store inside their nests.

from the mound and trails. Patrollers choose the direction foragers will take each day and also respond to damage to the nest or an invasion by alien ants. Midden workers build and sort the colony's refuse pile.

Gordon [150, 151] has shown that task allocation is a process of continual adjustment. The number of workers engaged in a specific task is appropriate to the current condition. When small piles of mixed seeds are placed outside the nest mound, away from the foraging trails but in front of scouting patrollers, early in the morning, active recruitment of foragers takes place. When toothpicks are placed near the nest entrance, early in the morning at the beginning of nest-maintenance activity, the number of nest-maintenance workers increases significantly.

The surprising fact is that task allocation is achieved without any central control. The queen does not decide which worker does what. No master ant could possibly oversee the entire colony and broadcast instructions to the individual workers. An individual ant can only perceive local information from the ants nearby through chemical and tactile communication. Each individual ant processes this partial information in order to decide which of the many possible functional roles it should play in the colony.

The cooperative behavior of an ant colony that results from local interactions between its members and not from the existence of a central controller is referred to as *emergent behavior*. Emergent properties are defined as large-scale effects of locally interacting agents that are often surprising and hard to predict even in the case of simple interactions. Such a definition is not very satisfying: what might be surprising to someone could be not so surprising to someone else.

A system such as an ant colony, which consists of large populations of connected agents (that is, collections of interacting elements), is said to be *complex* if there exists an emergent global dynamics resulting from the actions of its parts rather than being imposed by a central controller.

Ant colonies are not the only multiagent systems that exhibit coordinated behaviors without a centralized control.

Animal groups display a variety of remarkable coordinated behaviors [278, 64]. All the members in a school of fish change direction simultaneously without any obvious cue; while foraging, birds in a flock alternate feeding and scanning. No individual in these groups has a sense of the overall orderly pattern. There is no apparent leader. In a school of fish, the direction of each member is determined by the average direction of its neighbors [339, 331]. In a flock of birds, each individual chooses to scan for predators if a majority of its neighbors are eating and chooses to eat if a majority of its neighbors are already scanning [23]. The existence of sentinels in animal groups engaged in dangerous activities is a typical example of cooperation. Recent studies suggest that guarding may be an individual's optimal activity once its stomach is full and no other animal is on guard [90].

Self-organized motion in schools of fish, flocks of birds, or herds of ungulate mammals is not specific to animal groups. Vehicle traffic on a highway exhibits emergent behaviors such as the existence of traffic jams that propagate in the opposite direction of the traffic flow, keeping their structure and characteristic parameters for a long time [183], or the synchronization of average velocities in neighboring lanes in congested traffic [184]. Similarly, pedestrian crowds display self-organized spatiotemporal patterns that are not imposed by any regulation: on a crowded sidewalk, pedestrians walking in opposite directions tend to form lanes along which walkers move in the same direction.

A high degree of self-organization is also found in social networks that can be viewed as graphs.[2] The collection of scientific articles published in refereed journals is a directed graph, the vertices being the articles and the arcs being the links connecting an article to the papers cited in its list of references. A recent study [295] has shown that the citation distribution—that is, the number of papers $N(x)$ that have been cited a total of x times—has a power-law tail, $N(x) \sim x^{-\alpha}$ with $\alpha \approx 3$. Minimally cited papers are usually referenced by their authors and close associates, while heavily cited papers become known through collective effects.

Other social networks, such as the World Wide Web or the casting pattern of movie actors, exhibit a similar emergent behavior [37]. In the World Wide Web, the vertices are the HTML[3] documents, and the arcs are the links pointing from one document to another. In a movie database, the vertices are the actors, two of them being connected by an *undirected* edge if they have been cast in the same movie.

In October 1987, major indexes of stock market valuation in the United States declined by 30% or more. An analysis [321] of the time behavior of the U. S. stock exchange index S&P500 before the crash identifies precursory patterns suggesting that the crash may be viewed as a dynamical critical point. That is, as a function of time t, the S&P500 behaves as $(t - t_c)^{0.7}$, where t is the time in years and $t_c \approx 1987.65$. This result shows that the stock market is a complex system that exhibits self-organizing cooperative effects.

All the examples of complex systems above exhibit some common characteristics:

1. They consist of a large number of interacting *agents*.
2. They exhibit *emergence*; that is, a self-organizing collective behavior difficult to anticipate from the knowledge of the agents' behavior.
3. Their emergent behavior does not result from the existence of a central controller.

The appearance of emergent properties is the single most distinguishing feature of complex systems. Probably, the most famous example of a system that exhibits emergent properties as a result of simple interacting rules between its agents is the *game of life* invented by John H. Conway. This game is

---

[2] A directed graph (or digraph) G consists of a nonempty set of elements $V(G)$, called vertices, and a subset $E(G)$ of ordered pairs of distinct elements of $V(G)$, called directed edges or arcs.

[3] Hypertext Markup Language.

played on an (infinite) two-dimensional square lattice. Each cell of the lattice is either on (occupied by a living organism) or off (empty). If a cell is off, it turns on if exactly three of its eight neighboring cells (four adjacent orthogonally and four adjacent diagonally) are on (birth of a new organism). If a cell is on, it stays on if exactly two or three of its neighboring cells are on (survival), otherwise it turns off (death from isolation or overpopulation). These rules are applied simultaneously to all cells. Populations evolving according to these rules exhibit endless unusual and unexpected changing patterns [137].

"To help people explore and learn about decentralized systems and emergent phenomena," Mitchell Resnick[4] developed the StarLogo[5] modeling environment. Among the various sample projects consider, for example, the project inspired by the behavior of termites gathering wood chips into piles. Each cell of a 100×100 square lattice is either empty or occupied by a wood chip or/and a termite. Each termite starts wandering randomly. If it bumps into a wood chip, it picks the chip up and continues to wander randomly. When it bumps into another wood chip, it finds a nearby empty space and puts its wood chip down. With these simple rules, the wood chips eventually end up in a single pile (Figure 1.1). Although rather simple, this model is representative of a complex system. It is interesting to notice that while the gathering of all wood chips into a single pile may, at first sight, look surprising, on reflection it is no wonder. Actually it is clear that the number of piles cannot increase, and, since the probability for any pile to disappear is nonzero, this number has to decrease and ultimately become equal to one.[6]

1.2 What is a model?

A model is a simplified mathematical representation of a system. In the actual system, many features are likely to be important. Not all of them, however, should be included in the model. Only the few relevant features that are thought to play an essential role in the interpretation of the observed phenomena should be retained. Models should be distinguished from what is usually called a *simulation*. To clarify this distinction, it is probably best to quote John Maynard Smith [234]:

---

[4] See Mitchell Resnick's Web page: http://mres.www.media.mit.edu/people/mres. Resnick's research is described in his book [297].

[5] StarLogo is freeware that can be downloaded from: http://el.www.media.mit.edu/groups/el/Projects/starlogo.

[6] Here is a similar mathematical model that can be solved exactly. Consider a random distribution of N identical balls in B identical boxes, and assume that, at each time step, a ball is transferred from one box to another, not necessarily different, with a probability $P(n \to n \pm 1)$ of changing by one unit the number n of balls in a given box depending only on the number n of balls in that box. Moreover, if this probability is equal to zero for $n = 0$ (an empty box stays empty), then it can be shown that the probability for a given box to become empty is equal to $1 - n/N$. Hence, ultimately all balls end up in one unique box.

Fig. 1.1. *StarLogo sample project termites. Randomly distributed wood chips (left figure) eventually end up in a single pile (right figure). Density of wood: 0.25; number of termites: 75.*

If, for example, one wished to know how many fur seals can be culled annually from a population without threatening its future survival, it would be necessary to have a description of that population, in its particular environment, which includes as much relevant detail as possible. At a minimum, one would require age-specific birth and death rates, and knowledge of how these rates varied with the density of the population, and with other features of the environment likely to alter in the future. Such information could be built into a simulation of the population, which could be used to predict the effects of particular management policies.

The value of such simulations is obvious, but their utility lies mainly in analyzing particular cases. A theory of ecology must make statements about ecosystems as a whole, as well as about particular species at particular times, and it must make statements which are true for many different species and not for just one. Any actual ecosystem contains far too many species, which interact in far too many ways, for simulation to be a practical approach. The better a simulation is for its own purposes, by the inclusion of all relevant details, the more difficult it is to generalize its conclusions to other species. For the discovery of general ideas in ecology, therefore, different kinds of mathematical description, which may be called models, are called for. Whereas a good simulation should include as much detail as possible, a good model should include as little as possible.

A simple model, if it captures the key elements of a complex system, may elicit highly relevant questions.

For example, the growth of a population is often modeled by a differential equation of the form

$$\frac{dN}{dt} = f(N), \tag{1.1}$$

where the time-dependent function N is the number of inhabitants of a given area. It might seem paradoxical that such a model, which ignores the influence of sex ratios on reproduction, or age structure on mortality, would be of any help. But many populations have regular sex ratios and, in large populations near equilibrium, the number of old individuals is a function of the size of the population. Thus, taking into account these additional features maybe is not as essential as it seems.

To be more specific, in an isolated population (that is, if there is neither immigration nor emigration), what should be the form of a reasonable function f? According to Hutchinson [177] any equation describing the evolution of a population should take into account that:

1. Every living organism must have at least one parent of like kind.
2. In a finite space, due to the limiting effect of the environment, there is an upper limit to the number of organisms that can occupy that space.

The simplest model satisfying these two requirements is the so-called *logistic model*:

$$\frac{dN}{dt} = rN\left(1 - \frac{N}{K}\right). \tag{1.2}$$

The word "logistic" was coined by Pierre François Verhulst (1804–1849), who used this equation for the first time in 1838 to discuss population growth.[7] His paper [338] did not, at that time, arouse much interest. Verhulst's equation was rediscovered about 80 years later by Raymond Pearl and Lowell J. Reed. After the publication of their paper [280], the logistic model began to be used extensively.[8] Interesting details on Verhulst's ideas and the beginning of scientific demography can be found in the first chapter of Hutchinson's book [177].

In Equation (1.2), the constant r is referred to as the *intrinsic rate of increase* and K is called the *carrying capacity* because it represents the population size that the resources of the environment can just maintain (carry) without a tendency to either increase or decrease. The logistic equation is clearly a very crude model but, in spite of its obvious limitations,[9] it is often a good starting point.[10]

The logistic equation contains two parameters. This number can be reduced if we express the model in non-dimensional terms. Since r has the

---

[7] The French word "logistique" had, since 1840, the same meaning as the word "logistics" in English, but in old French, since 1611, it meant "l'art de compter"; *i.e.*, the art of calculation. See *Le nouveau petit Robert, dictionnaire de la langue française* (Paris: Dictionnaires Le Robert, 1993).

[8] For a critical review of experimental attempts to verify the validity of the logistic model, see Willy Feller [123].

[9] See, *e.g.*, Chapter 6 of Begon, Harper, and Townsend's book [39].

[10] On the history of the logistic model, see [185].

dimension of the inverse of a time and K has the dimension of a number of individuals, if we put

$$\tau = rt \quad \text{and} \quad n = \frac{N}{K},$$

Equation (1.2) becomes

$$\frac{dn}{d\tau} = n(1-n). \tag{1.3}$$

This equation contains no more parameters. That is, if the unit of time is r^{-1} and the unit of number of inhabitants is K, then the reduced logistic equation (1.3) is universal; it is system-independent.

Equation (1.3) is very simple and can be integrated exactly.[11] Most equations cannot be solved analytically. But, following ideas going back to Poincaré,[12] a geometrical approach, developed essentially during the second half of the twentieth century, gives, in many cases of interest, a description of the qualitative behavior of the solutions.

The reduction of equations to a dimensionless form simplifies the mathematics and, usually, leads to some insight even without solving the equation. Moreover, the value of a dimensionless variable carries more information than the value of the variable itself.

For simple models such as Equation (1.2) the definition of scaled variables is straightforward. If the model is not so simple, reduced variables may be defined using a systematic technique. To illustrate this technique, consider the following model of insect population outbreaks due to Ludwig, Jones, and Holling [217].

Certain insect populations exhibit outbreaks in abundance as they move from a low-density equilibrium to a high-density equilibrium and back again. This is the case, for instance, of the spruce budworm (*Choristoneura fumiferana*), which feeds on the needles of the terminal shoots of spruce, balsam fir, and other evergreen trees in eastern North America.

In an immature balsam fir and white spruce forest, the quantity of food for the budworms is low and their *rate of recruitment* (that is, the amount by which the population increases during one time unit) is low. It is then reasonable to assume that the budworm population is kept at a low-density equilibrium by its predators (essentially birds). However, as the forest gradually matures, more food becomes available, the rate of budworm recruitment increases, and the budworm density grows. Above a certain rate of recruitment threshold, avian predators can no longer contain the growth of the budworm density, which jumps to a high-level value. This outbreak of the budworm

---

[11] Its general solution reads:

$$n(\tau) = \frac{1}{1 + ae^{-\tau}},$$

where a is an integration constant whose value depends upon the initial value $n(0)$.

[12] See Chapter 3.

density quickly defoliates the mature trees; the forest then reverts to immaturity, the rate of recruitment decreases, and the budworm density jumps back to a low-level equilibrium.

The budworm can increase its density several hundredfold in a few years. Therefore, a characteristic time interval for the budworm is of the order of months. The trees, however, cannot put on foliage at a comparable rate. A characteristic time interval for trees to completely replace their foliage is of the order of 7 to 10 years. Moreover, in the absence of the budworm, the life span of the trees is of the order of 100 years. Therefore, in analyzing the dynamics of the budworm population, we may assume that the foliage quantity is held constant.[13]

The main limiting features of the budworm population are food supply and the effects of parasites and predators. In the absence of predation, we may assume that the budworm density B satisfies the logistic equation

$$\frac{dB}{dt} = r_B B \left(1 - \frac{B}{K_B}\right),$$

where r_B and K_B are, respectively, the intrinsic rate of increase of the spruce budworm and the carrying capacity of the environment.

Predation may be taken into account by subtracting a term $p(B)$ from the right-hand side of the logistic equation. What conditions should satisfy the function p?

1. At high prey density, predation usually saturates. Hence, when B becomes increasingly large, $p(B)$ should approach an upper limit a ($a > 0$).
2. At low prey density, predation is less effective. Birds are relatively unselective predators. If a prey becomes less common, they seek food elsewhere. Hence, when B tends to zero, $p(B)$ should tend to zero faster than B.

A simple form for $p(B)$ that has the properties of saturation at a level a and vanishes like B^2 is

$$p(B) = \frac{aB^2}{b^2 + B^2}.$$

The positive constant b is a critical budworm density. It determines the scale of budworm densities at which saturation begins to take place.

The dynamics of the budworm density B is then governed by

$$\frac{dB}{dt} = r_B B \left(1 - \frac{B}{K_B}\right) - \frac{aB^2}{b^2 + B^2}. \tag{1.4}$$

This equation, which is of the general form (1.1), contains four parameters: r_B, K_B, a, and b. Their dimensions are the same as, respectively, t^{-1}, B, Bt^{-1}, and B. Since the equation relates two variables B and t, we have to define two dimensionless variables

---

[13] This *adiabatic approximation* is familiar to physicists. For a nice discussion of its validity and its use in solid-state theory, see Weinreich's book [343].

$$\tau = \frac{t}{t_0} \quad \text{and} \quad x = \frac{B}{B_0}. \tag{1.5}$$

To reduce (1.4) to a dimensionless form, we have to define the constants t_0 and B_0 in terms of the parameters r_B, K_B, a, and b. Replacing (1.5) into (1.4), we obtain

$$\frac{dx}{d\tau} = r_B t_0 x \left(1 - \frac{x B_0}{K_B} \right) - \frac{a B_0 t_0 x^2}{b^2 + x^2 B_0^2}.$$

To reduce this equation to as simple a form as possible, we may choose either

$$t_0 = r_B^{-1} \quad \text{and} \quad B_0 = K_B,$$

or

$$t_0 = b a^{-1} \quad \text{and} \quad B_0 = b.$$

The first choice simplifies the logistic part of Equation (1.4), whereas the second one simplifies the predation part. The corresponding reduced forms of (1.4) are, respectively,

$$\frac{dx}{d\tau} = x(1 - x) - \frac{\alpha x^2}{\beta^2 + x^2} \tag{1.6}$$

and

$$\frac{dx}{d\tau} = rx \left(1 - \frac{x}{k} \right) - \frac{x^2}{1 + x^2}. \tag{1.7}$$

To study budworm outbreaks as a function of the available foliage per acre of forest, the second choice is better. To study the influence of the predator density, however, the first choice is preferable. Both reduced equations contain two parameters: the scaled upper limit of predation α and the scaled critical density β in the first case and the scaled rate of increase r and the scaled carrying capacity k in the second case.

It is not very difficult to prove that, if the evolution of a model is governed by a set of equations containing n parameters that relate variables involving d independent dimensions, the final reduced equations will contain $n - d$ scaled parameters.

1.3 What is a dynamical system?

The notion of a dynamical system includes the following ingredients: a *phase space* S whose elements represent possible states of the system[14]; *time t*, which may be discrete or continuous; and an *evolution law* (that is, a rule that allows determination of the state at time t from the knowledge of the

---

[14] S is also called the *state space*.

states at all previous times). In most examples, knowing the state at time t_0 allows determination of the state at any time $t > t_0$.

The two models of population growth presented in the previous section are examples of dynamical systems. In both cases, the phase space is the set of nonnegative real numbers, and the evolution law is given by the solution of a nonlinear first-order differential equation of the form (1.1).

The name *dynamical system* arose, by extension, after the name of the equations governing the motion of a system of particles. Today the expression dynamical system is used as a synonym of nonlinear system of equations.

Dynamical systems may be divided into two broad categories. According to whether the time variable may be considered as continuous or discrete, the dynamics of a given system is described by differential equations or finite-difference equations of the form[15]

$$\frac{d\mathbf{x}}{dt} = \dot{\mathbf{x}} = \mathbf{X}(\mathbf{x}), \tag{1.8}$$

$$\mathbf{x}_{t+1} = \mathbf{f}(\mathbf{x}_t), \tag{1.9}$$

where t belongs to the set of nonnegative real numbers $\mathbb{R}_+$ in (1.8) and the set of nonnegative integers $\mathbb{N}_0$ (that is, the union of the set $\mathbb{N}$ of positive integers and $\{0\}$) in (1.9). Such equations determine how the state $x \in S$ of the system varies with time.[16] To solve (1.8) or (1.9) we need to specify the initial state $x(0) \in S$. The *state of a system at time t* represents all the information characterizing the system at this particular time. Here are some illustrative examples.

Example 1. The simple pendulum. In the absence of friction, the equation of motion of a simple pendulum moving in a vertical plane is

$$\frac{d^2\theta}{dt^2} + \frac{g}{\ell}\sin\theta = 0, \tag{1.10}$$

where θ is the displacement angle from the stable equilibrium position, g the acceleration of gravity, and ℓ the length of the pendulum. If we put

$$x_1 = \theta, \quad x_2 = \dot{\theta},$$

---

[15] Here, we are considering *autonomous systems*; that is, we are assuming that the functions $\mathbf{X}$ and $\mathbf{f}$ do not depend explicitly on time. A *nonautonomous* system may always be written as an autonomous system of higher dimensionality (see Example 1).

[16] Assuming of course that, for a given initial state, the equations above have a unique solution. Since we are essentially interested in applications, we will not discuss problems of existence and uniqueness of solutions. These problems are important for the mathematician, and nonunicity is certainly an interesting phenomenon. But for someone interested in applications, nonunicity is an unpleasant feature indicating that the model has to be modified, since, according to experience, a *real* system has a unique evolution for any *realizable* initial state.

Equation (1.10) may be written

$$\frac{dx_1}{dt} = x_2$$

$$\frac{dx_2}{dt} = -\frac{g}{\ell}\sin x_1.$$

This type of transformation is general. Any system of differential equations of order higher than one can be written as a first-order system of higher dimensionality.

The state of the pendulum is represented by the ordered pair (x_1, x_2). Since $x_1 \in [-\pi, \pi[$ and $x_2 \in \mathbb{R}$, the phase space X is the cylinder $\mathbb{S}^1 \times \mathbb{R}$, where $\mathbb{S}^n$ denotes the unit sphere in $\mathbb{R}^{n+1}$. This surface is a two-dimensional *manifold*. A manifold is a locally Euclidean space that generalizes the idea of parametric representation of curves and surfaces in $\mathbb{R}^3$.[17]

Example 2. Nonlinear oscillators. Models of nonlinear oscillators have been the source of many important and interesting problems.[18] They are described by second-order differential equations of the form

$$\ddot{x} + g(x, \dot{x}) = 0.$$

While the dynamics of such systems is already nontrivial (see, for instance, the van der Pol oscillator discussed in Example 16), the addition of a periodic forcing term $f(t) = f(t + T)$ yields

$$\ddot{x} + g(x, \dot{x}) = f(t) \tag{1.11}$$

and can introduce completely new phenomena. If we put

$$x_1 = x, \quad x_2 = \dot{x}, \quad x_3 = t,$$

Equation (1.11) may be written

$$\dot{x}_1 = x_2,$$
$$\dot{x}_2 = -g(x_1, x_2) + f(x_3),$$
$$\dot{x}_3 = 1.$$

Here again, this type of transformation is general. Any nonautonomous system of differential equations of order higher than one can be written as a first-order system of higher dimensionality.

The state of the system is represented by the triplet (x_1, x_2, x_3). If the period T of the function f is, say, 2π, the phase space is $X = \mathbb{R} \times \mathbb{R} \times S^1$; that is, a three-dimensional manifold.

---

[17] See also Section 3.1.

[18] Refer, in particular, to [160].

Example 3. Age distribution. A one-species population may be characterized by its density ρ. Since ρ should be nonnegative and not greater than 1, the phase space is the interval [0,1]. The population density is a global variable that ignores, for instance, age structure. A more precise characterization of the population should take into account its age distribution. If $f(t, a)\, da$ represents the density of individuals whose age, at time t, lies between a and $a + da$, then the state of the system is represented by the *age distribution function* $a \mapsto f(t, a)$. The total population density at time t is

$$\rho(t) = \int_0^\infty f(t, a)\, da$$

and, in this case, the state space is a set of positive integrable functions on $\mathbb{R}_+$.

Example 4. Population growth with a time delay. In the logistic model the growth rate of a population at any time t depends on the number of individuals in the system at that time. This assumption is seldom justified, for reproduction is not an instantaneous process. If we assume that the growth rate $\dot{N}(t)/N(t)$ is a decreasing function of the number of individuals at time $t - T$, the simplest model is

$$\dot{N}(t) = \frac{dN}{dt} = rN(t)\left(1 - \frac{N(t - T)}{K}\right). \tag{1.12}$$

This logistic model with a time lag is due to Hutchinson [176, 177], who was the first ecologist to consider time-delayed responses.

To solve Equation (1.12), we need to know not only the value of an initial population but a *history function* h such that

$$(\forall u \in [0, T]) \qquad N(-u) = h(u).$$

If we put

$$x_1(t) = N(t), \quad x_2(t, u) = N(t - u),$$

we have

$$\frac{\partial x_2}{\partial t} = \frac{dN(t - u)}{dt} = -\frac{\partial x_2}{\partial u},$$

and we may, therefore, write the logistic equation with a time lag under the form

$$\frac{dx_1}{dt} = rx_1(t)\left(1 - \frac{x_2(t, u)}{K}\right),$$

$$\frac{\partial x_2}{\partial t} = -\frac{\partial x_2}{\partial u}.$$

Here, the state space is two-dimensional. x_1 is a nonnegative real and $u \mapsto x_2(t, u)$ a nonnegative function defined on the interval $[0, T]$. The boundary conditions are $x_1(0) = h(0)$ and $x_2(0, u) = h(u)$ for all $u \in [0, T]$.

In the more general case of a logistic equation of the form

$$\frac{dN}{dt} = rN(t)\left(1 - \frac{1}{K}\int_0^t N(t-u)Q(u)\,du\right),\qquad (1.13)$$

Q, called the *delay kernel*, is a positive integrable normalized function on $\mathbb{R}_+$; that is, a function defined for $u \geq 0$ such that

$$\int_0^\infty Q(u)\,du = 1.\qquad (1.14)$$

Here are two typical illustrative examples of delay kernels found in the literature[19]:

$$Q_1(u) = \frac{1}{T}\exp(-u/T),$$

$$Q_2(u) = \frac{1}{T^2}u\exp(-u/T).$$

Hutchinson's equation corresponds to the singular kernel $Q = \delta_T$, where δ_T denotes the Dirac distribution at T.[20] Taking into account (1.14), it is easy to verify that K is the only nontrivial equilibrium point of (1.13). The parameter K corresponds, therefore, to the carrying capacity of the standard logistic model.

The history function $h : u \mapsto N(-u)$ is defined on $\mathbb{R}_+$, and we have

$$\frac{dx_1}{dt} = rx_1(t)\left(1 - \frac{1}{K}\int_0^\infty x_2(t,u)Q(u)\,du\right),$$

$$\frac{\partial x_2}{\partial t} = -\frac{\partial x_2}{\partial u},$$

with the boundary conditions $x_1(0) = h(0)$ and $x_2(0,u) = h(u)$ for all $u \in \mathbb{R}_+$.

Example 5. Random walkers on a lattice. Let $\mathbb{Z}_L$ be a one-dimensional finite lattice of length L with periodic boundary conditions,[21] and denote by $n(t,i)$ the occupation number of site i at time t. $n(t,i) = 0$ if the site is vacant, and $n(t,i) = 1$ if the site is occupied by a random walker. The evolution rule of the system is such that, at each time step, a random walker selected at random—that is, the probability for a walker to be selected is uniform—will move with a probability $\frac{1}{2}$ either to the right or the left neighboring site if this site is vacant. If the randomly selected site is not vacant, then the walker will not move. The state of the system at time t is represented by the function $i \mapsto n(t,i)$, and the phase space is $X = \{0,1\}^{\mathbb{Z}_L}$. An element of such a phase space is called a *configuration*.

---

[19] On delay models in population ecology, consult [97].

[20] On distribution theory and its applications to differential and integral equations, see [46], Chapter 4.

[21] $\mathbb{Z}_L$ denotes the set of integers modulo L. Similarly, $\mathbb{Z}_L^d$ represents a finite d-dimensional lattice of volume L^d with periodic boundary conditions.

In most situations of interest, the phase space of a dynamical system possesses a certain structure that the evolution law respects. In applications, we are usually interested in lasting rather than transient phenomena and so in steady states. Therefore, steady solutions of the governing equations of evolution are of special importance. Consider, for instance, Equation (1.2); its steady solutions, which are such that

$$\frac{dN}{dt} = 0,$$

are

$$N = 0 \quad \text{and} \quad N = K.$$

In this simple case, it is not difficult to verify that, if the initial condition is $N(0) > 0$, $N(t)$ tends to K when t tends to infinity. The expression "$N(t)$ tends to K when t tends to infinity" is meaningful if, and only if, the phase space X has a topology. Roughly speaking, a topological space is a space in which the notion of neighborhood has been defined. A simple way to induce a topology is to define a distance, that is, to each ordered pair of points (x_1, x_2) in X we should be able to associate a nonnegative number $d(x_1, x_2)$, said to be the distance between x_1 and x_2, satisfying the following conditions:

1. $d(x_1, x_2) = 0 \Leftrightarrow x_1 = x_2$,
2. $d(x_1, x_2) = d(x_2, x_1)$,
3. $d(x_1, x_3) \leq d(x_1, x_2) + d(x_2, x_3)$.

In the Euclidean space $\mathbb{R}^n$, the distance is defined by

$$d(x_1, x_2) = \left(\sum_{\alpha=1}^{n} (x_1^\alpha - x_2^\alpha)^2 \right)^{1/2}.$$

If, as in Example 1, X is a manifold, we use a suitable coordinate system to define the distance.

In Example 3, if we assume that age distribution functions are Lebesgue integrable,[22] then the distance between two functions f_1 and f_2 may be defined by

$$d(f_1, f_2) = \left(\int_0^\infty |f_1(\xi) f_2(\xi)|^p \, d\xi \right)^{1/p},$$

where $p \geq 1$.

In Example 5, the *Hamming distance* $d_H(c_1, c_2)$ between two configurations c_1 and c_2 is defined by

$$d_H(c_1, c_2) = \frac{1}{L} \sum_{i=1}^{L} |n_1(i) - n_2(i)|,$$

---

[22] For an elementary presentation of the notion of measure and Lebesgue theory of integration, see [46], Chapter 1. In applications, this requirement is not restrictive, but it allows definition of complete metric spaces.

where $n_1(i)$ and $n_2(i)$ are, respectively, the occupation numbers of site i in configurations 1 and 2.

When the evolution of a system is not deterministic, as is the case for the random walkers of Example 5, it is necessary to introduce the notion of a *random process*. Since a random process is a family of measurable mappings on the space Ω of elementary events in the phase space X, the phase space has to be measurable.

To summarize the discussion above, we shall assume that, if the evolution is *deterministic*, the phase space is a *metric space*, whereas, if the evolution is *stochastic*, the phase space is a *measurable metric space*.

To conclude this section, we present two examples of dynamical systems that can be viewed as mathematical recreations.

Example 6. Bulgarian solitaire. Like many other mathematical recreations, Bulgarian solitaire has been made popular by Martin Gardner [138]. A pack of $N = \frac{1}{2}n(n+1)$ cards is divided into k packs of $n_1, n_2, \ldots, n_k$ cards, where $n_1 + n_2 + \cdots + n_k = N$. A move consists in taking exactly one card of each pack and forming a new pack. By repeating this operation a sufficiently large number of times any initial configuration eventually converges to a configuration that consists of n packs of, respectively, $1, 2, \ldots, n$ cards. For instance, if $N = 10$ (which corresponds to $n = 4$), starting from the partition $\{1, 2, 7\}$, we obtain the following sequence:

$$\{1, 3, 6\}, \{2, 3, 5\}, \{1, 2, 3, 4\}.$$

Numbers N of the form $\frac{1}{2}n(n+1)$ are known as *triangular numbers*. Then, what happens if the number of cards is not triangular? Since the number of partitions of a finite integer is finite, any initial partition leads into a cycle of partitions. For example, if $N = 8$, starting from $\{8\}$, we obtain the sequence:

$$\{7, 1\}, \{6, 2\}, \{5, 2, 1\}, \{4, 3, 1\}, \{3, 3, 2\}, \{3, 2, 2, 1\}, \{4, 2, 1, 1\}, \{4, 3, 1\}.$$

For any positive integer N, the convergence towards a cycle, which is of length 1 if N is triangular, has been proved by J. Brandt [71] (see also [1]). In the case of a triangular number, it has been shown that the number of moves before the final configuration is reached is at most equal to $n(n-1)$ [178, 118].

The Bulgarian solitaire is a time-discrete dynamical system. The phase space consists of all the partitions of the number N.

Example 7. The original Collatz problem. Many iteration problems are simple to state but often intractably hard to solve. Probably the most famous one is the so-called $3x + 1$ problem, also known as the Collatz conjecture, which asserts that, starting from any positive integer n, repeated iteration of the function f defined by

$$f(n) = \begin{cases} \frac{1}{2}n, & \text{if } n \text{ is even,} \\ \frac{1}{2}(3n+1), & \text{if } n \text{ is odd,} \end{cases}$$

always returns 1. In what follows, we shall present a less known conjecture that, like the $3x + 1$ problem, has not been solved. Consider the function f defined, for all positive integers, by

$$f(n) = \begin{cases} \frac{2}{3}n, & \text{if } n = 0 \mod 3, \\ \frac{4}{3}n - \frac{1}{3}, & \text{if } n = 1 \mod 3, \\ \frac{4}{3}n + \frac{1}{3}, & \text{if } n = 2 \mod 3. \end{cases}$$

Its inverse f^{-1} is defined by

$$f(n) = \begin{cases} \frac{3}{2}n, & \text{if } n = 0 \mod 2, \\ \frac{3}{4}n + \frac{1}{4}, & \text{if } n = 1 \mod 4, \\ \frac{3}{4}n - \frac{1}{4}, & \text{if } n = 3 \mod 4. \end{cases}$$

f, which is bijective, is a permutation of the natural numbers. The study of the iterates of f has been called the original Collatz problem [198]. If we consider the first natural numbers, we obtain the following permutation:

$$\begin{pmatrix} 1\ 2\ 3\ 4\ 5\ 6\ 7\ \ 8\ \ 9\ \cdots \\ 1\ 3\ 2\ 5\ 7\ 4\ 9\ 11\ 6\ \cdots \end{pmatrix}.$$

While some cycles are finite, e.g. $(3, 2, 3)$ or $(5, 7, 9, 6, 4, 5)$, it has been conjectured that there exist infinite cycles. For instance, none of the 200,000 successive iterates of 8 is equal to 8. This is also the case for 14 and 16. For this particular dynamical system, the phase space is the set $\mathbb{N}$ of all positive integers, and the evolution rule is reversible.

2

How to Build Up a Model

Nature offers a puzzling variety of interactions between species. *Predation* is one of them. According to the way predators feed on their prey, various categories of predators may be distinguished [330, 39]. *Parasites*, such as tapeworms or tuberculosis bacteria, live throughout a major period of their life in a single host. Their attack is harmful but rarely lethal in the short term. *Grazers*, such as sheep or biting flies that feed on the blood of mammals, also consume only parts of their prey without causing immediate death. However, unlike parasites, they attack large numbers of prey during their lifetime. *True predators*, such as wolves or plankton-eating aquatic animals, also attack many preys during their lifetime, but unlike grazers, they quickly kill their prey.

Our purpose in this chapter is to build up models to study the effects of true predation on the population dynamics of the predator and its prey. More precisely, among the various patterns of predator-prey abundance, we focus on two-species systems in which it appears that predator and prey populations exhibit coupled density oscillations. In order to give an idea of the variety of dynamical systems used in modeling, we describe different models of predator-prey systems.

2.1 Lotka-Volterra model

The simplest two-species predator-prey model has been proposed independently by Lotka [214] and Volterra [340].[1] Vito Volterra (1860–1940) was stimulated to study this problem by his future son-in-law, Umberto D'Ancona, who, analyzing market statistics of the Adriatic fisheries, found that, during the First World War, certain predacious species increased when fishing was severely limited. A year before, Alfred James Lotka (1880–1949) had come up with an almost identical solution to the predator-prey problem. His method was very general, but, probably because of that, his book did not receive the

---

[1] See also [99] and [310].

attention it deserved.[2] This model assumes that, in the absence of predators, the prey population, denoted by H for "herbivore," grows exponentially,
whereas, in the absence of prey, predators starve to death and their population, denoted by P, declines exponentially. As a result of the interaction
between the two species, H decreases and P increases at a rate proportional
to the frequency of predator–prey encounters. We have then

$$\dot{H} = bH - sHP, \tag{2.1}$$

$$\dot{P} = -dP + esHP, \tag{2.2}$$

where b is the birth rate of the prey, d the death rate of the predator, s the
searching efficiency of the predator, and e the efficiency with which extra food
is turned into extra predators.[3]

Lotka-Volterra equations contain four parameters. This number can be
reduced if we express the model in dimensionless form.

If we put

$$h = \frac{Hes}{d}, \quad p = \frac{Ps}{b}, \quad \tau = \sqrt{bd}\,t, \quad \rho = \sqrt{\frac{b}{d}}, \tag{2.3}$$

(2.1) and (2.2) become

$$\frac{dh}{d\tau} = \rho h\left(1 - p\right), \tag{2.4}$$

$$\frac{dp}{d\tau} = -\frac{1}{\rho}p\left(1 - h\right). \tag{2.5}$$

These equations contain only one parameter, which makes them much easier
to analyze.

Before analyzing this particular model, let us show how the asymptotic
stability of a given equilibrium state of a two-species model of the form

$$\frac{dx_1}{d\tau} = f_1(x_1, x_2),$$

$$\frac{dx_2}{d\tau} = f_2(x_1, x_2),$$

may be discussed. A more rigorous discussion of the stability of equilibrium
points of a system of differential equations is given in Section 3.2. Here, we
are just interested in the properties of the Lotka-Volterra model. Let (x_1^*, x_2^*)
be an equilibrium point of the differential system above, and put

---

[2] On the relations between Lotka and Volterra, and how ecologists in the 1920s
  perceived mathematical modeling, consult Kingsland [186].

[3] Chemists will note the similarity of these equations with the rate equations of
  chemical kinetics. For a treatment of chemical kinetics from the point of view of
  dynamical systems theory, see Gavalas [139].

$$u_1 = x_1 - x_1^* \quad \text{and} \quad u_2 = x_2 - x_2^*.$$

In the neighborhood of this equilibrium point, neglecting quadratic terms in u_1 and u_2, we have

$$\frac{du_1}{d\tau} = \frac{\partial f_1}{\partial x_1}(x_1^*, x_2^*)u_1 + \frac{\partial f_1}{\partial x_2}(x_1^*, x_2^*)u_2,$$

$$\frac{du_2}{d\tau} = \frac{\partial f_2}{\partial x_1}(x_1^*, x_2^*)u_1 + \frac{\partial f_2}{\partial x_2}(x_1^*, x_2^*)u_2.$$

The general solution of this linear system is

$$\mathbf{u}(\tau) = \exp\left(Df(\mathbf{x}^*)\tau\right)\mathbf{u}(0),$$

where $\mathbf{f} = (f_1, f_2)$, $\mathbf{x}^* = (x_1, x_2)$, $\mathbf{u} = (u_1, u_2)$, and $Df(\mathbf{x}^*)$ (called the *community matrix* in ecology) is the Jacobian matrix of $\mathbf{f}$ at $\mathbf{x} = (x_1^*, x_2^*)$, that is,

$$Df(\mathbf{x}^*) = \begin{bmatrix} \frac{\partial f_1}{\partial x_1}(x_1^*, x_2^*) & \frac{\partial f_1}{\partial x_2}(x_1^*, x_2^*) \\ \frac{\partial f_2}{\partial x_1}(x_1^*, x_2^*) & \frac{\partial f_2}{\partial x_2}(x_1^*, x_2^*) \end{bmatrix}.$$

Therefore, the equilibrium point $\mathbf{x}^*$ is *asymptotically stable*[4] if, and only if, the eigenvalues of the matrix $Df(x_1^*, x_2^*)$ have *negative real parts*.[5]

The Lotka-Volterra model has two equilibrium states $(0, 0)$ and $(1, 1)$. Since

$$Df(0, 0) = \begin{bmatrix} \rho & 0 \\ 0 & -\frac{1}{\rho} \end{bmatrix}, \quad Df(1, 1) = \begin{bmatrix} 0 & -\rho \\ \frac{1}{\rho} & 0 \end{bmatrix},$$

---

[4] The precise definition of asymptotic stability is given in Definition 5. Essentially, it means that any solution $\mathbf{x}(t, 0, \mathbf{x}_0)$ of the system of differential equations satisfying the initial condition $\mathbf{x} = \mathbf{x}_0$ at $t = 0$ tends to $\mathbf{x}^*$ as t tends to infinity.

[5] Given a 2×2 real matrix A, there exists a real invertible matrix M such that $J = MAM^{-1}$ may be written under one of the following canonical forms

$$\begin{bmatrix} \lambda_1 & 0 \\ 0 & \lambda_2 \end{bmatrix}, \quad \begin{bmatrix} \lambda & 1 \\ 0 & \lambda \end{bmatrix}, \quad \begin{bmatrix} a & -b \\ b & a \end{bmatrix}.$$

The corresponding forms of $e^{J\tau}$ are

$$\begin{bmatrix} e^{\lambda_1 \tau} & 0 \\ 0 & e^{\lambda_2 \tau} \end{bmatrix}, \quad e^{\lambda\tau}\begin{bmatrix} 1 & \tau \\ 0 & 1 \end{bmatrix}, \quad e^{a\tau}\begin{bmatrix} \cos b\tau & -\sin b\tau \\ \sin b\tau & \cos b\tau \end{bmatrix}.$$

$e^{A\tau}$ can then be determined from the relation

$$e^{A\tau} = M^{-1}e^{J\tau}M.$$

More details are given in Subsection 3.2.1.

$(0,0)$ is unstable and $(1,1)$ is stable but not asymptotically stable. The eigenvalues of the matrix $Df(1,1)$ being pure imaginary, if the system is in the neighborhood of $(1,1)$, it remains in this neighborhood. The equilibrium point $(1,1)$ is said to be *neutrally stable*.[6]

The set of all trajectories in the (h,p) phase space is called the *phase portrait* of the differential system (2.4–2.5). Typical phase-space trajectories are represented in Figure 2.1. Except the coordinate axes and the equilibrium point $(0,0)$, all the trajectories are closed orbits oriented counterclockwise.

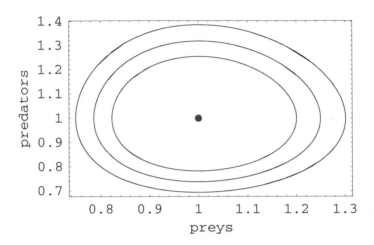

Fig. 2.1. *Lotka-Volterra model. Typical trajectories around the neutral fixed point for $\rho = 0.8$.*

Since the trajectories in the predator-prey phase space are closed orbits, the populations of the two species are periodic functions of time (Figure 2.2). This result is encouraging because it might point toward a simple relevant mechanism for predator-prey cycles.

There is an abundant literature on cyclic variations of animal populations.[7] They were first observed in the records of fur-trading companies. The classic example is the records of furs received by the Hudson Bay Company from 1821 to 1934. They show that the numbers of snowshoe hares[8] (*Lepus americanus*) and Canadian lynx (*Lynx canadensis*) trapped for the company vary periodically, the period being about 10 years. The hare feeds on a variety of herbs, shrubs, and other vegetable matter. The lynx is essentially single-prey oriented, and although it consumes other small animals if starving, it cannot

---

[6] For the exact meaning of neutrally stable, see Section 3.2, and, in particular, Example 13.

[7] See, in particular, Finerty [127].

[8] Also called *varying hares*. They have large, heavily furred hind feet and a coat that is brown in summer and white in winter.

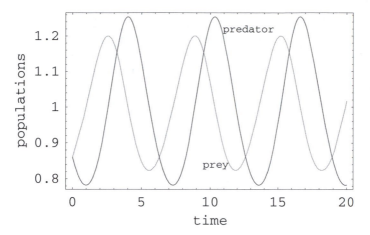

Fig. 2.2. *Lotka-Volterra model. Scaled predator and prey populations as functions of scaled time.*

live successfully without the snowshoe hare. This dependence is reflected in the variation of lynx numbers, which closely follows the cyclic peaks of abundance of the hare, usually lagging a year behind. The hare density may vary from one hare per square mile of woods to 1000 or even 10,000 per square mile.[9] In this particular case, however, the understanding of the coupled periodic variations of predator and prey populations seems to require a more elaborate model. The two species are actually parts of a multispecies system. In the boreal forests of North America, the snowshoe hare is the dominant herbivore, and the hare-plant interaction is probably the essential mechanism responsible for the observed cycles. When the hare density is not too high, moderate browsing removes the annual growth and has a pruning effect. But at high hare density, browsing may reduce all new growth for several years and, consequently, lower the carrying capacity for hares. The shortage in food supply causes a marked drop in the number of hares. It has also been suggested that when hares are numerous, the plants on which they feed respond to heavy grazing by producing shoots with high levels of toxins.[10] If this interpretation is correct, the hare cycles would be the result of the herbivore-forage interaction (in this case, hares are "preying" on vegetation), and the lynx, because they depend almost exclusively upon the snowshoe hares, track the hare cycles.

A careful study of the variations of the numbers of pelts sold by the Hudson Bay Company as a function of time poses a difficult problem of interpretation. Assuming that these numbers represent a fixed proportion of the total

---

[9] Many interesting facts concerning northern mammals may be found in Seton [312]. For a statistical analysis of the lynx-hare and other 10-year cycles in the Canadian forests, see Bulmer [73].

[10] See [39], pp. 356–357.

populations of a two-species system, they seem to indicate that *the hares are eating the lynx* [144] since the predator's oscillation precedes the prey's. It should be the opposite: an increase in the predator population should lead to a decrease of the prey population, as illustrated in Figure 2.2.

Although it accounts in a very simple way for the existence of coupled cyclic variations in animal populations, the Lotka-Volterra model exhibits some unsatisfactory features, however.

Since, in nature, the environment is continually changing, in phase space, the point representing the state of the system will continually jump from one orbit to another. From an ecological viewpoint, an adequate model should not yield an infinity of neutrally stable cycles but one *stable limit cycle*. That is, in the (h, p) phase space, there should exist a closed trajectory C such that any trajectory in the neighborhood of C should, as time increases, become closer and closer to C.

Furthermore, the Lotka-Volterra model assumes that, in the absence of predators, the prey population grows exponentially. This Malthusian growth[11] is not realistic. Hence, if we assume that, in the absence of predation, the growth of the prey population follows the logistic model, we have

$$\dot{H} = bH\left(1 - \frac{H}{K}\right) - sHP, \tag{2.6}$$

$$\dot{P} = -dP + esHP, \tag{2.7}$$

where K is the carrying capacity of the prey. For large K, this model is just a small perturbation of the Lotka-Volterra model. If, to the dimensionless variables defined in (2.3), we add the scaled carrying capacity

$$k = \frac{Kes}{d}, \tag{2.8}$$

Equations (2.6) and (2.7) become

$$\frac{dh}{d\tau} = \rho h \left(1 - \frac{h}{k} - p\right), \tag{2.9}$$

$$\frac{dp}{d\tau} = -\frac{1}{\rho} p \left(1 - h\right). \tag{2.10}$$

The equilibrium points are $(0,0)$, $(k,0)$, and $(1, 1 - 1/k)$. Note that the last equilibrium point exists if, and only if, $k > 1$; that is, if the carrying capacity of the prey is high enough to support the predator. Since

---

[11] After Thomas Robert Malthus (1766–1834), who, in his most influential book [221], stated that because a population grows much faster than its means of subsistence—the first increasing geometrically whereas the second increases only arithmetically—"vice and misery" will operate to restrain population growth. To avoid these disastrous results, many demographers, in the nineteenth century, were led to advocate birth control. More details on Malthus and his impact may be found in [177], pp. 11–18.

$$Df(0,0) = \begin{bmatrix} \rho & 0 \\ 0 & -1/\rho \end{bmatrix}, \quad Df(k,0) = \begin{bmatrix} -\rho & -\rho k \\ 0 & -(1-k)/\rho \end{bmatrix},$$

and

$$Df\left(1, 1 - \frac{1}{k}\right) = \begin{bmatrix} -\rho/k & -\rho \\ (k-1)/\rho k & 0 \end{bmatrix},$$

it follows that $(0,0)$ and $(k,0)$ are unstable, whereas $(1, 1 - 1/k)$ is stable. A finite carrying capacity for the prey transformed the neutrally stable equilibrium point of the Lotka-Volterra model into an asymptotically stable equilibrium point. It is easy to verify that, if k is large enough for the condition

$$\rho^2 < 4k(k-1)$$

to be satisfied, the eigenvalues are complex, and, in the neighborhood of the asymptotically stable equilibrium point, the trajectories are converging spirals oriented counterclockwise (Figure 2.3). The predator and prey populations are no longer periodic functions of time, they exhibit damped oscillations, the predator oscillations lagging in phase behind the prey (Figure 2.4). If $\rho^2 > 4k(k-1)$, the eigenvalues are real and the approach of the asymptotically stable equilibrium point is nonoscillatory.

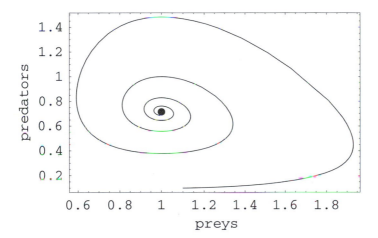

Fig. 2.3. *Modified Lotka-Volterra model. A typical trajectory around the stable fixed point (big dot) for $\rho = 0.8$ and $k = 3.5$.*

A small perturbation—corresponding to the existence of a finite carrying capacity for the prey—has qualitatively changed the phase portrait of the Lotka-Volterra model.[12] A model whose qualitative properties do not change

---

[12] A precise definition of what exactly is meant by *small perturbation* and *qualitative change of the phase portrait* will be given when we study structural stability (see Section 3.4).

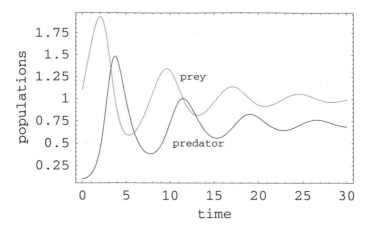

Fig. 2.4. *Modified Lotka–Volterra model. Scaled predator and prey populations as functions of scaled time.*

significantly when it is subjected to small perturbations is said to be *structurally stable*. Since a model is not a precise description of a system, qualitative predictions should not be altered by slight modifications. Satisfactory models should be structurally stable.

2.2 More realistic predator-prey models

If we limit our discussion of predation to two-species systems assuming, as we did so far, that

- time is a continuous variable,
- there is no time lag in the responses of either population to changes, and
- population densities are not space-dependent,

a somewhat realistic model, formulated in terms of ordinary differential equations, should at least take into account the following relevant features[13]:

1. *Intraspecific competition*; that is, competition between individuals belonging to the same species.
2. *Predator's functional response*; that is, the relation between the predator's consumption rate and prey density.
3. *Predator's numerical response*; that is, the efficiency with which extra food is transformed into extra predators.

Essential resources being, in general, limited, intraspecific competition reduces the growth rate, which eventually goes to zero. The simplest way to take

---

[13] See May [229], pp. 80–84, Pielou [283], pp. 91–95.

this feature into account is to introduce into the model carrying capacities for both the prey and the predator.

A predator has to devote a certain time to search, catch, and consume its prey. If the prey density increases, searching becomes easier, but consuming a prey takes the same amount of time. The functional response is, therefore, an increasing function of the prey density—obviously equal to zero at zero prey density—approaching a finite limit at high densities. In the Lotka-Volterra model the functional response, represented by the term sH, is not bounded. According to Holling [173, 174], the behavior of the functional response at low prey density depends upon the predator. If the predator eats essentially one type of prey, then the functional response should be linear at low prey density. If, on the contrary, the predator hunts different types of prey, the functional response should increase as a power greater than 1 (usually 2) of prey density.

In the Lotka-Volterra model, the predator's numerical response is a linear function of the prey density. As for the functional response, it can be argued that there should exist a saturation effect; that is, the predator's birth rate should tend to a finite limit at high prey densities.

Possible predator-prey models are

$$\dot{H} = r_H H \left(1 - \frac{H}{K}\right) - \frac{a_H P H}{b + H},$$

$$\dot{P} = \frac{a_P P H}{b + H} - cP,$$

or

$$\dot{H} = r_H H \left(1 - \frac{H}{K}\right) - \frac{a_H P H^2}{b + H^2},$$

$$\dot{P} = \frac{a_P P H^2}{b + H^2} - cP.$$

In these two models, the predator equation follows the usual assumption that the predator's numerical response is proportional to the rate of prey consumption. The efficiency of converting prey to predator is given by a_H/a_P. The term $-cP$ represents the rate at which the predators would decrease in the absence of the prey.

Following Leslie [201] (see Example 10), an alternative equation for the predator could be

$$\dot{P} = r_P P \left(1 - \frac{P}{cH}\right),$$

an equation that has the logistic form with a carrying capacity for the predator proportional to the prey density.

2.3 A model with a stable limit cycle

In a recent paper, Harrison [165] studied a variety of predator-prey models in order to find which model gives the best quantitative agreement with Luckinbill's data on *Didinium* and *Paramecium* [215]. Luckinbill grew *Paramecium aurelia* together with its predator *Didinium nasutum* and, under favorable experimental conditions aimed at reducing the searching effectiveness of the *Didinium*, he was able to observe oscillations of both populations for 33 days before they became extinct.

Harrison found that the predator-prey model

$$\dot{H} = r_H H \left(1 - \frac{H}{K}\right) - \frac{a_H PH}{b + H}, \tag{2.11}$$

$$\dot{P} = \frac{a_P PH}{b + H} - cP, \tag{2.12}$$

predicts the outcome of Luckinbill's experiment qualitatively.[14]

In order to simplify the discussion of this model, we first define reduced variables. Equations (2.11) and (2.12) have only one nontrivial equilibrium point corresponding to the coexistence of both populations. It is the solution of the system[15]

$$r_H \left(1 - \frac{H}{K}\right) - \frac{a_H P}{b + H} = 0, \qquad \frac{a_P H}{b + H} - c = 0.$$

Denote as (H^*, P^*) this nontrivial equilibrium point, and let

$$h = \frac{H}{H^*}, \quad p = \frac{P}{P^*}, \quad \tau = r_H t, \quad k = \frac{K}{H^*}, \quad \beta = \frac{b}{H^*}, \quad \gamma = \frac{c}{r}.$$

Equations (2.11) and (2.12) become

$$\frac{dh}{d\tau} = h \left(1 - \frac{h}{k}\right) - \frac{\alpha_h ph}{\beta + h}, \tag{2.13}$$

$$\frac{dp}{d\tau} = \frac{\alpha_p ph}{\beta + h} - \gamma p, \tag{2.14}$$

where[16]

$$\alpha_h = \left(1 - \frac{1}{k}\right)(\beta + 1) \quad \text{and} \quad \alpha_p = \gamma(\beta + 1).$$

Equations (2.13) and (2.14) contain only three independent parameters: k, β, and γ. In terms of the scaled populations, the nontrivial fixed point is $(1, 1)$, and the expression of the Jacobian matrix at this point is

---

[14] The reader interested in how Harrison modified this model to obtain a better quantitative fit should refer to Harrison's paper [165].

[15] Since the coordinates (H^*, P^*) of an acceptable equilibrium point have to be positive, the coefficients of Equations (2.11) and (2.12) have to satisfy certain conditions. Note that there exist two trivial equilibrium points, $(0, 0)$ and $(K, 0)$.

[16] These relations express that $(1, 1)$ is an equilibrium point of Equations (2.13) and (2.14).

$$Df(1,1) = \begin{bmatrix} \dfrac{k-2-\beta}{k(1+\beta)} & -1+\dfrac{1}{k} \\[2ex] \dfrac{\beta\gamma}{1+\beta} & 0 \end{bmatrix}.$$

For $(1,1)$ to be asymptotically stable, the determinant of the Jacobian has to be positive and its trace negative. Since $k > 1$, the determinant is positive, but the trace is negative if, and only if, $k < 2+\beta$. Below and above the threshold value $k_c = 2+\beta$, the phase portrait is qualitatively different. For $k < k_c$, the trajectories converge to the fixed point $(1,1)$, whereas for $k > k_c$ they converge to a limit cycle, as shown in Figure 2.5. The value k_c of the parameter k where this structural change occurs is called a *bifurcation point*.[17] This particular type of bifurcation is a Hopf bifurcation (see Chapter 3, Example 22).

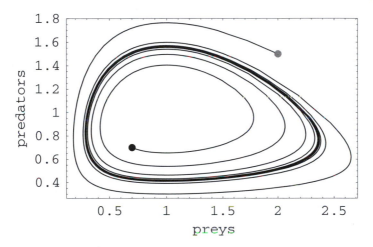

Fig. 2.5. *Two trajectories in the (h,p)-plane converging to a stable limit cycle of the predator-prey model described by Equations (2.13) and (2.14). Big dots represent initial points. Scaled parameters values are $k = 3.5$, $\beta = 1$, and $\gamma = 0.5$.*

2.4 Fluctuating environments

Population oscillations may be driven by fluctuating environments. Consider, for example, the logistic equation

$$\dot{N} = rN\left(1 - \frac{N}{K(t)}\right), \tag{2.15}$$

---

[17] See Section 3.5.

where $K(t)$ represents a time-dependent carrying capacity. If $K(t) = K_0(1 + a \cos \Omega t)$, where a is a real such that $|a| < 1$, (2.15) takes the reduced form

$$\frac{dn}{d\tau} = n \left(1 - \frac{n}{1 + a \cos \omega \tau} \right), \qquad (2.16)$$

where

$$n = \frac{N}{K_0}, \quad \omega = \frac{\Omega}{r}, \quad \tau = rt.$$

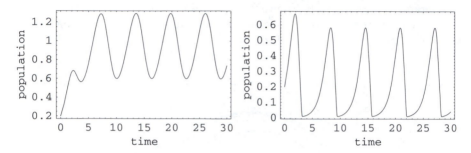

Fig. 2.6. *Scaled population evolving according to the logistic equation with a periodic carrying capacity.* $\omega = 1$ *and* $a = 0.5$ *(left);* $\omega = 1$ *and* $a = 0.999$ *(right).*

Numerical solutions of Equation (2.16), represented in Figure 2.6, show the existence of periodic solutions. Note that this nonlinear differential equation is a Riccati equation and, therefore, linearizable. If we put $n = k \, \dot{u}/u$, where $k(\tau) = 1 + a \cos \omega \tau$, (2.16) becomes

$$\ddot{u} + \left(\frac{\dot{k}}{k} - 1 \right) \dot{u} = 0.$$

Hence,

$$u(\tau) = c_1 \int \frac{e^\tau}{k(\tau)} \, d\tau + c_2,$$

where c_1 and c_2 are constants to be determined by the initial conditions.

2.5 Hutchinson's time-delay model

Since reproduction is not an instantaneous process, Hutchinson suggested (see Chapter 1, Example 4) that he logistic equation modeling population growth should be replaced by the following time-delay logistic equation

$$\dot{N}(t) = \frac{dN}{dt} = rN(t) \left(1 - \frac{N(t-T)}{K} \right). \qquad (2.17)$$

Mathematically, this model is not trivial. Its behavior may, however, be understood as follows. K is clearly an equilibrium point (steady solution). If $N(0)$ is less than K, as t increases, $N(t)$ does not necessarily tend monotonically to K from below. The time delay allows $N(t)$ to be momentarily greater than K. In fact, if at time t the population reaches the value K, it may still grow, for the growth rate, equal to $r(1 - N(t - T)/K)$, is positive. When $N(t - T)$ exceeds K, the growth rate becomes negative and the population declines. If T is large enough, the model will, therefore, exhibit oscillations.

In this problem, there exist two time scales: $1/r$ and T. As a result, the stability depends on the relative sizes of these time scales measured by the dimensionless parameter rT. Qualitatively, we can say that if rT is small, no oscillations—or damped oscillations—are observed, and, as for the standard logistic model, the equilibrium point K is asymptotically stable. If rT is large, K is no more stable, and the population oscillates. Hence, there exists a bifurcation point; that is, a threshold value for the parameter rT above which $N(t)$ tends to a periodic function of time.

This model is instructive for it proves that single-species populations may exhibit oscillatory behaviors even in stable environments.

The following discussion shows how to analyze the stability of time-delay models and find bifurcation points.

Consider the general time-delay equation (1.13)

$$\frac{dN}{dt} = rN(t)\left(1 - \frac{1}{K}\int_0^t N(t - u)Q(u)\,du\right),$$

where Q is the delay kernel whose integral over $[0, \infty[$ is equal to 1, and replace $N(t)$ by $K + n(t)$. Neglecting quadratic terms in n, we obtain

$$\frac{dn(t)}{dt} = -r\int_0^t n(t - u)Q(u)\,du. \qquad (2.18)$$

This type of equation can be solved using the Laplace transform.[18] The Laplace transform of n is defined by

$$L(z, n) = \int_0^\infty e^{-zt}n(t)\,dt.$$

Substituting in (2.18) gives[19]

$$zL(z, n) - n_0 = -rL(z, n)L(z, Q),$$

---

[18] On using the Laplace transform to solve differential equations, see Boccara [46], pp. 94–103 and 226–231.

[19] If $L(z, n)$ is the Laplace transform of n, the Laplace transform of $\dot{n}$ is $L(z, \dot{n}) = zL(z, n) - n(0)$, and the Laplace transform of the convolution of n and Q, defined by $\int_0^t n(t - u)Q(u)\,du$ (since the support of n and Q is $\mathbb{R}_+$), is the product $L(z, n)L(z, Q)$ of the Laplace transforms of n and Q.

where n_0 is the initial value $n(0)$. Thus,

$$L(z,n) = \frac{n_0}{z + rL(z,Q)}.$$ (2.19)

If, as in Equation (2.17), the delay kernel is δ_T, the Dirac distribution at T, its Laplace transform is e^{-zT}, and we have

$$L(z,n) = \frac{n_0}{z + re^{-zT}}.$$

$n(t)$ is, therefore, the inverse Laplace transform of $L(z,n)$, that is

$$n(t) = \frac{n_0}{2i\pi} \int_{c-i\infty}^{c+i\infty} \frac{dz}{z + re^{-zT}},$$

where c is such that all the singularities of the function $z \mapsto (z+re^{-zT})^{-1}$ are in the half-plane $\{z \mid \operatorname{Re} z < c\}$. The only singularities of the integrand are the zeros of $z + re^{-zT}$. As a simplification, define the dimensionless variable $\zeta = zT$ and parameter $\rho = rT$. We then have to find the solutions of

$$\zeta + \rho e^{-\zeta} = 0.$$ (2.20)

If $\zeta = \xi + i\eta$, where ξ and η are real, we therefore have to solve the system

$$\xi + \rho e^{-\xi} \cos \eta = 0,$$ (2.21)
$$\eta - \rho e^{-\xi} \sin \eta = 0.$$ (2.22)

(i) *If ζ is real,* $\eta = 0$, and ξ is the solution of $\xi + \rho e^{-\xi} = 0$. This equation has no real root for $\rho > 1/e$, a double root equal to -1 for $\rho = 1/e$, and two negative real roots for $\rho < 1/e$.

(ii) *If ζ is complex,* eliminating $\rho e^{-\xi}$ between Equations (2.21) and (2.22) yields

$$\xi = -\eta \cot \eta,$$ (2.23)

which, when substituted back into (2.22), gives

$$\frac{\eta}{\rho} = e^{\eta \cot \eta} \sin \eta.$$ (2.24)

Complex roots of Equation (2.20) are found by solving (2.22) for η and obtaining ξ from η by (2.21). Since complex roots appear in complex conjugate pairs, it is sufficient to consider the case $\eta > 0$.

The function $f : \eta \mapsto e^{\eta \cot \eta} \sin \eta$ has the following properties (k is a positive integer):

$$\lim_{\eta \uparrow 2k\pi} f(\eta) = 0,$$

$$\lim_{\eta \downarrow (2k-1)\pi} f(\eta) = -\infty,$$

$$\lim_{\eta \uparrow (2k-1)\pi} f(\eta) = \infty.$$

In each open interval $](2k-1)\pi, 2k\pi[$, f is increasing, and in each open interval $]2k\pi, (2k+1)\pi[$, f is decreasing. Finally, in the interval $[0, \pi[$, the graph of f is represented in Figure 2.7.

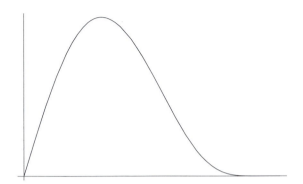

Fig. 2.7. *Graph of the function* $f : \eta \mapsto e^{\eta \cot \eta} \sin \eta$ *for* $\eta \in]0, \pi[$.

Since the slope at the origin $f'(0) = e$, if $\rho < 1/e$, Equation (2.24) has no root in $[0, \pi[$ but a root in each interval of the form $[2k\pi, \eta_k]$, where η_k is the solution of the equation $e\eta = e^{\eta \cot \eta} \sin \eta$ in the interval $[2k\pi, (2k+1)\pi[$. From (2.23) it could be verified that the corresponding value of ξ is negative.

If $1/e < \rho < \pi/2$, we find, as above, that there are no real roots and that there is an additional complex root with $0 < \eta < \pi/2$ and, from (2.23), a negative corresponding ξ.

Finally, if $\rho > \pi/2$, there is a complex root with $\pi/2 < \eta < \pi$ and, from (2.23), a positive corresponding ξ.

Consequently, if $\rho < \pi/2$, the equilibrium point $N^* = K$ is asymptotically stable; but, for $\rho > \pi/2$, this point is unstable. $\rho = \pi/2$ is a bifurcation point.

The method we have just described is applicable to any other delay kernel Q. Since each pole z_* of the Laplace transform of n contributes in the expression of $n(t)$ with a term of the form Ae^{z_*t}, the problem is to find the poles of the Laplace transform of n given by (2.19) and determine the sign of their real part. These poles are the solutions of the equation $z + rL(z, Q) = 0$.

2.6 Discrete-time models

When a species may breed only at a specific time, the growth process occurs in discrete time steps. To a continuous-time model, we may always associate a discrete-time one. If τ_0 represents some characteristic time interval, the following finite difference equations

$$\frac{1}{\tau_0}\big(x_1(\tau+\tau_0)-x_1(\tau)\big)=f_1\big(x_1(\tau),x_2(\tau)\big), \qquad (2.25)$$

$$\frac{1}{\tau_0}\big(x_2(\tau+\tau_0)-x_2(\tau)\big)=f_2\big(x_1(\tau),x_2(\tau)\big), \qquad (2.26)$$

coincide, when τ_0 tends to zero, with the differential equations

$$\frac{dx_1}{d\tau}=f_1(x_1,x_2), \qquad (2.27)$$

$$\frac{dx_2}{d\tau}=f_2(x_1,x_2). \qquad (2.28)$$

If τ_0 is taken equal to 1, Equations (2.25) and (2.26) become

$$x_1(\tau+1)=x_1(\tau)+f_1\big(x_1(\tau),x_2(\tau)\big), \qquad (2.29)$$
$$x_2(\tau+1)=x_2(\tau)+f_2\big(x_1(\tau),x_2(\tau)\big). \qquad (2.30)$$

Equations (2.29) and (2.30) are called the *time-discrete analogues* of Equations (2.27) and (2.28).

Let (x_1^*,x_2^*) be an equilibrium point of the difference equations (2.29) and (2.30),[20] and put

$$u_1=x_1-x_1^* \quad \text{and} \quad u_2=x_2-x_2^*.$$

In the neighborhood of this equilibrium point, we have

$$u_1(\tau+1)=u_1(\tau)+\frac{\partial f_1}{\partial x_1}(x_1^*,x_2^*)u_1(\tau)+\frac{\partial f_1}{\partial x_2}(x_1^*,x_2^*)u_2(\tau),$$

$$u_2(\tau+1)=u_2(\tau)+\frac{\partial f_2}{\partial x_1}(x_1^*,x_2^*)u_1(\tau)+\frac{\partial f_2}{\partial x_2}(x_1^*,x_2^*)u_2(\tau).$$

The general solution of this linear system is

$$u(\tau)=\big(I+Df(x_1^*,x_2^*)\big)^\tau u(0),$$

where I is the 2×2 identity matrix. Here τ is an integer. The equilibrium point (x_1^*,x_2^*) is asymptotically stable if the absolute values of the eigenvalues of $I+Df(x_1^*,x_2^*)$ are less than 1. That is, in the complex plane, the eigenvalues of $Df(x_1^*,x_2^*)$ must belong to the open disk $\{z \mid |z-1|<1\}$.

The stability criterion for discrete models is more stringent than for continuous models. The existence of a finite time interval between generations is a destabilizing factor.

---

[20] That is, a solution of the system

$$f_1(x_1^*,x_2^*)=0$$
$$f_2(x_1^*,x_2^*)=0.$$

For example, the time-discrete analogue of the reduced Lotka-Volterra equations (2.4) and (2.5) are

$$h(\tau + 1) = h(\tau) + \rho\, h(\tau)\big(1 - p(\tau)\big), \tag{2.31}$$

$$p(\tau + 1) = p(\tau) - \frac{1}{\rho}\, p(\tau)\big(1 - h(\tau)\big). \tag{2.32}$$

Since the eigenvalues of the Jacobian matrix at the fixed point $(1, 1)$, which are i and $-i$, are outside the open disk $\{z \mid |z - 1| < 1\}$, the neutrally stable fixed point of the time-continuous model is unstable for the time-discrete model.

2.7 Lattice models

To catch a prey, a predator has to be in the immediate neighborhood of the prey. In predator-prey models formulated in terms of either differential equations (ordinary or partial) or difference equations, the short-range character of the predation process is not correctly taken into account.[21] One way to correctly take into account the short-range character of the predation process is to discretize space; that is, to consider lattice models.

Dynamical systems in which states, space, and time are discrete are called *automata networks*. If the network is periodic, the dynamical system is a cellular automaton. More precisely, a *one-dimensional cellular automaton* is a dynamical system whose state $s(i, t) \in \{0, 1, 2, \ldots, q - 1\}$ at position $i \in \mathbb{Z}$ and time $t \in \mathbb{N}$ evolves according to a *local rule* f such that

$$s(i, t + 1) = f\big(s(i - r_\ell, t), s(i - r_\ell + 1, t), \ldots, s(i + r_r, t)\big).$$

The numbers r_ℓ and r_r, called the *left* and *right radii* of rule f, are positive integers.

In what follows, we briefly describe a lattice predator-prey model studied by Boccara, Roblin, and Roger [55]. Consider a finite two-dimensional lattice $\mathbb{Z}_L^2$ with periodic boundary conditions. The total number of vertices is equal to L^2. Each vertex of the lattice is either empty or occupied by a prey or a predator. According to the process under consideration, for a given vertex, we consider two different neighborhoods (Figure 2.8): a von Neumann *predation neighborhood* (4 vertices) and a Moore *pursuit and evasion neighborhood* (8 vertices).

The evolution of preys and predators is governed by the following set of rules.

---

[21] This will be manifest when we discuss systems that exhibit bifurcations. In phase transition theory, for instance, it is well-known that in the vicinity of a bifurcation point—*i.e.*, a second-order transition point—certain physical quantities have a singular behavior (see Boccara [45], pp. 155–189). It is only above a certain spatial dimensionality—known as the upper critical dimensionality—that the behavior of the system is correctly described by a partial differential equation.

Fig. 2.8. *Von Neumann predation neighborhood (left) and Moore pursuit and evasion neighborhood (right).*

1. A prey has a probability d_h of being captured and eaten by a predator in its predation neighborhood.
2. If there is no predator in its predation neighborhood, a prey has a probability b_h of giving birth to a prey at an empty site of this neighborhood.
3. After having eaten a prey, a predator has a probability b_p of giving birth to a predator at the site previously occupied by the prey.
4. A predator has a probability d_p of dying.
5. Predators move to catch prey and prey move to evade predators. Predators and prey move to a site of their von Neumann neighborhood according to the occupation of the pursuit and evasion Moore neighborhood. This neighborhood is divided into four quarters along its diagonals, and prey and predator densities are evaluated in each quarter. Predators move to a neighboring site in the direction of highest local prey density of the Moore neighborhood. In case of equal highest density in two or four directions, one of them is chosen at random. If three directions correspond to the same highest density, predators select the middle one. Preys move to a neighboring site in the opposite direction of highest local predator density. If the four directions are equivalent, one is selected at random. If three directions correspond to the same maximum density, preys choose the remaining one. If two directions correspond to the same maximum density, preys choose at random one of the other two. If H and P denote, respectively, the prey and predator densities, then mHL^2 preys and mPL^2 predators are sequentially selected at random to perform a move. This sequential process allows some individuals to move more than others. The parameter m, which is a positive number, represents the *average number of tentative moves per individual* during a unit of time.

Rules 1, 2, 3, and 4 are applied *simultaneously*. Predation, birth, and death processes are modeled by a *three-state two-dimensional cellular automaton rule*. Rule 5 is applied *sequentially*.

At each time step, the evolution results from the application of the synchronous cellular automaton subrule followed by the sequential one.

To study such a lattice model, it is usually useful to start with the mean-field approximation that ignores space dependence and neglects correlations. In lattice models with local interactions, quantitative predictions of such an approximation are not very good. However, it gives interesting information about the qualitative behavior of the system in the limit $m \to \infty$.

If H_t and P_t denote the densities at time t of preys and predators, respectively, we have

$$H_{t+1} = H_t - H_t f(1, d_h P_t) + (1 - H_t - P_t)f(1 - P_t, b_h H_t),$$
$$P_{t+1} = P_t - d_p P_t + b_p H_t f(1, d_h P_t),$$

where

$$f(p_1, p_2) = p_1^4 - (p_1 - p_2)^4.$$

Within the mean-field approximation, the evolution of our predator-prey model is governed by two coupled finite-difference equations.

The model has three fixed points that are solutions of the system:

$$H f(1, d_h P) - (1 - H - P)f(1 - P, b_h t) = 0,$$
$$d_p P_t - b_p H_t f(1, d_h P_t) = 0.$$

In the prey-predator phase space ((H, P)-plane), $(0, 0)$ is always unstable. $(0, 1)$ is stable if $d_p - 4b_p d_h > 0$. When $d_p - 4b_p d_h$ changes sign, a nontrivial fixed point (H^*, P^*), whose coordinates depend upon the numerical values of the parameters, becomes stable, as illustrated in Figure 2.9.

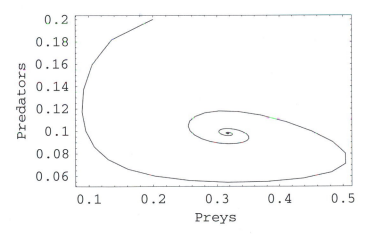

Fig. 2.9. *Orbit in the prey-predator phase space approaching the stable fixed point* $(0.288, 0.103)$ *for* $b_h = 0.2, d_h = 0.9, b_p = 0.2, d_p = 0.2$.

Increasing the probability b_p for a predator to give birth, we observe a *Hopf bifurcation* and the system exhibits a stable limit cycle, as illustrated in Figure 2.10.

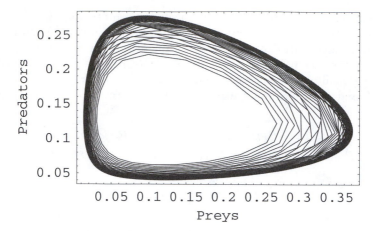

Fig. 2.10. *Orbit in the prey-predator phase space approaching the stable limit cycle for $b_h = 0.2, d_h = 0.9, b_p = 0.6, d_p = 0.2$.*

We will not describe in detail the results of numerical simulations, which can be found in [55]. For large m values, the mean-field approximation provides useful qualitative—although not exact—information on the general temporal behavior as a function of the different parameters.

If we examine the influence of the pursuit and evasion process (Rule 5)— *i.e.*, if we neglect the birth, death, and predation processes ($b_h = b_d = d_h = d_p = 0$)—we observe the formation of small clusters of preys surrounded by predators preventing these preys from escaping. Preys that are not trapped by predators move more or less randomly avoiding predators.

For $b_h = 0.2, d_h = 0.9, b_p = 0.2, d_p = 0.2$, the mean-field approximation exhibits a stable fixed point in the prey-predator phase space located at $(H^*, P^*) = (0.288, 0.103)$. Simulations show that, for these parameter values, a nontrivial fixed point exists only if $m > m_0 = 0.350$. Below m_0, the stable fixed point is $(1, 0)$. This result is quite intuitive; if the average number of tentative moves is too small, all predators eventually die and prey density grows to reach its maximum value. As m increases from m_0 to $m = 500$, the location of the nontrivial fixed point approaches the mean-field fixed point.

For $b_h = 0.2, d_h = 0.9, b_p = 0.6, d_p = 0.2$, the mean-field approximation exhibits a stable limit cycle in the prey-predator phase space. This oscillatory behavior *is not* observed for two-dimensional large lattices [84]. A quasicyclic behavior of the predator and prey densities on a scale of the order of the mean displacements of the individuals may, however, be observed. As mentioned above, cyclic behaviors observed in population dynamics have received a variety of interpretations. This automata network predator-prey model suggests another possible explanation: approximate cyclic behaviors could result as a consequence of a not too large habitat; *i.e.*, when the size of the habitat is of the order of magnitude of the mean displacements of the individuals.

Part I

Mean-Field Type Models

Mean-field type models deal with average quantities. As models of complex systems, it is important to emphasize that they ignore space correlations between the elements of the system and replace local interactions by uniform long-range ones. These models are, therefore, rather crude representations of complex systems, but, as a first attempt to understand the behavior of multi-agent systems, they are often very useful.

The mean-field type models we shall study are formulated in terms of either differential equations or recurrence equations whose theory is well-developed. Since the purpose of this book is to show how models are built up and studied, we shall use the results of dynamical systems theory without giving proofs of most theorems, especially when these proofs are rather technical. We shall, however, make every effort to give precise definitions and state theorems with accuracy—avoiding unnecessary and overly general statements—and indicate references to mathematical texts where the interested reader will find correct proofs.

3

Differential Equations

The study of dynamical models formulated in terms of ordinary differential equations began with Newton's attempts to explain the motion of bodies in the solar system. Except in very simple cases, such as the two-body problem, most problems in celestial mechanics proved extremely difficult. At the end of the nineteenth century, Poincaré developed new methods to analyze the qualitative behavior of solutions to nonlinear differential equations. In a paper [287][1] devoted to functions defined as solutions of differential equations, he explains:

> *Malheureusement, il est évident que, dans la grande généralité des cas qui se présentent, on ne peut intégrer ces équations à l'aide des fonctions déjà connues, ...*
>
> *Il est donc nécessaire d'étudier les fonctions définies par des équations différentielles en elles-mêmes et sans chercher à les ramener à des fonctions plus simples, ...[2]*

The modern qualitative theory of differential equations has its origin in this work.

There exists a wide variety of models formulated in terms of differential equations, and some of them are presented in this chapter.

3.1 Flows

Consider a system whose dynamics is described by the differential equation

---

[1] This paper is a revision of the work for which Poincaré was awarded a prize offered by the king of Sweden in 1889.

[2] Unfortunately, it is clear that in most cases we cannot solve these equations using known functions, ...

It is therefore necessary to study functions defined by differential equations for themselves without trying to reduce them to simpler functions, ...

$$\frac{d\mathbf{x}}{dt} = \dot{\mathbf{x}} = \mathbf{X}(\mathbf{x}), \tag{3.1}$$

where $\mathbf{x}$, which represents the state of the system, belongs to the *state* or *phase space* $\mathcal{S}$, and $\mathbf{X}$ is a given *vector field*. Figure 3.1 shows an example of a vector field. We have seen (Chapter 1, Example 1) that a differential equation of order higher than one, autonomous or nonautonomous, can always be written under the above form.

To present the theory, we need to recall some definitions.

Definition 1. *A function* $\mathbf{f} : \mathbb{R}^n \to \mathbb{R}^n$ *is differentiable at* $\mathbf{x}_0 \in \mathbb{R}^n$ *if there exists a linear transformation* $D\mathbf{f}(\mathbf{x}_0)$ *that satisfies*

$$\lim_{\|\mathbf{h}\| \to 0} \frac{\|\mathbf{f}(\mathbf{x}_0 + \mathbf{h}) - \mathbf{f}(\mathbf{x}_0) - D\mathbf{f}(\mathbf{x}_0)\mathbf{h}\|}{\|\mathbf{h}\|} = 0. \tag{3.2}$$

The linear transformation $D\mathbf{f}(\mathbf{x}_0)$ *is called the* derivative *of* $\mathbf{f}$ *at* $\mathbf{x}_0$.

Instead of the word *function*, many authors use the words *map* or *mapping*. In this text, we shall indifferently use any of these terms.

It is easily verified that the derivative $D\mathbf{f}$ is given by the $n \times n$ Jacobian matrix

$$\frac{\partial(f_1, f_2, \ldots, f_n)}{\partial(x_1, x_2, \ldots, x_n)}.$$

Let U be an open subset of $\mathbb{R}^n$; the function $\mathbf{f} : U \to \mathbb{R}^n$ is continuously differentiable, or of *class* C^1, in U if all the partial derivatives

$$\frac{\partial f_i}{\partial x_j} \quad (1 \le i \le n, \ 1 \le j \le n)$$

are continuous in U. More generally, $\mathbf{f}$ is of *class* C^k in U if all the partial derivatives

$$\frac{\partial^k f_i}{\partial x_{j_1} \partial x_{j_2} \cdots \partial x_{j_k}} \quad (1 \le i \le n, \ 1 \le j_1 \le n, \ 1 \le j_2 \le n, \ \ldots, \ 1 \le j_k \le n)$$

exist and are continuous in U. Continuous but not differentiable functions are referred to as C^0 functions. A function is said to be *smooth* if it is differentiable a sufficient number of times.

Definition 2. *A function* $\mathbf{f} : U \to V$, *where* U *and* V *are open subsets of* $\mathbb{R}^n$, *is said to be a* C^k diffeomorphism *if it is a bijection, and both* $\mathbf{f}$ *and* $\mathbf{f}^{-1}$ *are* C^k *functions. If* $\mathbf{f}$ *and* $\mathbf{f}^{-1}$ *are* C^0, $\mathbf{f}$ *is called a* homeomorphism.

In many models, the phase space is not Euclidean. It may have, for instance, the structure of a circle or a sphere. But if, as in both these cases, we can define *local coordinates*, the notions of derivative and diffeomorphism can be easily extended. Phase spaces that have a structure similar to the structure of a circle or a sphere are called *manifolds*. More precisely, $\mathcal{M}$ is a manifold of dimension n if, for any $\mathbf{x} \in \mathcal{M}$, there exist a neighborhood $N(\mathbf{x}) \subseteq \mathcal{M}$ containing $\mathbf{x}$ and a homeomorphism $\mathbf{h} : N(\mathbf{x}) \to \mathbb{R}^n$ that maps $N(\mathbf{x})$ onto a neighborhood of $\mathbf{h}(\mathbf{x}) \in \mathbb{R}^n$. Since we can define coordinates in $\mathbf{h}\big(N(\mathbf{x})\big) \subseteq \mathbb{R}^n$, $\mathbf{h}$ defines local coordinates on $N(\mathbf{x})$. The pair $\big(\mathbf{h}(N(\mathbf{x})), \mathbf{h}\big)$ is called a *chart*. In order to obtain a global description of $\mathcal{M}$, we cover it with a family of open sets N_i, each associated with a chart $(\mathbf{h}_i(N_i), \mathbf{h}_i)$. The set of all these charts is called an *atlas*. In all the models we shall study, even if the phase space is a manifold, the functions $\mathbf{f}$ will be given in terms of local coordinates. We shall, therefore, never be really involved with charts and atlases. In all definitions and theorems involving "differential manifolds of dimension n," the reader could replace this expression by "open sets of $\mathbb{R}^n$."

If, in Equation (3.1), the vector field $\mathbf{X}$ defined on an open subset U of $\mathbb{R}^n$ is C^k, then given $\mathbf{x}_0 \in U$ and $t_0 \in \mathbb{R}$, for $|t - t_0|$ sufficiently small, there exists a solution of Equation (3.1) through the point $\mathbf{x}_0$ at t_0, denoted $\mathbf{x}(t, t_0, \mathbf{x}_0)$ with $\mathbf{x}(t_0, t_0, \mathbf{x}_0) = \mathbf{x}_0$. This solution is unique and is a C^k function of t, t_0, and $\mathbf{x}_0$. As we already pointed out (page 10, Footnote 16), we shall not give a proof of this fundamental theorem since a differential equation modeling a real system should have a unique evolution for any realizable initial state.[3]

The solution of Equation (3.1) being unique, we have

$$\mathbf{x}(t + s, t_0, \mathbf{x}_0) = \mathbf{x}(s, t + t_0, \mathbf{x}(t, t_0, \mathbf{x}_0)).$$

This property shows that the solutions of (3.1) form a one-parameter group of C^k diffeomorphisms of the phase space. These diffeomorphisms are referred to as a *phase flow* or just a *flow*. The common notation for flows is $\varphi(t, \mathbf{x})$ or $\varphi_t(\mathbf{x})$, and we have

$$\varphi_t \circ \varphi_s = \varphi_{t+s} \tag{3.3}$$

for all t and s in $\mathbb{R}$. Note that φ_0 is the identity and that $(\varphi_t)^{-1}$ exists and is given by φ_{-t}.[4]

In most cases $t_0 = 0$ and, from the definition of φ_t above we have

$$\mathbf{X}(\mathbf{x}) = \frac{d}{dt} \varphi_t(\mathbf{x}) \Big|_{t=0}. \tag{3.4}$$

Given a point $\mathbf{x} \in U \subseteq \mathbb{R}^n$, the *orbit* or *trajectory* of φ passing through $\mathbf{x} \in U$ is the set $\{\varphi_t(\mathbf{x}) \mid t \in \mathbb{R}\}$ oriented in the sense of increasing t. There is only one trajectory of φ passing through any given point $\mathbf{x} \in U$. That is, if two trajectories intersect, they must coincide.

The set of all trajectories of a flow is called its *phase portrait*. A helpful representation of a flow is obtained by plotting typical trajectories (Figure 3.1).

---

[3] The mathematically oriented reader may consult Hale [163] or Hirsch and Smale [171].

[4] The mapping $t \mapsto \varphi_t$ is an *isomorphism* from the group of real numbers $\mathbb{R}$ to the group $\{\varphi_t \mid t \in \mathbf{R}\}$. This group is called an *action of the group $\mathbb{R}$ on the state space U*.

Definition 3. *Let* $\mathbf{X}$ *be a vector field defined on an open set* U *of* $\mathbb{R}^n$; *a point* $\mathbf{x}^* \in U$ *is an* equilibrium point *of the differential equation* $\dot{\mathbf{x}} = \mathbf{X}(\mathbf{x})$ *if* $\mathbf{X}(\mathbf{x}^*) = 0$.

Note that if $\mathbf{x}^*$ is an equilibrium point, then $\varphi_t(\mathbf{x}^*) = \mathbf{x}^*$ for all $t \in \mathbb{R}$. Thus $\mathbf{x}^*$ is also called a *fixed point* of the flow φ. The orbit of a fixed point is the fixed point itself.

A *closed orbit* of a flow φ is a trajectory that is not a fixed point but is such that $\varphi_\tau(\mathbf{x}) = \mathbf{x}$ for some $\mathbf{x}$ on the trajectory and a nonzero τ. The smallest nonzero value of τ is usually denoted by T and is called the *period of the orbit*. That is, we have $\varphi_T(\mathbf{x}) = \mathbf{x}$, but $\varphi_t(\mathbf{x}) \neq \mathbf{x}$ for $0 < t < T$.

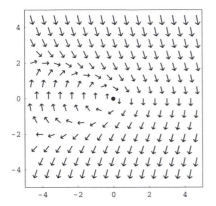 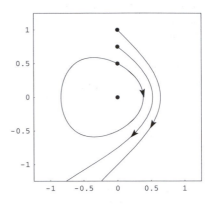

Fig. 3.1. *Vector field (left) and phase portrait (right) of the two-dimensional system* $\dot{x}_1 = x_2$, $\dot{x}_2 = -x_1 - x_2^2$. $(0,0)$ *is a nonhyperbolic equilibrium point.*

Definition 4. *An equilibrium point* $\mathbf{x}^*$ *of the differential equation* $\dot{\mathbf{x}} = \mathbf{X}(\mathbf{x})$ *is said to be* Lyapunov stable *(or L-stable) if, for any given positive* ε, *there exists a positive* δ *(which depends on* ε *only) such that, for all* $\mathbf{x}_0$ *in the neighborhood of* $\mathbf{x}^*$ *defined by* $\|\mathbf{x}_0 - \mathbf{x}^*\| < \delta$, *the solution* $\mathbf{x}(t, 0, \mathbf{x}_0)$ *of the differential equation above satisfying the initial condition* $\mathbf{x}(0, 0, \mathbf{x}_0) = \mathbf{x}_0$ *is such that* $\|\mathbf{x}(t, 0, \mathbf{x}_0) - \mathbf{x}^*\| < \varepsilon$ *for all* $t > 0$. *The equilibrium point is said to be* unstable *if it is not stable.*

An equilibrium point $\mathbf{x}^*$ of a differential equation is stable if the trajectory in the phase space going through a point sufficiently close to the equilibrium point at $t = 0$ remains close to the equilibrium point as t increases. Lyapunov stability does not imply that, as t tends to infinity, the point $\mathbf{x}(t, 0, \mathbf{x}_0)$ tends to $\mathbf{x}^*$. But:

Definition 5. *An equilibrium point* $\mathbf{x}^*$ *of the differential equation* $\dot{\mathbf{x}} = \mathbf{X}(\mathbf{x})$ *is said to be* asymptotically stable *if it is Lyapunov stable and*

$$\lim_{t \to \infty} \mathbf{x}(t, 0, \mathbf{x}_0) = \mathbf{x}^*.$$

Example 8. Kermack-McKendrick epidemic model. To discuss the spread of an infection within a population, Kermack and McKendrick [182] divide the population into three disjoint groups.

1. *Susceptible* individuals are capable of contracting the disease and becoming infective.
2. *Infective* individuals are capable of transmitting the disease to others.
3. *Removed* individuals have had the disease and are dead, have recovered and are permanently immune, or are isolated until recovery and permanent immunity occur.[5]

Infection and removal are governed by the following rules.

1. The rate of change in the susceptible population is proportional to the number of contacts between susceptible and infective individuals, where the number of contacts is taken to be proportional to the product of the numbers S and I of, respectively, susceptible and infective individuals. The model ignores incubation periods.
2. Infective individuals are removed at a rate proportional to their number I.
3. The total number of individuals $S + I + R$, where R is the number of removed individuals, is constant, that is, the model ignores births, deaths by other causes, immigration, emigration, etc.[6]

Taking into account the rules above yields

$$\begin{aligned} \dot{S} &= -iSI, \\ \dot{I} &= iSI - rI, \\ \dot{R} &= rI, \end{aligned} \tag{3.5}$$

where i and r are positive constants representing infection and the removal rates.

From the first equation, it is clear that S is a nonincreasing function, whereas the second equation implies that $I(t)$ increases with t, if $S(t) > r/i$, and decreases otherwise. Therefore, if, at $t = 0$, the initial number of susceptible individuals S_0 is less than r/i, since $S(t) \leq S_0$, the infection dies out, that is, no epidemic occurs. If, on the contrary, S_0 is greater than the critical value r/i, the epidemic occurs; that is, the number of infective individuals first increases and then decreases when $S(t)$ becomes less than r/i.[7]

Remark 1. This *threshold phenomenon* shows that an epidemic occurs if, and only if, the initial number of susceptible individuals S_0 is greater than a threshold value S_{th}. For this model, $S_{\mathrm{th}} = r/i$; *i.e.*, in the case of a deadly disease, an epidemic has less chance to occur if the death rate due to the disease is high!

---

[5] Models of this type are called *SIR models*.
[6] Or we could say that birth, death, and migration are in exact balance.
[7] That is, *by definition*, an epidemic occurs if the time derivative of the number of infective individuals $\dot{I}$ *is positive at* $t = 0$.

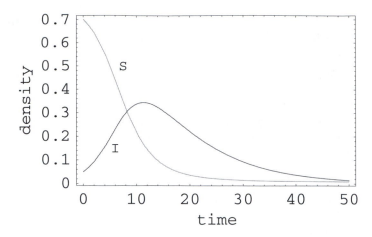

Fig. 3.2. *Kermack-McKendrick epidemic model. Susceptible and infective densities as functions of time for $i = 0.6$, $r = 0.1$, $S(0) = 0.7$, and $I(0) = 0.05$. $S(0)$ is above the threshold value, and, as a consequence of the epidemic, the total population is seriously reduced.*

Note that $\dot{S}$ is nonincreasing and positive, and $\dot{R}$ is positive and less than or equal to the total population N, therefore, $\lim_{t \to \infty} S(t)$ and $\lim_{t \to \infty} R(t)$ exist. Since $I(t) = N - S(t) - R(t)$, $\lim_{t \to \infty} I(t)$ also exists.[8] Moreover, from the first and third equations of (3.5), it follows that

$$\frac{dS}{dR} = -\frac{i}{r} S;$$

that is,

$$S = S_0 \exp\left(-\frac{iR}{r}\right)$$

$$\geq S_0 \exp\left(-\frac{iN}{r}\right) > 0. \tag{3.6}$$

In this model, even in the case of a very serious epidemic, some individuals are not infected. The spread of the disease does not stop for lack of susceptible individuals (see Figure 3.2).

Example 9. Hethcote-York model for the spread of gonorrhea [170]. Gonorrhea is a sexually transmitted disease that presents the following important characteristics that differ from other infections such as measles or mumps:

---

[8] Equations (3.5) show that in the steady state $I(\infty) = 0$. Then, $R(\infty) = N - S(\infty)$, and Relation (3.6) shows that $S(\infty)$ is the only positive root of $x = S_0 \exp(-i(N - x)/r)$. More details can be found in Waltman [341].

- Gonococcal infection does not confer protective immunity, so individuals are susceptible again as soon as they recover from infection.[9]
- The latent period is very short: 2 days, compared to 12 days for measles.
- The seasonal oscillations in gonorrhea incidence[10] are very small (less than 10%), while the incidence of influenza or measles often varies by a factor of 5 to 50.

If we assume that the infection is transmitted only through heterosexual intercourse, we divide the population into two groups, N_f females and N_m males at risk, each group being divided into two subgroups, $N_f S_f$ (resp. $N_m S_m$) susceptible females (resp. males) and $N_f I_f$ (resp. $N_m I_m$) infective females (resp. males). N_f and N_m are assumed to be constant. The dynamics of gonorrhea is then modeled by the four-dimensional system

$$N_f \dot{S}_f = -\lambda_f S_f N_m I_m + N_f I_f/d_f,$$

$$N_f \dot{I}_f = \lambda_f S_f N_m I_m - N_f I_f/d_f,$$

$$N_m \dot{S}_m = -\lambda_m S_m N_f I_f + N_m I_m/d_m,$$

$$N_m \dot{I}_m - \lambda_m S_m N_f I_f - N_m I_m/d_m,$$

where λ_f (resp. λ_m) is the rate of infection of susceptible females (resp. males), and d_f (resp. d_m) is the average duration of infection for females (resp. males). The rates λ_f and λ_m are different since the probability of transmission of gonococcal infection during a single sexual exposure from an infectious woman to a susceptible man is estimated to be about 0.2–0.3, while the probability of transmission from an infectious man to a susceptible woman is about 0.5–0.7. The average durations of infection d_f and d_m are also different since 90% of all men who have a gonococcal infection notice symptoms within a few days after exposure and promptly seek medical treatment, while up to 75% of women with gonorrhea fail to have symptoms and remain untreated for some time.

Since $S_f + I_f = 1$ and $S_m + I_m = 1$, the four-dimensional system reduces to the two-dimensional system

$$\dot{I}_f = \frac{\lambda_f}{r}(1 - I_f)I_m - \frac{I_f}{d_f},$$

$$\dot{I}_m = r\lambda_m(1 - I_m)I_f - \frac{I_m}{d_m},$$

where $r = N_f/N_m$. The system has two equilibrium points $(I_f, I_m) = (0,0)$ and

$$(I_f,\ I_m) = \left(\frac{d_f d_m \lambda_f \lambda_m - 1}{d_m \lambda_m (r + d_f \lambda_f)},\ \frac{r(d_f d_m \lambda_f \lambda_m - 1)}{d_f \lambda_f (1 + r d_m \lambda_m)} \right).$$

---

[9] Models of this type are called *SIS models*.
[10] *Incidence* is the number of new cases in a time interval.

Since acceptable solutions should not be negative, we find that the nontrivial equilibrium point exists if

$$d_f d_m \lambda_f \lambda_m > 1.$$

The coefficient λ_f/r (resp. $\lambda_m r$) represents the average fraction of females (resp. males) being infected by one male (resp. female) per unit of time. Since males (resp. females) are infectious during the period d_m (resp. d_f), then the average fraction of females (resp. males) being infected by one infective male (female) during his (resp. her) period of infection is $\lambda_f d_m/r$ (resp. $\lambda_m d_f r$). The condition $\lambda_m d_f \lambda_f d_m > 1$ therefore expresses that the average fraction of females infected by one male will infect, during their period of infection, more than one male. In this case, gonorrhea remains endemic. If the condition is not satisfied, then gonorrhea dies out. As a consequence, for this model, if the nontrivial equilibrium point exists, it is asymptotically stable; if it does not, then the trivial fixed point is asymptotically stable.

Example 10. Leslie's predator-prey model. After the publication of the Lotka-Volterra model (Equations (2.1) and (2.1)), many other predator-prey models were proposed. In 1948, Leslie [201] suggested the system

$$\dot{H} = r_H H \left(1 - \frac{H}{K}\right) - sHP,$$
$$\dot{P} = r_P P \left(1 - \frac{P}{cH}\right), \tag{3.7}$$

where H and P denote, respectively, the prey and predator populations. The equation for the preys is similar to Lotka-Volterra equation (2.1) except that, in the absence of predators, the growth of the preys is modeled by the logistic equation. The equation for the predators is a logistic equation in which the carrying capacity is proportional to the prey population. This model contains five parameters. There is only one nontrivial equilibrium point (H^*, P^*), which is the unique solution of the linear system

$$r_H \left(1 - \frac{H}{K}\right) = sP, \quad P = cH.$$

If we put

$$h = \frac{H}{H^*}, \quad p = \frac{P}{P^*}, \quad \rho = \sqrt{\frac{r_H}{r_P}}, \quad k = \frac{K}{H^*}, \quad \tau = \sqrt{r_H r_P}\, t, \tag{3.8}$$

Equations (3.7) become

$$\frac{dh}{d\tau} = \rho h \left(1 - \frac{h}{k}\right) - \alpha p h,$$
$$\frac{dp}{d\tau} = \frac{1}{\rho} p \left(1 - \frac{p}{h}\right). \tag{3.9}$$

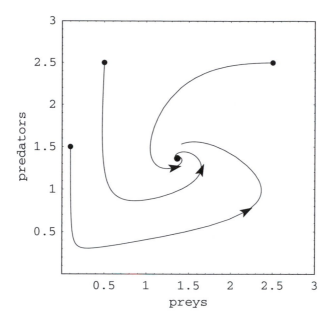

Fig. 3.3. *Phase portrait of the scaled Leslie model for $k = 5$ and $\rho = 1.5$.*

These equations contain two independent parameters, ρ and k, the extra parameter α being given in terms of these by[11]

$$\alpha = \rho \left(1 - \frac{1}{k} \right).$$

The equilibrium points are $(0,0)$, $(k,0)$, and $(1,1)$. A few trajectories converging to the asymptotically stable equilibrium point $(1,1)$ are shown in Figure 3.3.

Definition 6. *Let $\boldsymbol{\varphi}_t : U \to U$ and $\boldsymbol{\psi}_t : W \to W$ be two flows; if there exists a diffeomorphism $\mathbf{h} : U \to W$ such that, for all $t \in \mathbb{R}$,*

$$\mathbf{h} \circ \boldsymbol{\varphi}_t = \boldsymbol{\psi}_t \circ \mathbf{h}, \tag{3.10}$$

the flows $\boldsymbol{\varphi}_t$ and $\boldsymbol{\psi}_t$ are said to be conjugate.

In other words, the diagram

$$
\begin{array}{ccc}
U & \xrightarrow{\;\varphi_t\;} & U \\
\mathbf{h} \downarrow & & \downarrow \mathbf{h} \\
W & \xrightarrow{\;\psi_t\;} & W
\end{array}
$$

---

[11] This result follows from the first of the two equations (3.9) when we put
$$\frac{dh}{d\tau} = 0, \quad h = 1, \quad p = 1.$$

is commutative. The purpose of Definition 6 is to provide with a way of characterizing when two flows have qualitatively the same dynamics. Equation (3.10) can also be written

$$\psi_t = \mathbf{h} \circ \varphi_t \circ \mathbf{h}^{-1}.$$

That is, $\mathbf{h}$ takes the orbits of the flow φ_t into the orbits of the flow ψ_t. In other words, the flow φ_t becomes the flow ψ_t under the change of coordinates $\mathbf{h}$.

It is readily verified that conjugacy is an equivalence relation, *i.e.*,

$$(\varphi_t \sim \varphi_t), \ (\varphi_t \sim \psi_t) \Rightarrow (\psi_t \sim \varphi_t), \ \text{and} \ (\varphi_t \sim \psi_t, \psi_t \sim \chi_t) \Rightarrow (\varphi_t \sim \chi_t).$$

If $\mathbf{h}$ is a C^1 function such that $\mathbf{h} \circ \varphi_t = \psi_t \circ \mathbf{h}$, then differentiating both sides with respect to t and evaluating at $t = 0$ yields

$$D\mathbf{h}(\varphi_t(\mathbf{x})) \frac{d}{dt} \varphi_t(\mathbf{x}) \Big|_{t=0} = \frac{d}{dt} \psi_t(\mathbf{h}(\mathbf{x})) \Big|_{t=0},$$

and taking into account Relation (3.4), we obtain

$$D\mathbf{h}(\mathbf{x})\mathbf{X}(\mathbf{x}) = \mathbf{Y}(\mathbf{h}(\mathbf{x})). \tag{3.11}$$

That is, if $\mathbf{h}$ is a differentiable flow conjugacy of the flows φ_t and ψ_t, the derivative $D\mathbf{h}(\mathbf{x})$ transforms $\mathbf{X}(\mathbf{x})$ into $\mathbf{Y}(\mathbf{h}(\mathbf{x}))$.

Remark 2. Definition 6 requires conjugacy to preserve the parameter t. If we are required to preserve only the *orientation* along the orbits of φ_t and ψ_t, we obtain more satisfactory equivalence classes for flows. In that case, Relation (3.10) would have to be replaced by

$$\mathbf{h} \circ \varphi_t = \psi_{\tau(t,\mathbf{x})} \circ \mathbf{h}, \tag{3.12}$$

where, for all $\mathbf{x}$, the function $t \mapsto \tau(t,\mathbf{x})$ is strictly increasing; *i.e.*, its derivative with respect to t has to be positive for all $\mathbf{x}$. If there exist a homeomorphism $\mathbf{h}$ and a differentiable function τ such that Relation (3.12) is satisfied, it is said that the flows φ_t and ψ_t are *topologically equivalent*. Here is a simple example. Consider the two one-dimensional flows:

$$\varphi_1(t,x) = e^{-\lambda_1 t}x \quad \text{and} \quad \varphi_2(t,x) = e^{-\lambda_2 t}x,$$

where λ_1 and λ_2 are different positive numbers. Let h be such that

$$h(e^{-\lambda_1 t}x) = e^{-\lambda_2 t}h(x). \tag{3.13}$$

If h is a *diffeomorphism*, then taking the derivative with respect to x of each side of this relation yields

$$e^{-\lambda_1 t}h'(e^{-\lambda_1 t}x) = e^{-\lambda_2 t}h'(x).$$

A diffeomorphism being invertible by definition, $h'(0) \neq 0$, and we obtain $e^{-\lambda_1 t} = e^{-\lambda_2 t}$, which contradicts the assumption $\lambda_1 \neq \lambda_2$. If, on the contrary, we assume that h is not differentiable everywhere (*i.e.*, h is not a diffeomorphism but only a homeomorphism), then the homeomorphism[12]

---

[12] h is continuous and invertible.

$$h \mapsto \begin{cases} -|x|^{\lambda_2/\lambda_1}, & \text{if } x < 0 \\ 0, & \text{if } x = 0 \\ x^{\lambda_2/\lambda_1}, & \text{if } x > 0 \end{cases}$$

satisfies (3.13) and shows that φ_1 and φ_2 are topologically equivalent.

3.2 Linearization and stability

In order to analyze a model described by a nonlinear differential equation of the form (3.1), we first have to determine its equilibrium points and study the behavior of the system near these points. Under certain conditions, this behavior is qualitatively the same as the behavior of a linear system. We therefore begin this section with a brief study of linear differential equations.

3.2.1 Linear systems

The solution of the one-dimensional linear differential equation

$$\dot{x} = ax,$$

which satisfies the initial condition $x(0) = x_0$, is $x(t) = x_0 e^{at}$.

Question. Let $\mathbf{A}$ be a time-independent linear operator defined on $\mathbb{R}^n$; is it possible to generalize the result above and say that, if $\mathbf{x} \in \mathbb{R}^n$, the solution to the linear differential equation

$$\dot{\mathbf{x}} = \mathbf{A}\mathbf{x}, \tag{3.14}$$

which satisfies the initial condition $\mathbf{x}(0) = \mathbf{x}_0$, is $\mathbf{x}(t) = e^{\mathbf{A}t}\mathbf{x}_0$?

The answer is yes, provided we define and show how to express the linear operator $e^{\mathbf{A}t}$.

Definition 7. *Let $\mathbf{A}$ be a linear operator defined on $\mathbb{R}^n$; the exponential of $\mathbf{A}$ is the linear operator defined on $\mathbf{R}^n$ by*[13]

$$e^{\mathbf{A}} = \sum_{k=0}^{\infty} \frac{\mathbf{A}^k}{k!}. \tag{3.15}$$

---

[13] For this definition to make sense, it is necessary to show that the series converges and, therefore, to first define a metric on the space of linear operators on $\mathbb{R}^n$ in order to be able to introduce the notion of limit of a sequence of linear operators. If $\|\mathbf{x}\|$ is the norm of $\mathbf{x} \in \mathbb{R}^n$, the norm of a linear operator $\mathbf{A}$ may be defined as $\|\mathbf{A}\| = \sup_{\|\mathbf{x}\| \le 1} \|\mathbf{A}\mathbf{x}\|$. The distance d between two linear operators $\mathbf{A}$ and $\mathbf{B}$ on $\mathbb{R}^n$ is then defined as $d(\mathbf{A}, \mathbf{B}) = \|\mathbf{A} - \mathbf{B}\|$.

Depending on whether the real linear operator $\mathbf{A}$ has real or complex distinct or multiple eigenvalues, the real linear operator $e^{\mathbf{A}t}$ may take different forms, as described below.[14]

1. *All the eigenvalues of $\mathbf{A}$ are distinct.*

If $\mathbf{A}$ has distinct real eigenvalues λ_i and corresponding eigenvectors $\mathbf{u}_i$, where $i = 1, 2, \ldots, k$ and distinct complex eigenvalues $\lambda_j = \alpha_j + i\beta_j$ and $\overline{\lambda}_j = \alpha_j - i\beta_j$ and corresponding complex eigenvectors $\mathbf{w}_j = \mathbf{u}_j + i\mathbf{v}_j$ and $\overline{\mathbf{w}}_j = \mathbf{u}_j - i\mathbf{v}_j$, where $j = k+1, k+2, \ldots, \ell$, then the matrix[15]

$$\mathbf{M} = [\mathbf{u}_1, \ldots, \mathbf{u}_k, \mathbf{u}_{k+1}, \mathbf{v}_{k+1}, \ldots, \mathbf{u}_\ell, \mathbf{v}_\ell]$$

is invertible, and

$$\mathbf{M}^{-1}\mathbf{A}\mathbf{M} = \operatorname{diag}[\lambda_1, \ldots, \lambda_k, B_{k+1}, \ldots, B_\ell].$$

The right-hand side denotes the matrix

$$\begin{bmatrix} \lambda_1 & 0 & \cdots & \cdots & \cdots & \cdots & \cdots & 0 \\ 0 & \lambda_2 & 0 & \cdots & \cdots & \cdots & \cdots & 0 \\ \cdots & \cdots & \cdots & \cdots & \cdots & \cdots & \cdots & \cdots \\ \cdots & \cdots & \cdots & \cdots & \cdots & \cdots & \cdots & \cdots \\ 0 & 0 & \cdots & 0 & \lambda_k & 0 & \cdots & 0 \\ 0 & 0 & \cdots & \cdots & 0 & B_{k+1} & \cdots & 0 \\ \cdots & \cdots & \cdots & \cdots & \cdots & \cdots & \cdots & \cdots \\ 0 & 0 & \cdots & \cdots & \cdots & \cdots & \cdots & B_\ell \end{bmatrix},$$

where, for $j = k+1, k+2, \ldots, \ell$, the B_j are 2×2 blocks given by

$$B_j = \begin{bmatrix} \alpha_j & -\beta_j \\ \beta_j & \alpha_j \end{bmatrix}.$$

In this case, we have

$$e^{\mathbf{A}t} = \mathbf{M}\operatorname{diag}[e^{\lambda_1 t}, \ldots, e^{\lambda_k t}, E_{k+1}, \ldots, E_\ell]\mathbf{M}^{-1}, \qquad (3.16)$$

the 2×2 block E_j being given by

$$E_j = e^{\alpha_j t}\begin{bmatrix} \cos\beta_j t & -\sin\beta_j t \\ \sin\beta_j t & \cos\beta_j t \end{bmatrix}.$$

----

[14] For a simple and rigorous treatment of linear differential systems, see Hirsch and Smale [171].

[15] If $\mathbf{a}_1, \mathbf{a}_2, \ldots, \mathbf{a}_n$ are n independent vectors of $\mathbb{R}^n$, the matrix $[\mathbf{a}_1, \mathbf{a}_2, \ldots, \mathbf{a}_n]$ denotes the matrix

$$\begin{bmatrix} a_{11} & a_{21} & \cdots & a_{n1} \\ a_{12} & a_{22} & \cdots & a_{n2} \\ \cdots & \cdots & \cdots & \cdots \\ a_{1n} & a_{2n} & \cdots & a_{nn} \end{bmatrix}.$$

Note that the dimension of the vector space on which $\mathbf{A}$ and $e^{\mathbf{A}t}$ are defined is $n = 2\ell - k$.

2. $\mathbf{A}$ *has real multiple eigenvalues.*

If $\mathbf{A}$ has real eigenvalues $\lambda_1, \lambda_2, \ldots, \lambda_n$ repeated according to their multiplicity and if $(\mathbf{u}_1, \mathbf{u}_2, \ldots, \mathbf{u}_n)$ is a basis of *generalized eigenvectors*,[16] then the matrix

$$\mathbf{M} = [\mathbf{u}_1, \mathbf{u}_2, \ldots, \mathbf{u}_n]$$

is invertible, and the operator $\mathbf{A}$ can be written as the sum of two matrices $\mathbf{S} + \mathbf{N}$, where

$$\mathbf{SN} = \mathbf{NS}, \quad \mathbf{M}^{-1}\mathbf{SM} = \operatorname{diag}[\lambda_1, \lambda_2, \ldots, \lambda_n],$$

and $\mathbf{N}$ is *nilpotent of order* $k \leq n$.[17]

In this case, we have[18]

$$e^{\mathbf{A}t} = \mathbf{M} \operatorname{diag}[e^{\lambda_1 t}, e^{\lambda_2 t}, \ldots, e^{\lambda_n t}] \, \mathbf{M}^{-1}$$

$$\times \left(\mathbf{I} + \mathbf{N}t + \cdots + \mathbf{N}^{k-1} \frac{t^{k-1}}{(k-1)!} \right). \tag{3.17}$$

3. $\mathbf{A}$ *has complex multiple eigenvalues.*

If a real linear operator $\mathbf{A}$, represented by a $2n \times 2n$ matrix, has complex eigenvalues $\lambda_j = \alpha_j + i\beta_j$ and $\overline{\lambda}_j = \alpha_j - i\beta_j$, where $j = 1, 2, \ldots, n$, there exists a basis of generalized eigenvectors $\mathbf{w}_j = \mathbf{u}_j + i\mathbf{v}_j$ and $\overline{\mathbf{w}}_j = \mathbf{u}_j - i\mathbf{v}_j$ for $\mathbb{C}^n$, $(\mathbf{u}_1, \mathbf{v}_1, \ldots, \mathbf{u}_n, \mathbf{v}_n)$ is a basis for $\mathbb{R}^{2n}$, the $2n \times 2n$ matrix $\mathbf{M} = [\mathbf{u}_1, \mathbf{v}_1, \ldots, \mathbf{u}_n, \mathbf{v}_n]$ is invertible, and the operator $\mathbf{A}$ can be written as the sum of two matrices $\mathbf{S} + \mathbf{N}$, where

$$\mathbf{SN} = \mathbf{NS}, \quad \mathbf{M}^{-1}\mathbf{SM} = \operatorname{diag}\left[\begin{bmatrix} \alpha_1 & -\beta_1 \\ \beta_1 & \alpha_1 \end{bmatrix}, \begin{bmatrix} \alpha_2 & -\beta_2 \\ \beta_2 & \alpha_2 \end{bmatrix}, \cdots, \begin{bmatrix} \alpha_n & -\beta_n \\ \beta_n & \alpha_n \end{bmatrix} \right],$$

and $\mathbf{N}$ is nilpotent of order $k \leq 2n$.

In this case, we have

$$e^{\mathbf{A}t} = \mathbf{M} \operatorname{diag}\left[e^{\alpha_1 t} \begin{bmatrix} \cos\beta_1 & -\sin\beta_1 \\ \sin\beta_1 & \cos\beta_1 \end{bmatrix}, \cdots, e^{\alpha_n t} \begin{bmatrix} \cos\beta_n & -\sin\beta_n \\ \sin\beta_n & \cos\beta_n \end{bmatrix} \right] \mathbf{M}^{-1}$$

$$\times \left(\mathbf{I} + \mathbf{N}t + \cdots + \mathbf{N}^{k-1} \frac{t^{k-1}}{(k-1)!} \right). \tag{3.18}$$

---

[16] If the real eigenvalue λ has multiplicity $m < n$, then for $k = 1, 2, \ldots, m$, any nonzero solution of $(\mathbf{A} - \lambda\mathbf{I})^k\mathbf{u} = 0$ is called a *generalized eigenvector*.

[17] A linear operator $\mathbf{N}$ is *nilpotent of order* k if $\mathbf{N}^{k-1} \neq \mathbf{0}$ and $\mathbf{N}^k = \mathbf{0}$.

[18] If the linear operators $\mathbf{A}$ and $\mathbf{B}$ commute, then $e^{\mathbf{A}+\mathbf{B}} = e^{\mathbf{A}}e^{\mathbf{B}}$. The series defining the exponential of a nilpotent linear operator of order k is a polynomial of degree $k - 1$.

4. **A** *has both real and complex multiple eigenvalues.*

In this case, we use a combination of the results above to find the expression of linear operator $e^{\mathbf{A}t}$.

Example 11. Classification of two-dimensional linear flows. Let $\operatorname{tr}\mathbf{A}$ and $\det\mathbf{A}$ denote, respectively, the *trace* and the *determinant* of the 2×2 time-independent real matrix **A**. The eigenvalues of **A** are the roots of its *characteristic polynomial*:

$$\det(\mathbf{A} - \lambda\mathbf{I}) = \lambda^2 - \operatorname{tr}\mathbf{A}\,\lambda + \det\mathbf{A}. \tag{3.19}$$

We may distinguish the following cases:

1. If $\det\mathbf{A} < 0$, the eigenvalues of **A** are real and have opposite sign. The origin is said to be a *saddle*.
2. If $\det\mathbf{A} > 0$ and $(\operatorname{tr}\mathbf{A})^2 \geq 4\det\mathbf{A}$, the eigenvalues of **A** are real and have the same sign. The origin is said to be an *attractive node* if $\operatorname{tr}\mathbf{A} < 0$ and a *repulsive node* if $\operatorname{tr}\mathbf{A} > 0$.
3. If $\operatorname{tr}\mathbf{A} \neq 0$ and $(\operatorname{tr}\mathbf{A})^2 < 4\det\mathbf{A}$, the eigenvalues of **A** are complex. The origin is said to be an *attractive focus* if $\operatorname{tr}\mathbf{A} < 0$ and a *repulsive focus* if $\operatorname{tr}\mathbf{A} > 0$.
4. If $\operatorname{tr}\mathbf{A} = 0$ and $\det\mathbf{A} < 0$, the eigenvalues of **A** are complex with a zero real part. The origin is said to be a *center*.

The phase portraits corresponding to the different cases described above are represented in Figure 3.4. Attractive nodes and attractive foci are asymptotically stable equilibrium points for linear two-dimensional systems. Centers are Lyapunov stable but not asymptotically stable equilibrium points.

To the classification above, we have to add two *degenerate* cases illustrated in Figure 3.5.

1. If $\operatorname{tr}\mathbf{A} \neq 0$ and $\det\mathbf{A} = 0$, one eigenvalue is equal to zero and the other one equal to $\operatorname{tr}\mathbf{A}$. The origin is said to be a *saddle node*.[19] In this case, there exists a basis in which $e^{\mathbf{A}t} = \operatorname{diag}[1, e^{\lambda t}]$, showing that the equations of the orbits are $x = a, y > 0$ and $x = a, y < 0$, with $a \in \mathbf{R}$.
2. If $\operatorname{tr}\mathbf{A} = 0$ and $\det\mathbf{A} = 0$ with $\mathbf{A} \neq \mathbf{0}$, there exists a basis in which **A** is of the form $\begin{bmatrix} 0 & a \\ 0 & 0 \end{bmatrix}$. All the orbits are the straight lines $y = y_0$ traveled at constant velocity ay_0.

We mentioned (Remark 2) that, if we require only the orientation along the orbits to be preserved, we obtain more satisfactory flow equivalence classes. In the case of two-dimensional linear systems it can be shown [164][20] that if the eigenvalues of the two matrices **A** and **B** have nonzero real parts, then the two linear systems $\dot{\mathbf{x}} = \mathbf{A}\mathbf{x}$ and $\dot{\mathbf{x}} = \mathbf{B}\mathbf{x}$ are topologically equivalent if,

---

[19] Also called a *fold* or a *tangent bifurcation point*. See Section 3.5.
[20] See pp. 238–246.

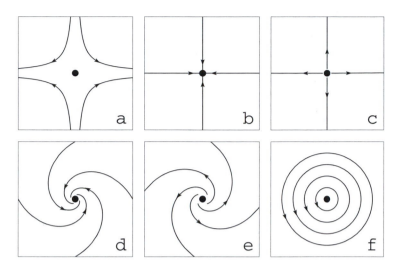

Fig. 3.4. *Two-dimensional linear flows. Phase portraits of the nondegenerate cases. (a) saddle, (b) attractive node, (c) repulsive node, (d) attractive focus, (e) repulsive focus, (f) center.*

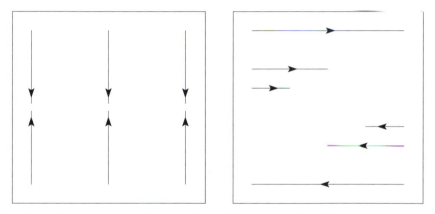

Fig. 3.5. *Two-dimensional linear flows. Phase portraits of the degenerate cases. Left:* $\operatorname{tr} \mathbf{A} \neq 0$ *and* $\det \mathbf{A} = 0$. *Right:* $\operatorname{tr} \mathbf{A} = 0$ *and* $\det \mathbf{A} = 0$ *with* $\mathbf{A} \neq \mathbf{0}$.

and only if, $\mathbf{A}$ and $\mathbf{B}$ have the same number of eigenvalues with negative (and hence positive) real parts. Consequently, up to topological equivalence, there are three distinct equivalence classes of hyperbolic[21] two-dimensional linear systems; that is, cases (a), (b), and (c) in Figure 3.4.

If the origin is a hyperbolic equilibrium point of a linear system $\dot{\mathbf{x}} = \mathbf{A}\mathbf{x}$, the subspace spanned by the (generalized) eigenvectors corresponding to the

---

[21] That is, whose eigenvalues have nonzero real parts. See Definition 8.

eigenvalues with negative (resp. positive) real parts is called the *stable* (resp. *unstable*) *manifold* of the hyperbolic equilibrium point.

3.2.2 Nonlinear systems

Definition 8. *Let* $\mathbf{x}^* \in U \subseteq \mathbb{R}^n$ *be an equilibrium point of the differential equation* $\dot{\mathbf{x}} = \mathbf{X}(\mathbf{x})$; $\mathbf{x}^*$ *is said to be* hyperbolic *if all the eigenvalues of the Jacobian matrix* $D\mathbf{X}(\mathbf{x}^*)$ *have nonzero real part. The linear function*

$$\mathbf{x} \mapsto D\mathbf{X}(\mathbf{x}^*)\mathbf{x} \tag{3.20}$$

is called the linear part *of* $\mathbf{X}$ *at* $\mathbf{x}^*$.

Let $\mathbf{x}^*$ be an equilibrium point of Equation (3.1). In order to determine the stability of $\mathbf{x}^*$, we have to understand the nature of the solutions near $\mathbf{x}^*$. Let

$$\mathbf{x} = \mathbf{x}^* + \mathbf{y}.$$

substituting in (3.1) and Taylor expanding about $\mathbf{x}^*$ yields

$$\dot{\mathbf{y}} = D\mathbf{X}(\mathbf{x}^*)\mathbf{y} + O(\|\mathbf{y}\|^2). \tag{3.21}$$

Since the stability of $\mathbf{x}^*$ is determined by the behavior of orbits through points arbitrarily close to $\mathbf{x}^*$, we might think that the stability could be determined by studying the stability of the equilibrium point $\mathbf{y} = \mathbf{0}$ of the linear system

$$\dot{\mathbf{y}} = D\mathbf{X}(\mathbf{x}^*)\mathbf{y}. \tag{3.22}$$

The solution of (3.22) through the point $\mathbf{y}_0 \in \mathbb{R}^n$ at $t = 0$ is

$$\mathbf{y}(t) = \exp\left(D\mathbf{X}(\mathbf{x}^*)t\right)\mathbf{y}_0. \tag{3.23}$$

Thus, the equilibrium point $\mathbf{y} = \mathbf{0}$ is asymptotically stable if all the eigenvalues of $D\mathbf{X}(\mathbf{x}^*)$ have negative real parts.[22]

Question: If all the eigenvalues of $D\mathbf{X}(\mathbf{x}^*)$ have negative real parts, is the equilibrium point $\mathbf{x}^*$ of Equation (3.1) asymptotically stable? The answer is yes. More precisely:

Theorem 1. *If* $\mathbf{x}^*$ *is a hyperbolic equilibrium point of the differential equation* $\dot{\mathbf{x}} = \mathbf{X}(\mathbf{x})$, *the flow generated by the vector field* $\mathbf{X}$ *in the neighborhood of* $\mathbf{x}^*$ *is* C^0 *conjugate to the flow generated by* $D\mathbf{X}(\mathbf{x}^*)(\mathbf{x} - \mathbf{x}^*)$.

This result is known as the *Hartman-Grobman theorem*.[23] Hence, if $D\mathbf{X}(\mathbf{x}^*)$ has no purely imaginary eigenvalues, the stability of the equilibrium point $\mathbf{x}^*$ of the nonlinear differential equation $\dot{\mathbf{x}} = \mathbf{X}(\mathbf{x})$ can be determined from the study of the linear differential equation $\dot{\mathbf{y}} = D\mathbf{X}(\mathbf{x}^*)\mathbf{y}$, where $\mathbf{y} = \mathbf{x} - \mathbf{x}^*$. If $D\mathbf{X}(\mathbf{x}^*)$ has purely imaginary eigenvalues, this is not the case. The following examples illustrate the various possibilities.

---

[22] See Subsection 3.2.1.

[23] For a proof, consult Palis and de Melo [276].

Example 12. The damped pendulum. The equation for the damped pendulum is

$$\ddot{\theta} + 2a\dot{\theta} + \omega^2 \sin\theta = 0, \qquad (3.24)$$

where θ is the displacement angle from the stable equilibrium position, $a > 0$ is the friction coefficient, and ω^2 is equal to the acceleration of gravity g divided by the pendulum length ℓ. If we put

$$x_1 = \theta, \quad x_2 = \dot{\theta},$$

Equation (3.24) may be written

$$\frac{dx_1}{dt} = x_2,$$
$$\frac{dx_2}{dt} = -\omega^2 \sin x_1 - 2ax_2.$$

The equilibrium points of this system are $(n\pi, 0)$, where n is any integer $(n \in \mathbb{Z})$. The Jacobian of the vector field $\mathbf{X} = (x_2, -\omega^2 \sin x_1 - 2ax_2)$ is

$$D\mathbf{X}(x_1, x_2) = \begin{bmatrix} 0 & 1 \\ -\omega^2 \cos x_1 & -2a \end{bmatrix}.$$

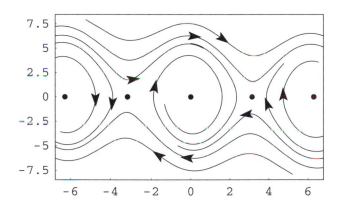

Fig. 3.6. *Phase portrait of a damped pendulum $\dot{x}_1 = x_2$, $\dot{x}_2 = -\omega^2 \sin x_1 - 2ax_2$ for $\omega = 2.8$ and $a = 0.1$. The equilibrium points 0 and $\pm 2\pi$ are asymptotically stable, while the equilibrium points $\pm\pi$ are unstable (saddle points).*

If n is even, the eigenvalues of the Jacobian matrix at $(n\pi, 0)$ are $\lambda_{1,2} = -a \pm \sqrt{a^2 - \omega^2}$. If $\omega \le a$, both eigenvalues are real and negative; if $\omega > a$, the eigenvalues are complex conjugate and their real part is negative (it is equal to $-a$). Therefore, if n is an even integer, the equilibrium point $(n\pi, 0)$ is asymptotically stable (see Figure 3.6).

If n is odd, the eigenvalues of the Jacobian matrix are $\lambda_{1,2} = -a \pm \sqrt{a^2 + \omega^2}$, that is, real and of opposite signs. The equilibrium point $(n\pi, 0)$ is therefore unstable (see Figure 3.6).

Remark 3. A hyperbolic equilibrium point $\mathbf{x}^*$ of the differential equation $\dot{\mathbf{x}} = \mathbf{X}(\mathbf{x})$ is called a *sink* if all the eigenvalues of $D\mathbf{X}(\mathbf{x}^*)$ have negative real parts; it is called a *source* if all the eigenvalues of $D\mathbf{X}(\mathbf{x}^*)$ have positive real parts; and it is called a *saddle* if it is hyperbolic and $D\mathbf{X}(\mathbf{x}^*)$ has at least one eigenvalue with a negative real part and at least one eigenvalue with a positive real part.

Example 13. A perturbed harmonic oscillator. Consider the system

$$\dot{x}_1 = x_2 + \lambda x_1(x_1^2 + x_2^2),$$
$$\dot{x}_2 = -x_1 + \lambda x_2(x_1^2 + x_2^2),$$

where λ is a parameter. These equations describe the dynamics of a perturbed harmonic oscillator. For all values of λ, the origin is an equilibrium point. The Jacobian at the origin of the vector field $\mathbf{X} = (x_2 + \lambda x_1(x_1^2 + x_2^2), -x_1 + \lambda x_2(x_1^2 + x_2^2))$ is

$$D\mathbf{X}(0,0) = \begin{bmatrix} 0 & 1 \\ -1 & -0 \end{bmatrix}.$$

Its eigenvalues are $\pm i$. The origin is a nonhyperbolic point, and to study its stability we have to analyze the behavior of the orbits close to the origin. If $(x_1(t), x_2(t))$ are the coordinates of a phase point at time t, its distance from the origin will increase or decrease according to the sign of the time derivative

$$\frac{d}{dt}\left(x_1^2(t) + x_2^2(t)\right) = 2x_1(t)\dot{x}_1(t) + 2x_2(t)\dot{x}_2(t) = 2\lambda(x_1^2 + x_2^2)^2.$$

Thus, as t tends to infinity, $\|\mathbf{x}(t)\|^2$ tends to zero if $\lambda < 0$, and the origin is an asymptotically stable equilibrium; if $\lambda > 0$ with $\mathbf{x}_0 \neq \mathbf{0}$, $\|\mathbf{x}(t)\|^2$ tends to infinity, showing that, in this case, the origin is an unstable equilibrium.

We could have reached the same conclusion using polar coordinates defined by

$$x_1 = r\cos\theta, \quad x_2 = r\sin\theta.$$

In terms of the coordinates (r, θ), the system becomes

$$\dot{r} = \lambda r^3, \quad \dot{\theta} = -1.$$

Since $\dot{\theta} = -1$, the orbits spiral monotonically clockwise around the origin, and the stability of the origin is the same as the equilibrium point $r = 0$ of the one-dimensional system $\dot{r} = \lambda r^3$. That is, $r = 0$ is asymptotically stable if $\lambda < 0$ and unstable if $\lambda > 0$ with $r_0 \neq 0$.

Near a hyperbolic equilibrium point $\mathbf{x}^*$ of a nonlinear system $\dot{\mathbf{x}} = \mathbf{X}(\mathbf{x})$ we can define *(local) stable* and *unstable manifolds*. They are tangent to the respective stable and unstable manifolds of the linear system $\dot{\mathbf{y}} = D\mathbf{X}(\mathbf{x}^*)\mathbf{y}$.[24]

If the equilibrium point $\mathbf{x}^*$ is nonhyperbolic, we can define in a similar manner a *center manifold* tangent to the center subspace spanned by the n_z (generalized) eigenvectors corresponding to the n_z eigenvalues of the linear operator $D\mathbf{X}(\mathbf{x}^*)$ with zero real parts. Note that while the stable and unstable manifolds are unique, the center manifold is not. The essential interest of the center manifold is that it contains all the complicated dynamics in the neighborhood of a nonhyperbolic point. The following classical example illustrates the characteristic features of the center manifold.[25]

Example 14. Consider the system

$$\dot{x}_1 = x_1^2, \quad \dot{x}_2 = -x_2.$$

The origin is a nonhyperbolic fixed point. The solutions are

$$x_1(t) = \frac{x_1(0)}{1 - x_1(0)t} \quad \text{and} \quad x_2(t) = x_2(0)e^{-t}.$$

Eliminating t, we find that the equations of the trajectories in the (x_1, x_2)-space are

$$x_2(0) \exp\left(\frac{1}{x_1} - \frac{1}{x_1(0)}\right) - x_2 = 0.$$

For $x_1 < 0$, all the trajectories approach the origin with all the derivatives of x_2 with respect to x_1 equal to zero at the origin. For $x_1 \geq 0$, the only trajectory that goes through the origin is $x_2 = 0$. The center manifold, tangent to the eigenvector directed along the x_1-axis, which corresponds to the zero eigenvalue of the linear part of the vector field at the origin is therefore not unique. Except for $x_2 = 0$, all the center manifolds are not C^∞. This example shows that the invariant center manifold, unlike the invariant stable and unstable manifolds, is not necessarily unique and as smooth as the vector field.

---

[24] More precisely: *Let* $\mathbf{x}^*$ *be a hyperbolic equilibrium point of the nonlinear system* $\dot{\mathbf{x}} = \mathbf{X}(\mathbf{x})$, *where* $\mathbf{X}$ *is a* C^k *($k \geq 1$) vector field on* $\mathbb{R}^n$. *If the linear operator* $D\mathbf{X}(\mathbf{x}^*)$ *has* n_n *eigenvalues with negative real parts and* $n_p = n - n_n$ *eigenvalues with positive real parts, there exists an* n_n-*dimensional differentiable manifold* W_{loc}^s *tangent to the stable subspace of the linear system* $\dot{\mathbf{y}} = D\mathbf{X}(\mathbf{x}^*)\mathbf{y}$ *at* $\mathbf{x}^*$ *such that, for all* $\mathbf{x}_0 \in W_{\text{loc}}^s$, *and all* $t > 0$, $\lim_{t\to\infty} \varphi_t(\mathbf{x}_0) = \mathbf{x}^*$, *and there exists an* n_p-*dimensional differentiable manifold* W_{loc}^u *tangent to the unstable subspace of the linear system* $\dot{\mathbf{y}} = D\mathbf{X}(\mathbf{x}^*)\mathbf{y}$ *at* $\mathbf{x}^*$ *such that, for all* $\mathbf{x}_0 \in W_{\text{loc}}^u$ *and all* $t < 0$, $\lim_{t\to\infty} \varphi_t(\mathbf{x}_0) = \mathbf{x}^*$, *where* φ_t *is the flow generated by the vector field* $\mathbf{X}$. *This rather intuitive result is known as the* stable manifold theorem. For more details on invariant manifolds consult Hirsch, Pugh, and Shub [172].

[25] On center manifold theory, see Carr [79].

In order to determine if an equilibrium point $\mathbf{x}^*$ of the differential equation $\dot{\mathbf{x}} = \mathbf{X}(\mathbf{x})$ is stable we have to study the behavior of the function $\mathbf{x} \mapsto \|\mathbf{x} - \mathbf{x}^*\|$ in a neighborhood $N(\mathbf{x}^*)$ of $\mathbf{x}^*$. The Lyapunov method introduces more general functions. The essential idea on which the method rests is to determine how an adequately chosen real function varies along the trajectories of the flow φ_t generated by the vector field $\mathbf{X}$.

Definition 9. *Let $\mathbf{x}^*$ be an equilibrium point of the differential equation $\dot{\mathbf{x}} = \mathbf{X}(\mathbf{x})$ on $U \subseteq \mathbb{R}^n$. A C^1 function $V : U \to \mathbb{R}$ is called a* strong Lyapunov *function for the flow φ_t on an open neighborhood $N(\mathbf{x}^*)$ of $\mathbf{x}^*$ provided $V(\mathbf{x}) > V(\mathbf{x}^*)$ and*

$$\dot{V}(\mathbf{x}) = \frac{d}{dt} V(\varphi_t(\mathbf{x}))\bigg|_{t=0} < 0$$

for all $\mathbf{x} \in N(\mathbf{x}^*) \setminus \{\mathbf{x}^*\}$. If the condition $\dot{V}(\mathbf{x}) < 0$ is replaced by $\dot{V}(\mathbf{x}) \leq 0$, V is called a weak Lyapunov *function.*

It is not difficult to prove that [171], if $\mathbf{x}^*$ is an equilibrium point of the differential equation $\dot{\mathbf{x}} = \mathbf{X}(\mathbf{x})$ on $U \subseteq \mathbb{R}^n$ and there exists a weak Lyapunov function V defined on a neighborhood of $\mathbf{x}^*$, then $\mathbf{x}^*$ is Lyapunov stable. If there exists a strong Lyapunov function V defined on a neighborhood of $\mathbf{x}^*$, then $\mathbf{x}^*$ is asymptotically stable. This result is known as the *Lyapunov theorem.*[26]

The interesting feature of the Lyapunov method is that it is possible to calculate $\dot{V}(\mathbf{x})$ without actually knowing the solutions to the differential equation. To emphasize this particular feature, it is often called the *direct* method of Lyapunov. The inconvenience of the method is that finding a Lyapunov function is a matter of trial and error. In some cases, there are, however, some natural functions to try. As illustrated by the following example, in the case of a mechanical system, energy is often a Lyapunov function.

Example 15. Simple and damped pendulums. We have seen that the Equation (1.10) of the simple pendulum may be written

$$\dot{x}_1 = x_2,$$
$$\dot{x}_2 = -\frac{g}{\ell} \sin x_1.$$

The function

---

[26] Note that $\dot{V}(\mathbf{x})$ is equal to the dot product $\nabla V(\mathbf{x}) \cdot \mathbf{X}(\mathbf{x})$ of the gradient of V with the vector field $\mathbf{X}$ at $\mathbf{x}$. For two-dimensional systems, if $\dot{V}(\mathbf{x}) < 0$, the angle between $\nabla V(\mathbf{x})$ and $\mathbf{X}(\mathbf{x})$ is obtuse. Since the gradient is the outward normal vector to the curve $V(\mathbf{x}) = $ constant at $\mathbf{x}$, this implies that the orbit is crossing the curve from the outside to the inside. Similarly, it could be shown that, if $\dot{V}(\mathbf{x}) = 0$, the orbit is tangent to the curve, and, if $\dot{V}(\mathbf{x}) > 0$, the orbit is crossing the curve from the inside to the outside. These remarks make the Lyapunov theorem quite intuitive.

$$V : (x_1, x_2) \mapsto \frac{1}{2} \ell^2 x_2^2 + g\ell(1 - \cos x_1),$$

which represents the energy of the pendulum when the mass of the bob is equal to unity, is a weak Lyapunov function since

$$V(x_1, x_2) > 0 \text{ for } (x_1, x_2) \neq (0, 0) \text{ and } \dot{V}(x_1, x_2) = \frac{\partial V}{\partial x_1} \dot{x}_1 + \frac{\partial V}{\partial x_2} \dot{x}_2 \equiv 0.$$

The equilibrium point $(0, 0)$ is Lyapunov stable. In the case of the damped pendulum (see Example 12), we have

$$\dot{x}_1 = x_2,$$
$$\dot{x}_2 = -\frac{g}{\ell} \sin x_1 - 2ax_2, \quad (a > 0).$$

If here again we consider the function V defined by $V(x_1, x_2) = \frac{1}{2} \ell^2 x_2^2 + g\ell(1 - \cos x_1)$, we find that $\dot{V}(x_1, x_2) = -2a\ell^2 x_2^2$. For the damped pendulum, $\dot{V}(x_1, x_2)$ is a strong Lyapunov function in a neighborhood of the origin. The origin is, therefore, asymptotically stable.

Example 16. The van der Pol oscillator. The differential equation

$$\ddot{x} + \lambda(x^2 - 1)\dot{x} + x = 0 \tag{3.25}$$

describes the dynamics of the *van der Pol oscillator* [291], which arises in electric circuit theory. It is a harmonic oscillator that includes a nonlinear friction term: $\lambda(x^2 - 1)\dot{x}$. If the amplitude of the oscillations is large, the amplitude-dependent "coefficient" of friction is positive, and the oscillations are damped. As a result, the amplitude of the oscillations decreases, and the amplitude-dependent "coefficient" of friction eventually becomes negative, corresponding to a sort of antidamping.

If we put

$$x_1 = x \quad \text{and} \quad x_2 = \dot{x},$$

Equation (3.25) takes the form

$$\dot{x}_1 = x_2, \qquad \dot{x}_2 = -x_1 - \lambda(x_1^2 - 1)x_2. \tag{3.26}$$

The equilibrium point $(x_1, x_2) = (0, 0)$ is nonhyperbolic, but we may study its stability using the Lyapunov method. Consider the function

$$V : (x_1, x_2) \mapsto \frac{1}{2}(x_1^2 + x_2^2).$$

It is positive for $(x_1, x_2) \neq (0, 0,)$, and its time derivative

$$\dot{V}(x_1, x_2) = x_1 \dot{x}_1 + x_2 \dot{x}_2$$
$$= -\lambda(x_1^2 - 1)x_2^2$$

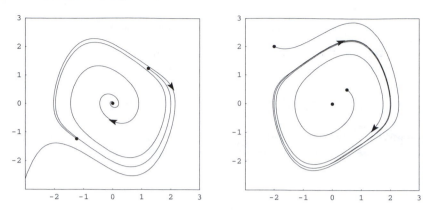

Fig. 3.7. *Phase portraits of the van der Pol equation:* $\ddot{x} + \lambda(x^2 - 1)\dot{x} + x = 0$. *For* $\lambda = -0.5$, *the origin is an asymptotically stable equilibrium point (left), while for* $\lambda = 0.5$ *the orbits converge to a stable limit cycle (right).*

is negative in a neighborhood of the equilibrium point $(0,0)$ if λ is negative. Hence, V is a strong Lyapunov function, which proves that $(0,0)$ is asymptotically stable if $\lambda < 0$. It is an attractive focus. If λ is positive, $(0,0)$ is unstable. It is a repulsive focus, and, as illustrated in Figure 3.7, trajectories converge to a stable limit cycle.

Arnol'd gives the following simple proof of the existence of a stable limit cycle.[27] Note first that the Lyapunov function V represents the energy of the system, which is a conserved quantity in the case of the harmonic oscillator, *i.e.*, for $\lambda = 0$. If the parameter λ is very small, trajectories are spirals in which the distance between adjacent coils is of the order of λ. To determine if these spirals either approach the origin or recede from it, we may compute an approximate value of the increment ΔV over one revolution around the origin. Since $\dot{V}(x_1, x_2) = -\lambda(x_1^2 - 1)x_2^2$, and, to first order in λ, $x_1(t) = A\cos(t - t_0)$ and $x_2(t) = -A\sin(t - t_0)$, we obtain

$$\Delta V = -\lambda \int_0^{2\pi} \Delta V\big(A\cos(t - t_0), -A\sin(t - t_0)\big)$$
$$= \pi\lambda\left(A^2 - \frac{A^4}{4}\right).$$

- If $\lambda < 0$ and the amplitude A of the oscillations is small, $\Delta V < 0$, *i.e.*, the system gives energy to the external world: the trajectory is a contracting spiral.
- If $\lambda > 0$ and the amplitude A of the oscillations is small, $\Delta V > 0$, *i.e.*, the system receives energy from the external world: the trajectory is an expanding spiral.

---

[27] See [9], pp. 150–151.

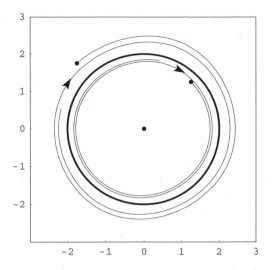

Fig. 3.8. *Phase portrait of the van der Pol equation: $\ddot{x} + \lambda(x^2 - 1)\dot{x} + x = 0$ for $\lambda = 0.05$. As explained in the text, for a small positive value of λ, the stable limit cycle (thick line) is close to the circle of radius 2 centered at the origin.*

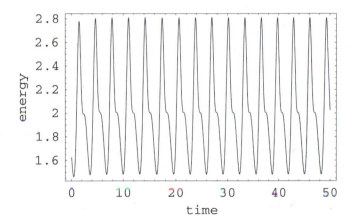

Fig. 3.9. *Energy of the van der Pol oscillator as a function of time along the limit cycle for $\lambda = 0.5$.*

- If $\Delta V = 0$, the energy is conserved, and, for a small positive λ, the trajectory is a cycle close to the circle $x_1^2 + x_2^2 = A^2$, where A is the positive root of $A^2(1 - \frac{1}{4}A^2) = 0$, *i.e.*, $A = 2$.

This result is illustrated in Figure 3.8. Note that for a finite positive value of λ, the energy is not conserved along a stable limit cycle but it varies periodically. As shown in Figure 3.9, the system receives energy from the external world during a part of the cycle and gives it back during the other part.

If $\mathbf{x}^*$ is an asymptotically stable equilibrium point of the differential equation $\dot{\mathbf{x}} = \mathbf{X}(\mathbf{x})$ on $U \subseteq \mathbb{R}^n$, it is of practical importance to determine its *basin of attraction*, that is, the set

$$\{\mathbf{x} \in U \mid \lim_{t \to \infty} \boldsymbol{\varphi}_t(\mathbf{x}) = \mathbf{x}^*\}. \tag{3.27}$$

The method of Lyapunov may be used to obtain estimates of the basin of attraction. The problem is to find the largest subset of U in which $\dot{V}(\mathbf{x})$ is negative.

Remark 4. In the case of the van der Pol equation, the symmetry of the equation, which is invariant under the transformation $t \mapsto -t, \lambda \mapsto -\lambda$, shows that, when $\lambda < 0$, the basin of attraction of the origin is the interior of the closed curve symmetrical, with respect to the $0x_1$-axis, to the stable limit cycle obtained for $|\lambda|$ (see Figure 3.7).

Definition 10. *A point* $\mathbf{y} \in \mathbb{R}^n$ *is an* ω-limit point *for the trajectory* $\{\boldsymbol{\varphi}_t(\mathbf{x}) \mid t \in \mathbb{R}\}$ *through* $\mathbf{x}$ *if there exists a sequence* (t_k) *going to infinity such that* $\lim_{k \to \infty} \boldsymbol{\varphi}_{t_k}(\mathbf{x}) = \mathbf{y}$. *The set of all* ω-limit points of $\mathbf{x}$ *is called the* ω-limit set *of* $\mathbf{x}$, *and is denoted by* $L_\omega(\mathbf{x})$.

α-limit points and the α-limit set are defined in the same way but with a sequence (t_k) going to $-\infty$. The α-limit set of $\mathbf{x}$ is denoted $L_\alpha(\mathbf{x})$.[28]

Let $\mathbf{y} \in L_\omega(\mathbf{x})$ and $\mathbf{z} = \boldsymbol{\varphi}_{t_k}(\mathbf{y})$; then $\lim_{k \to \infty} \boldsymbol{\varphi}_{t+t_k}(\mathbf{x}) = \mathbf{z}$, showing that $\mathbf{y}$ and $\mathbf{z}$ belong to the ω-limit set $L_\omega(\mathbf{x})$ of $\mathbf{x}$. The ω-limit set of $\mathbf{x}$ is, therefore, *invariant* under the flow $\boldsymbol{\varphi}_t$.[29] Similarly, we could have shown that the α-limit set of $\mathbf{x}$ is invariant under the flow.

If $\mathbf{x}^*$ is an asymptotically stable equilibrium point, it is the ω-limit set of every point in its basin of attraction. A closed orbit is the α-limit and ω-limit set of every point on it. While, in general, limit sets can be quite complicated, for two-dimensional systems the situation is much simpler. The following result known as the *Poincaré-Bendixson theorem* gives a criterion to detect limit cycles (see Definition 11 below) in systems modeled by a two-dimensional differential equation:

Theorem 2. *A nonempty compact limit set[30] of a two-dimensional flow defined by a C^1 vector field, which contains no fixed point, is a closed orbit.[31]*

Definition 11. *A limit cycle is a closed orbit γ such that either $\gamma \subset L_\omega(\mathbf{x})$ or $\gamma \subset L_\alpha(\mathbf{x})$ for some $\mathbf{x} \notin \gamma$. In the first case, γ is an ω-limit cycle; in the second case, it is an α-limit cycle.*

---

[28] The reason for this terminology is that α and ω are, respectively, the first and last letters of the Greek alphabet.

[29] A set M is also invariant under the flow $\boldsymbol{\varphi}_t$ if, for all $\mathbf{x} \in M$ and all $t \in \mathbb{R}$, $\boldsymbol{\varphi}_t(\mathbf{x}) \in M$. For instance, fixed points and closed orbits are invariant sets.

[30] A set is *compact* if, from any covering by open sets, it is possible to extract a finite covering. Any closed bounded subset of a finite-dimensional metric space is compact.

[31] See Hirsch and Smale [171].

Closed orbits around a center are not limit cycles. A limit cycle is an isolated closed orbit in the sense that there exists an annular neighborhood of the limit cycle that contains no other closed orbits. The stability of limit cycles is studied in the next chapter Section 4.3.

In order to prove the existence of a limit cycle using Theorem 2, one has to find a bounded subset D of $\mathbb{R}^2$ such that, for all $\mathbf{x} \in D$, the trajectories $\{\boldsymbol{\varphi}_t(\mathbf{x}) \mid t > 0\}$ remain in D^{32} and show that D does not contain an equilibrium point.

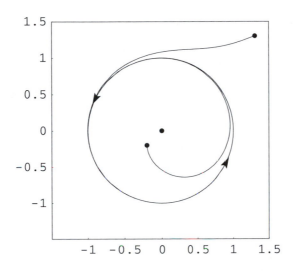

Fig. 3.10. *Phase portrait of the two-dimensional system of Example 17*

Example 17. Perturbed center. Consider the two-dimensional system

$$\dot{x}_1 = x_1 - x_2 - x_1(x_1^2 + x_2^2), \qquad \dot{x}_2 = x_1 + x_2 - x_2(x_1^2 + x_2^2).$$

Using polar coordinates, this system takes the following particularly simple form:

$$\dot{r} = r - r^3, \qquad \dot{\theta} = 1.$$

For $r = 1$, $\dot{r} = 0$; therefore, if r_1 and r_2 are two real numbers such that $0 < r_1 < 1 < r_2$, we verify that, for $r_1 \le r < 1$, $\dot{r} > 0$, and, for $1 < r \le r_2$, $\dot{r} < 0$. Thus, the closed annular set $\{\mathbf{x} \mid r_1 \le (x_1^2 + x_2^2)^{1/2} \le r_2\}$ is positively invariant (see Footnote 32) and does not contain an equilibrium point. According to the Poincaré-Bendixson theorem, this bounded subset contains a limit cycle (see Figure 3.10).

---

[32] Such a subset D is said to be *positively invariant*.

3.3 Graphical study of two-dimensional systems

There exists a wide variety of models describing the interactions between two populations. If one seeks to incorporate in such a model a minimum of broadly relevant features, the equations describing the model might become difficult to analyze. It is, however, frequently possible to analyze graphically the behavior of the system without entering into specific mathematical details.

The system

$$\dot{N}_1 = N_1 f_1(N_1, N_2), \qquad \dot{N}_2 = N_2 f_2(N_1, N_2), \qquad (3.28)$$

represents a model of two interacting populations, whose growth rates $\dot{N}_1/N_1$ and $\dot{N}_2/N_2$ are, respectively, equal to $f_1(N_1, N_2)$ and $f_2(N_1, N_2)$.

The general idea on which rests the qualitative graphical analysis of Equations (3.28) is to:

(i) divide the positive quadrant of the (N_1, N_2)-plane in domains bounded by the sets $\{(N_1, N_2) \mid f_1(N_1, N_2) = 0\}$ and $\{(N_1, N_2) \mid f_2(N_1, N_2) = 0\}$,[33]

(ii) find, in each domain, the sign of both growth rates, which determine the direction of the vector field, and

(iii) represent, in each domain, the direction of the vector field (*i.e.*, the flow) by an arrow.

As shown in the following example, a schematic phase portrait can then be easily obtained.

Example 18. Lotka-Volterra competition model. Assuming that two species compete for a common food supply, their growth could be described by the following simple two-dimensional system:

$$\dot{N}_1 = r_1 N_1 \left(1 - \frac{N_1}{K_1}\right) - \lambda_1 N_1 N_2, \quad \dot{N}_2 = r_2 N_2 \left(1 - \frac{N_2}{K_2}\right) - \lambda_2 N_1 N_2. \quad (3.29)$$

This competition model is usually associated with the names of Lotka and Volterra.

Each population (N_1 and N_2) has a logistic growth but the presence of each reduces the growth rate of the other. The constants r_1, r_2, K_1, K_2, λ_1, and λ_2 are positive. As usual, we define reduced variables writing

$$\tau = \sqrt{r_1 r_2}\, t, \quad \rho = \sqrt{\frac{r_1}{r_2}}, \quad n_1 = \frac{N_1}{K_1}, \quad n_2 = \frac{N_2}{K_2}, \quad \alpha_1 = \frac{\lambda_1 K_2}{\sqrt{r_1 r_2}}, \quad \alpha_2 = \frac{\lambda_2 K_1}{\sqrt{r_1 r_2}},$$

and Equations (3.29) become

$$\frac{dn_1}{d\tau} = \rho n_1(1 - n_1) - \alpha_1 n_1 n_2, \quad \frac{dn_2}{d\tau} = \frac{1}{\rho} n_2(1 - n_2) - \alpha_2 n_1 n_2. \quad (3.30)$$

---

[33] These sets, which are the preimages of the point $(0,0)$ by f_1 and f_2, respectively, are called *null clines*.

To determine under which conditions the two species can coexist, the two straight lines

$$\rho(1 - n_1) - \alpha_1 n_2 = 0, \quad \frac{1}{\rho}(1 - n_2) - \alpha_2 n_1 = 0,$$

should intersect in the positive quadrant. There are two possibilities represented in Figure 3.11. We find that if $a_1 = \rho/\alpha_1$ and $a_2 = 1/\rho\alpha_2$ are both

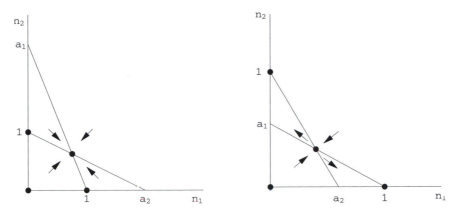

Fig. 3.11. *Lotka-Volterra competition model. Intersecting null clines. Dots show equilibrium points. In each domain, arrows represent the direction of the vector* $(\dot{n}_1, \dot{n}_2)$, *with time derivatives taken with respect to reduced time* τ. *Left:* $a_1 = \rho/\alpha_1 > 1$ *and* $a_2 = 1/\rho\alpha_2 > 1$. *Right:* $a_1 = \rho/\alpha_1 < 1$ *and* $a_2 = 1/\rho\alpha_2 < 1$.

greater than 1, the nontrivial equilibrium point is asymptotically stable: the two populations will coexist. If, on the contrary, $a_1 = \rho/\alpha_1$ and $a_2 = 1/\rho\alpha_2$ are both less than 1, the nontrivial equilibrium point is a saddle. The equilibrium points $(0, 1)$ and $(1, 0)$ are stable steady states. The population that will eventually survive depends upon the initial state.

When the null clines do not intersect, as in Figure 3.12, only one population will eventually survive. It is population 1 if $a_1 = \rho/\alpha_1 > 1$ and $a_2 = 1/\rho\alpha_2 < 1$, and population 2 if $a_1 = \rho/\alpha_1 < 1$ and $a_2 = 1/\rho\alpha_2 > 1$. The equilibrium point $(0, 0)$ is, in all cases, unstable.

These results illustrate the so-called *competitive exclusion principle* whereby *two species competing for the same limited resource cannot, in general, coexist.* Note that the species that will eventually survive is the species whose growth rate is less perturbed by the presence of the other.

It is possible to use the graphical method to study more complex models. For instance, Hirsch and Smale [171][34] discuss a large class of competition models. Their equations are of the form

---

[34] In a much older paper, Kolmogorov [193] had already presented a qualitative study of a general predator-prey system.

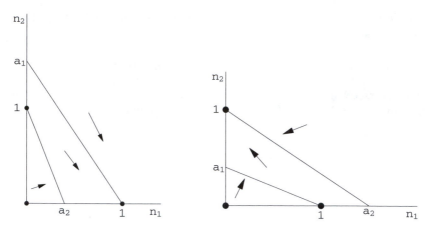

Fig. 3.12. *Lotka-Volterra competition model. Nonintersecting null clines. Dots show equilibrium points. In each domain, arrows represent the direction of the vector* $(\dot{n}_1, \dot{n}_2)$, *with time derivatives taken with respect to reduced time* τ. *Left:* $a_1 = \rho/\alpha_1 > 1$ *and* $a_2 = 1/\rho\alpha_2 < 1$. *Right:* $a_1 = \rho/\alpha_1 < 1$ *and* $a_2 = 1/\rho\alpha_2 > 1$.

$$\dot{N}_1 = N_1 f_1(N_1, N_2), \qquad \dot{N}_2 = N_2 f_2(N_1, N_2),$$

with the following assumptions:

1. If either species increases, the growth rate of the other decreases. Hence,

$$\frac{\partial f_1}{\partial N_2} < 0 \ \text{ and } \ \frac{\partial f_2}{\partial N_1} < 0.$$

2. If either population is very large, neither species can multiply. Hence,

$$f_1(N_1, N_2) \le 0 \ \text{ and } \ f_2(N_1, N_2) \le 0 \quad \text{if either } N_1 > K \text{ or } N_2 > K.$$

3. In the absence of either species, the other has a positive growth rate up to a certain population and a negative growth rate beyond that. There are, therefore, constants $a_1 > 0$ and $a_2 > 0$ such that

$$f_1(N_1, N_2) > 0 \ \text{ for } \ N_1 < a_1 \ \text{ and } \ f_1(N_1, N_2) < 0 \ \text{ for } \ N_1 > a_1$$

and

$$f_2(N_1, N_2) > 0 \ \text{ for } \ N_2 < a_2 \ \text{ and } \ f_2(N_1, N_2) < 0 \ \text{ for } \ N_2 > a_2.$$

Analyzing different possible *generic* shapes of the null clines, Hirsch and Smale show that the ω-limit set of any point in the positive quadrant of the (N_1, N_2)-plane exists and is one of a finite number of equilibria; that is, there are no closed orbits.

3.4 Structural stability

The qualitative properties of a model should not change significantly when the model is slightly modified: a model should be *robust*. To be precise we have to give a definition of what is a "slight modification." That is, in the space $\mathcal{V}(U)$ of all vector fields defined on an open set $U \subseteq \mathbb{R}^n$, we have to define an *appropriate* metric.[35] The metric, we said, has to be appropriate in the sense that, if two vector fields are close for this metric, then the dynamics they generate have the same qualitative properties. Actually, on the space $\mathcal{V}(U)$, we shall first define an appropriate *norm* and associate a metric with that norm.[36]

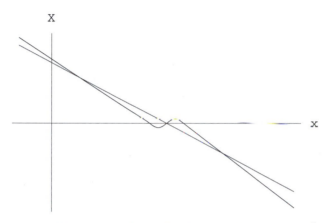

Fig. 3.13. *Two neighboring one-dimensional vector fields in the C^0 topology that do not have the same number of equilibrium points.*

If $\mathbf{X} \in \mathcal{V}(U)$, its C^0 norm is defined by

---

[35] We have already defined the notion of *distance* (or *metric*) (see page 14). A distance d on a space X is a mapping $d : X \times X \to \mathbb{R}_+$ satisfying the following conditions:

1. $d(x, y) = 0 \iff x = y$.
2. $d(x, y) = d(y, x)$.
3. $d(x, z) \leq d(x, y) + d(y, z)$.

The pair (X, d) is called a *metric space*.

[36] A mapping $x \mapsto \|x\|$ defined on a vector space X into $\mathbb{R}_+$ is a *norm* if it satisfies the following conditions:

1. $\|x\| = 0 \iff x = 0$.
2. For any $x \in X$ and any scalar λ, $\|\lambda x\| = |\lambda| \|x\|$.
3. For all x and y in X, $\|x + y\| \leq \|x\| + \|y\|$.

A vector space equipped with a norm is called a *normed vector space*.

$$\|\mathbf{X}\|_0 = \sup_{\mathbf{x}\in U} \|\mathbf{X}(\mathbf{x})\|,$$

where $\|\mathbf{X}(\mathbf{x})\|$ is the usual norm of the vector $\mathbf{X}(\mathbf{x})$ in $\mathbb{R}^n$. The C^0 distance between two vector fields $\mathbf{X}$ and $\mathbf{Y}$, which belong to $\mathcal{V}(U)$, is then defined by

$$d_0(\mathbf{X},\mathbf{Y}) = \|\mathbf{X}-\mathbf{Y}\|_0.$$

As shown in Figure 3.13, two vector fields that are close in the C^0 topology[37] may not have the same number of hyperbolic equilibrium points. To avoid this undesirable situation, we should define a distance requiring that the vector fields as well as their derivatives be close at all points of U. The C^1 distance between two vector fields $\mathbf{X}$ and $\mathbf{Y}$ belonging to $\mathcal{V}(U)$ is then defined by

$$d_1(\mathbf{X},\mathbf{Y}) = \sup_{\mathbf{x}\in U}\{\|\mathbf{X}(\mathbf{x}) - \mathbf{Y}(\mathbf{x})\|, \|D\mathbf{X}(\mathbf{x}) - D\mathbf{Y}(\mathbf{x})\|\}.$$

Similarly we could define C^k distances for k greater than 1.

In the space $\mathcal{V}(U)$ of all vector fields defined on an open set $U \subseteq \mathbf{R}^n$, an ε-*neighborhood* of $\mathbf{X} \in \mathcal{V}(U)$ is defined by

$$N_\varepsilon(\mathbf{X}) = \{\mathbf{Y} \in \mathcal{V}(U) \mid \|\mathbf{X}-\mathbf{Y}\|_1 < \varepsilon\}.$$

A vector field $\mathbf{Y}$ that belongs to an ε-*neighborhood* of $\mathbf{X}$ is said to be ε-C^1-*close* to $\mathbf{X}$ or an ε-C^1-*perturbation* of $\mathbf{X}$. In this case, the components $\big(X_1(\mathbf{x}), X_2(\mathbf{x}), \ldots, X_n(\mathbf{x})\big)$ and $\big(Y_1(\mathbf{x}), Y_2(\mathbf{x}), \ldots, Y_n(\mathbf{x})\big)$ of, respectively, $\mathbf{X}$ and $\mathbf{Y}$ and their first derivatives are close throughout U.

Theorem 3. *Let* $\mathbf{x}^*$ *be a hyperbolic equilibrium point of the flow* φ_t *generated by the vector field* $\mathbf{X} \in \mathcal{V}(U)$. *Then, there exists a neighborhood* V *of* $\mathbf{x}^*$ *and a neighborhood* N *of* $\mathbf{X}$ *such that each* $\mathbf{Y} \in N$ *generates a flow* ψ_t *that has a unique hyperbolic equilibrium point* $\mathbf{y}^* \in V$. *Moreover, the linear operators* $D\mathbf{X}(\mathbf{x}^*)$ *and* $D\mathbf{Y}(\mathbf{y}^*)$ *have the same number of eigenvalues with positive and negative real parts. In this case, the flow* φ_t *generated by the vector field* $\mathbf{X}$ *is said to be* locally structurally stable *at* $\mathbf{x}^*$.

---

[37] A collection $\mathcal{T}$ of subsets of a set X is said to be a *topology* in X if $\mathcal{T}$ has the following properties:

1. X and $\emptyset$ belong to $\mathcal{T}$.
2. If $\{O_i \mid i \in I\}$ is an arbitrary collection of elements of $\mathcal{T}$, then $\cup_{i\in I}O_i$ belongs to $\mathcal{T}$.
3. If O_1 and O_2 belong to $\mathcal{T}$, then $O_1 \cap O_2$ belongs to $\mathcal{T}$.

The ordered pair $(X, \mathcal{T})$ is called a *topological space*, and the elements of $\mathcal{T}$ are called *open sets* in X. When no ambiguity is possible, one may speak of the "topological space X."

Since the flows $\exp\left(D\mathbf{X}(\mathbf{x}^*)t\right)$ and $\exp\left(D\mathbf{X}(\mathbf{x}^*)t\right)$ are equivalent, this result follows directly from Theorem 1, which proves that the flows φ_t and ψ_t are conjugate.[38]

The vector field $(x_2, -\omega^2 \sin x_1)$ of the undamped pendulum generates a flow that is not structurally stable. Its equilibrium points are nonhyperbolic. The damped pendulum (Example 12), which is obtained by adding to the vector field of the undamped pendulum the perturbation $(0, -2ax_2)$, has hyperbolic equilibrium points, which are either asymptotically stable or unstable (see Figure 3.6).

Similarly, a linear harmonic oscillator whose vector field is $(x_2, -x_1)$ has only one equilibrium point $(0,0)$, which is a center. The flow generated by the vector field is not structurally stable. As shown in Example 13, the perturbation $(\lambda x_1(x_1^2 + x_2^2), \lambda x_2(x_1^2 + x_2^2))$ generates a qualitatively different flow: its ω-limit set is a stable limit cycle.

3.5 Local bifurcations of vector fields

In Section 2.3 we studied the predator-prey model used by Harrison to explain Luckinbill's experiment with *Didinium* and *Paramecium*. This model, whose dimensionless equations are

$$\frac{dh}{d\tau} = h\left(1 - \frac{h}{k}\right) - \frac{\alpha_h ph}{\beta + h},$$
$$\frac{dp}{d\tau} = \frac{\alpha_p ph}{\beta + h} - \gamma p,$$

where

$$\alpha_h = \left(1 - \frac{1}{k}\right)(\beta + 1) \quad \text{and} \quad \alpha_p = \gamma(\beta + 1),$$

exhibits two qualitatively different behaviors. For $k < \beta + 2$, the equilibrium point $(1,1)$ is asymptotically stable, but, for $k > \beta + 2$, $(1,1)$ is unstable and the ω-limit set is a limit cycle. The change of behavior occurs for $k = \beta + 2$; that is, when $(1,1)$ is nonhyperbolic. Such a change in a family of vector fields, which depends upon a finite number of parameters, is referred to as a *bifurcation*.

The van der Pol equation (Example 16) exhibits a similar bifurcation. Here again, at the bifurcation point, the equilibrium point $(0,0)$ is nonhyperbolic.

Like many concepts of the qualitative theory of differential equations, the theory of bifurcations has its origins in the work of Poincaré.[39] Let

---

[38] The flows have to be restricted on neighborhoods of the respective hyperbolic equilibrium points $\mathbf{x}^*$ and $\mathbf{y}^*$ on which Theorem 1 is valid. For more details, see Arrowsmith and Place [11].

[39] Poincaré was the first to use the French word *bifurcation* in this context [288].

$$(\mathbf{x}, \boldsymbol{\mu}) \mapsto \mathbf{X}(\mathbf{x}, \boldsymbol{\mu}) \qquad (\mathbf{x} \in \mathbb{R}^n, \boldsymbol{\mu} \in \mathbb{R}^r)$$

be a C^k r-parameter family of vector fields.[40] If $(\mathbf{x}^*, \boldsymbol{\mu}^*)$ is an equilibrium point of the flow generated by $\mathbf{X}(\mathbf{x}, \boldsymbol{\mu})$ (*i.e.*, $\mathbf{X}(\mathbf{x}^*, \boldsymbol{\mu}^*) = 0$), we should be able to answer the question: *Is the stability of the equilibrium point affected as $\boldsymbol{\mu}$ is varied?*

If the equilibrium point is hyperbolic (*i.e.*, if all the eigenvalues of the Jacobian matrix[41] $D\mathbf{X}(\mathbf{x}^*, \boldsymbol{\mu}^*)$ have nonzero real part), then, from Theorem 3, it follows that the flow generated by $\mathbf{X}(\mathbf{x}, \boldsymbol{\mu})$, is locally structurally stable at $(\mathbf{x}^*, \boldsymbol{\mu}^*)$. As a result, for values of $\boldsymbol{\mu}$ sufficiently close to $\boldsymbol{\mu}^*$, the stability of the equilibrium point is not affected.

More precisely, since all the eigenvalues of the Jacobian $D\mathbf{X}(\mathbf{x}^*, \boldsymbol{\mu}^*)$ have nonzero real part, the Jacobian is invertible, and, from the implicit function theorem,[42] it follows that there exists a unique C^k function $\mathbf{x} : \boldsymbol{\mu} \mapsto \mathbf{x}(\boldsymbol{\mu})$ such that, for $\boldsymbol{\mu}$ sufficiently close to $\boldsymbol{\mu}^*$,

$$\mathbf{X}(\mathbf{x}(\boldsymbol{\mu}), \boldsymbol{\mu}) = \mathbf{0} \quad \text{with} \quad \mathbf{x}(\boldsymbol{\mu}^*) = \boldsymbol{x}^*.$$

Since the eigenvalues of the Jacobian $D\mathbf{X}(\mathbf{x}(\boldsymbol{\mu}), \boldsymbol{\mu})$ are continuous functions of $\boldsymbol{\mu}$, for $\boldsymbol{\mu}$ sufficiently close to $\boldsymbol{\mu}^*$, all the eigenvalues of this Jacobian have nonzero real part. Hence, equilibrium points close to $(\mathbf{x}^*, \boldsymbol{\mu}^*)$ are hyperbolic and have the same type of stability as $(\mathbf{x}^*, \boldsymbol{\mu}^*)$.

If the equilibrium point $(\mathbf{x}^*, \boldsymbol{\mu}^*)$ is nonhyperbolic (*i.e.*, if some eigenvalues of $D\mathbf{X}(\mathbf{x}^*, \boldsymbol{\mu}^*)$ have zero real part), $\mathbf{X}(\mathbf{x}, \boldsymbol{\mu})$ is not structurally stable at $(\mathbf{x}^*, \boldsymbol{\mu}^*)$. In this case, for values of $\boldsymbol{\mu}$ close to $\boldsymbol{\mu}^*$, a totally new dynamical behavior can occur.

Our aim in this section is to investigate the simplest bifurcations that occur at nonhyperbolic equilibrium points in one- and two-dimensional systems.

---

[40] The degree of differentiability has to be as high as needed in order to satisfy the conditions for the family of vector fields to exhibit a given type of bifurcation. See the necessary and sufficient conditions below.

[41] The notation $D\mathbf{X}(\mathbf{x}^*, \boldsymbol{\mu}^*)$ means that the derivative is taken with respect to $\mathbf{x}$ at the point $(\mathbf{x}^*, \boldsymbol{\mu}^*)$. If there is a risk of confusion with respect to which variable the derivative is taken, we will write $D_{\mathbf{x}}\mathbf{X}(\mathbf{x}^*, \boldsymbol{\mu}^*)$.

[42] We shall often use this theorem, in particular in bifurcation theory, where it plays an essential role. Here is a simplified version that is sufficient in most cases: Let $f : I_1 \times I_2 \to \mathbb{R}$ be a C^k function of two real variables; if $(x_0, y_0) \in I_1 \times I_2$ and

$$f(x_0, y_0) = 0, \qquad \frac{\partial f}{\partial y}(x_0, y_0) \neq 0,$$

then there exists an open interval I, containing x_0, and a C^k function $\varphi : I \to \mathbb{R}$ such that

$$\varphi(x_0) = y_0, \quad \text{and} \quad f(x, \varphi(x)) = 0, \quad \text{for all } x \in I.$$

For a proof, see Lang [200], pp. 425–429, in which a proof of a more general version of the implicit function theorem is also given.

3.5.1 One-dimensional vector fields

In a one-dimensional system, a nonhyperbolic equilibrium point is necessarily associated with a zero eigenvalue of the derivative of the vector field at the equilibrium point. In this section, we describe the most important types of bifurcations that occur in one-dimensional systems

$$\dot{x} = X(x, \mu) \quad (x \in \mathbb{R}, \ \mu \in \mathbb{R}). \tag{3.31}$$

As a simplification, we assume that the bifurcation point (x^*, μ^*) is $(0, 0)$, that is,

$$X(0,0) = 0 \quad \text{and} \quad \frac{\partial X}{\partial x}(0,0) = 0. \tag{3.32}$$

In a neighborhood of a bifurcation point, the essential information concerning the bifurcation is captured by the *bifurcation diagram*, which consists of different curves. The locus of stable points is usually represented by a solid curve, while a broken curve represents the locus of unstable points (Figure 3.17).

Saddle-node bifurcation

Consider the equation

$$\dot{x} = \mu - x^2. \tag{3.33}$$

For $\mu = 0$, $x^* = 0$ is the only equilibrium point, and it is nonhyperbolic since $DX(0,0) = 0$. The vector field $X(x,0)$ is not structurally stable and $\mu = 0$ is a bifurcation value. For $\mu < 0$, there are no equilibrium points, while, for $\mu > 0$, there are two hyperbolic equilibrium points $x^* = \pm\sqrt{\mu}$. Since $DX(\pm\sqrt{\mu}, \mu) = \mp 2\sqrt{\mu}$, $\sqrt{\mu}$ is asymptotically stable, and $-\sqrt{\mu}$ is unstable. The phase portraits for Equation (3.33) are shown in Figure 3.14. This type of bifurcation is called a *saddle-node bifurcation*.[43]

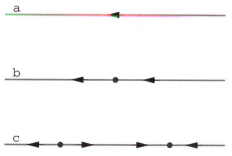

Fig. 3.14. *Saddle-node bifurcation. Phase portraits for the differential equation (3.33). (a) $\mu < 0$, (b) $\mu = 0$, (c) $\mu > 0$.*

---

[43] Also called *tangent bifurcation*.

Transcritical bifurcation

Consider the equation

$$\dot{x} = \mu x - x^2. \tag{3.34}$$

For $\mu = 0$, $x^* = 0$ is the only equilibrium point, and it is nonhyperbolic since $DX(0,0) = 0$. The vector field $X(x,0)$ is not structurally stable, and $\mu = 0$ is a bifurcation value. For $\mu \neq 0$, there are two equilibrium points, 0 and μ. At the bifurcation point, these two equilibrium points exchange their stability (see Figure 3.15). This type of bifurcation is called a *transcritical bifurcation*.

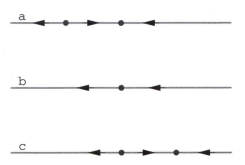

Fig. 3.15. *Transcritical bifurcation. Phase portraits for the differential equation (3.34). (a) $\mu < 0$, (b) $\mu = 0$, (c) $\mu > 0$.*

Pitchfork bifurcation

Consider the equation

$$\dot{x} = \mu x - x^3. \tag{3.35}$$

For $\mu = 0$, $x^* = 0$ is the only equilibrium point. It is nonhyperbolic since $DX(0,0) = 0$. The vector field $X(x,0)$ is not structurally stable and $\mu = 0$ is a bifurcation value. For $\mu \leq 0$, 0 is the only equilibrium point, and it is asymptotically stable. For $\mu > 0$, there are three equilibrium points, 0 is unstable, and $\pm\sqrt{\mu}$ are both asymptotically stable. The phase portraits for Equation (3.35) are shown in Figure 3.16. This type of bifurcation is called a *pitchfork bifurcation*.[44]

---

[44] Sometimes called *symmetry breaking bifurcation* since it is the bifurcation that characterizes the broken symmetry associated with a second-order phase transition in statistical physics.

Fig. 3.16. *Pitchfork bifurcation. Phase portraits for the differential equation (3.35).* *(a)* $\mu \leq 0$, *(b)* $\mu > 0$.

Necessary and sufficient conditions

It is possible to derive necessary and sufficient conditions under which a one-parameter family of one-dimensional vector fields exhibits a bifurcation of one of the types just described. These conditions involve derivatives of the vector field at the bifurcation point.

Saddle-node bifurcation. If the family $X(x,\mu)$ undergoes a saddle-node bifurcation, in the (μ, x)-plane there exists a unique curve of fixed points (see the top panel of Figure 3.17). This curve is tangent to the line $\mu = 0$ at $x = 0$, and it lies entirely to one side of $\mu = 0$. These two properties imply that

$$\frac{d\mu}{dx}(0) = 0, \qquad \frac{d^2\mu}{dx^2}(0) \neq 0. \tag{3.36}$$

The bifurcation point is a nonhyperbolic equilibrium; *i.e.*,

$$X(0,0) = 0, \qquad \frac{\partial X}{\partial x}(0,0) = 0.$$

If we assume

$$\frac{\partial X}{\partial \mu}(0,0) \neq 0,$$

then, by the implicit function theorem, there exists a unique function

$$\mu : x \mapsto \mu(x), \quad \text{such that} \quad \mu(0) = 0,$$

defined in a neighborhood of $x = 0$ that satisfies the relation $X(x, \mu(x)) = 0$. To express Conditions (3.36), which imply that $(0,0)$ is a nonhyperbolic equilibrium point at which a saddle-node bifurcation occurs, in terms of derivatives of X, we have to differentiate the relation $X(x, \mu(x)) = 0$ with respect to x. We obtain

$$\frac{dX}{dx}(x, \mu(x)) = \frac{\partial X}{\partial x}(x, \mu(x)) + \frac{\partial X}{\partial \mu}(x, \mu(x))\frac{d\mu}{dx}(x) = 0.$$

Hence,

$$\frac{d\mu}{dx}(0) = -\frac{\dfrac{\partial X}{\partial x}(0,0)}{\dfrac{\partial X}{\partial \mu}(0,0)},$$

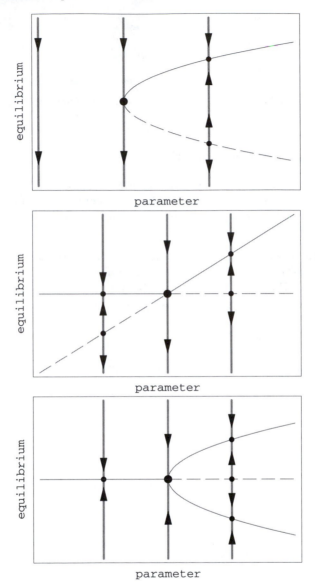

Fig. 3.17. *Bifurcation diagrams with phase portraits. Bifurcation points are represented by ● and hyperbolic equilibrium points by •. Stable equilibrium points are on solid curves. Top: saddle-node; middle: transcritical; bottom: pitchfork.*

which shows that

$$\frac{\partial X}{\partial x}(0,0) = 0 \quad \text{and} \quad \frac{\partial X}{\partial \mu}(0,0) \neq 0,$$

implies

$$\frac{d\mu}{dx}(0) = 0.$$

If we differentiate $X(x, \mu(x)) = 0$ once more with respect to x, we obtain

$$\frac{d^2 X}{dx^2}(x, \mu(x)) = \frac{\partial^2 X}{\partial^2 x}(x, \mu(x)) + 2\frac{\partial^2 X}{\partial x \partial \mu}(x, \mu(x))\frac{d\mu}{dx}(x)$$

$$+ \frac{\partial^2 X}{\partial^2 \mu}(x, \mu(x))\left(\frac{d\mu}{dx}(x)\right)^2 + \frac{\partial X}{\partial \mu}(x, \mu(x))\frac{d^2\mu}{dx^2}(x) = 0.$$

Hence, taking into account the expression of $d\mu/dx(0)$ found above,

$$\frac{d^2 X}{dx^2}(0) + \frac{\partial X}{\partial \mu}(0, 0)\frac{d^2\mu}{dx^2}(0) = 0,$$

which yields

$$\frac{d^2\mu}{d^2 x}(0) = -\frac{\dfrac{\partial^2 X}{\partial x^2}(0, 0)}{\dfrac{\partial X}{\partial \mu}(0, 0)}.$$

That is,

$$\frac{d^2\mu}{dx^2}(0) \neq 0 \quad \text{if} \quad \frac{\partial^2 X}{\partial x^2}(0, 0) \neq 0.$$

In short, the one-parameter family

$$\dot{x} = X(x, \mu)$$

undergoes a saddle-node bifurcation if

$$X(0, 0) = 0 \quad \text{and} \quad \frac{\partial X}{\partial x}(0, 0) = 0,$$

showing that $(0, 0)$ is a nonhyperbolic equilibrium point, and

$$\frac{\partial X}{\partial \mu}(0, 0) \neq 0 \quad \text{and} \quad \frac{\partial^2 X}{\partial^2 x}(0, 0) \neq 0,$$

which imply the existence of a unique curve of equilibrium points that passes through $(0, 0)$ and lies, in a neighborhood of $(0, 0)$, on one side of the line $\mu = 0$. In order to exhibit a saddle-node bifurcation, the one-parameter family X has to be C^k with $k \geq 2$.

Transcritical bifurcation. As for the saddle-node bifurcation, the implicit function theorem can be used to characterize the geometry of the bifurcation diagram in the neighborhood of the bifurcation point, assumed to be located at $(0, 0)$ in the (μ, x)-plane, in terms of the derivatives of the vector field X. In the case of the transcritical bifurcation, in the neighborhood of the bifurcation point, the bifurcation diagram is characterized (see the middle panel of

Figure 3.17) by the existence of two curves of equilibrium points: $x = \mu$ and $x = 0$. Both curves exist on both sides of $(0,0)$, and the equilibrium points on these curves exchange their stabilities on passing through the bifurcation point.

The bifurcation point $(0,0)$ being nonhyperbolic, we must have

$$X(0,0) = 0, \qquad \frac{\partial X}{\partial x}(0,0) = 0.$$

Since *two* curves of equilibrium points pass through this point,

$$\frac{\partial X}{\partial \mu}(0,0) = 0,$$

otherwise the implicit function theorem would imply the existence of *only one* curve passing through $(0,0)$. To be able to use the implicit function theorem, the result obtained in the discussion of the transcritical bifurcation will guide us. If the vector field X is assumed to be of the form $X(x,\mu) = x\,\Xi(x,\mu)$, then $x = 0$ is a curve of equilibrium points passing through $(0,0)$. To obtain the additional curve, Ξ has to satisfy $\Xi(0,0)$. The values of the derivatives of Ξ at $(0,0)$ are determined using the definition of Ξ; *i.e.*,

$$\Xi(x,\mu) = \begin{cases} \dfrac{X(x,\mu)}{x}, & \text{if } x \neq 0, \\[2ex] \dfrac{\partial X}{\partial x}(0,\mu), & \text{if } x = 0. \end{cases}$$

Hence,

$$\frac{\partial \Xi}{\partial x}(0,0) = \frac{\partial^2 X}{\partial^2 x}(0,0), \qquad \frac{\partial^2 \Xi}{\partial^2 x}(0,0) = \frac{\partial^3 X}{\partial^3 x}(0,0),$$

and

$$\frac{\partial \Xi}{\partial \mu}(0,0) = \frac{\partial^2 X}{\partial \mu \partial x}(0,0).$$

If

$$\frac{\partial \Xi}{\partial \mu}(0,0) \neq 0,$$

from the implicit function theorem, it follows that there exists a function $\mu : x \mapsto \mu(x)$, defined in the neighborhood of $x = 0$, that satisfies the relation $\Xi(x,\mu(x)) = 0$. For the curve $\mu = \mu(x)$ not to coincide with the curve $x = 0$ and to be defined on both sides of $(0,0)$, the function μ has to be such that

$$0 < \left| \frac{d\mu}{dx}(0) \right| < \infty.$$

Differentiating with respect to x the relation $\Xi(x,\mu(x)) = 0$, we obtain

$$\frac{d\mu}{dx}(0) = -\frac{\dfrac{\partial \Xi}{\partial x}(0,0)}{\dfrac{\partial \Xi}{\partial \mu}(0,0)} = -\frac{\dfrac{\partial^2 X}{\partial^2 x}(0,0)}{\dfrac{\partial^2 X}{\partial \mu \partial x}(0,0)}.$$

In short, the one-parameter family

$$\dot{x} = X(x,\mu)$$

undergoes a transcritical bifurcation if

$$X(0,0) = 0 \quad \text{and} \quad \frac{\partial X}{\partial x}(0,0) = 0,$$

showing that $(0,0)$ is a nonhyperbolic equilibrium point, and

$$\frac{\partial X}{\partial \mu}(0,0) = 0, \quad \frac{\partial^2 X}{\partial^2 x}(0,0) \neq 0, \quad \text{and} \quad \frac{\partial^2 X}{\partial \mu \partial x}(0,0) \neq 0,$$

which imply the existence of two curves of equilibrium points on both sides $(0,0)$ passing through that point. In order to exhibit a transcritical bifurcation, the one-parameter family X has to be C^k with $k \geq 2$.

Pitchfork bifurcation. The derivation of the necessary and sufficient conditions for a one-parameter family of vector fields $X(x,\mu)$ to exhibit a pitchfork bifurcation is similar to the derivation for such a family to exhibit a transcritical bifurcation.

In the case of a pitchfork bifurcation, the bifurcation diagram is characterized (see the bottom panel of Figure 3.17) by the existence of two curves of equilibrium points: $x = 0$ and $\mu = x^2$. The curve $x = 0$ exists on both sides of the bifurcation point $(0,0)$, and the equilibrium points on this curve change their stabilities on passing through this point. The curve $\mu = x^2$ exists only on one side of $(0,0)$ and at this point is tangent to the line $\mu = 0$.

The point $(0,0)$ being nonhyperbolic, we must have

$$X(0,0) = 0, \quad \frac{\partial X}{\partial x}(0,0) = 0.$$

Since *more than one curve* of equilibrium points passes through this point,

$$\frac{\partial X}{\partial \mu}(0,0) = 0.$$

As in the case of a transcritical bifurcation, we assume the vector field X to be of the form $X(x,\mu) = x\,\Xi(x,\mu)$ to ensure that $x = 0$ is a curve of equilibrium points passing through $(0,0)$. To obtain the additional curve, Ξ has to satisfy $\Xi(0,0)$. The values of the derivatives of Ξ at $(0,0)$ may be determined using the definition of Ξ, *i.e.*,

$$
\Xi(x,\mu) = \begin{cases} \dfrac{X(x,\mu)}{x}, & \text{if } x \neq 0, \\[3mm] \dfrac{\partial X}{\partial x}(0,\mu), & \text{if } x = 0. \end{cases}
$$

Then,

$$
\frac{\partial \Xi}{\partial x}(0,0) = \frac{\partial^2 X}{\partial^2 x}(0,0), \qquad \frac{\partial^2 \Xi}{\partial^2 x}(0,0) = \frac{\partial^3 X}{\partial^3 x}(0,0),
$$

and

$$
\frac{\partial \Xi}{\partial \mu}(0,0) = \frac{\partial^2 X}{\partial \mu \partial x}(0,0).
$$

If

$$
\frac{\partial \Xi}{\partial \mu}(0,0) \neq 0,
$$

from the implicit function theorem, it follows that there exists a function $\mu : x \mapsto \mu(x)$, defined in the neighborhood of $x = 0$, that satisfies the relation $\Xi(x, \mu(x)) = 0$. For the curve $\mu = \mu(x)$ to have, close to $x = 0$, the geometric properties of $\mu = x^2$, it suffices to have

$$
\frac{d\mu}{dx}(0) = 0 \quad \text{and} \quad \frac{d^2\mu}{dx^2}(0) \neq 0.
$$

Differentiating the relation $\Xi(x, \mu(x)) = 0$, we therefore obtain

$$
\frac{d\mu}{dx}(0) = -\frac{\dfrac{\partial \Xi}{\partial x}(0,0)}{\dfrac{\partial \Xi}{\partial \mu}(0,0)} = -\frac{\dfrac{\partial^2 X}{\partial^2 x}(0,0)}{\dfrac{\partial^2 X}{\partial \mu \partial x}(0,0)} = 0,
$$

$$
\frac{d^2\mu}{dx^2}(0) = -\frac{\dfrac{\partial^\Xi}{\partial^2 x}(0,0)}{\dfrac{\partial \Xi}{\partial \mu}(0,0)} = -\frac{\dfrac{\partial^3 X}{\partial^3 x}(0,0)}{\dfrac{\partial^2 X}{\partial \mu \partial x}(0,0)} \neq 0.
$$

In short, the one-parameter family

$$
\dot{x} = X(x,\mu)
$$

undergoes a pitchfork bifurcation if

$$
X(0,0) = 0 \quad \text{and} \quad \frac{\partial X}{\partial x}(0,0) = 0,
$$

showing that $(0,0)$ is a nonhyperbolic equilibrium point, and

$$
\frac{\partial X}{\partial \mu}(0,0) = 0 \quad \frac{\partial^2 X}{\partial^2 x}(0,0) = 0, \quad \frac{\partial^2 X}{\partial \mu \partial x}(0,0) \neq 0, \quad \text{and} \quad \frac{\partial^3 X}{\partial^3 x}(0,0) \neq 0,
$$

which imply the existence of two curves of equilibrium points passing through $(0,0)$, one being defined on both sides and the other one only on one side of this point. In order to exhibit a pitchfork bifurcation, the one-parameter family X has to be C^k with $k \geq 3$.

Example 19. Bead sliding on a rotating hoop. Consider a bead of mass m sliding without friction on a hoop of radius R (Figure 3.18). The hoop rotates with angular velocity ω about a vertical axis. The acceleration due to gravity is g. If x is the angle between the vertical downward direction and the position vector of the bead, the Lagrangian of the system is

$$L(x, \dot{x}) = \frac{1}{2} mR^2 \dot{x}^2 + \frac{1}{2} m\omega^2 R^2 \sin^2 x + mgR \cos x,$$

and the equation of motion is

$$mR^2 \ddot{x} = m\omega^2 R^2 \sin x \cos x - mgR \sin x$$

or

$$R\ddot{x} = \omega^2 R \sin x \cos x - g \sin x.$$

Introducing the reduced variables

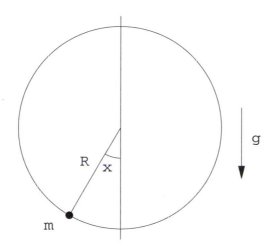

Fig. 3.18. *Bead sliding on a vertical rotating hoop without friction.*

$$\tau = \sqrt{\frac{g}{R}}\, t, \quad \mu = \frac{R}{g} \omega^2,$$

the equation of motion takes the form

$$\frac{d^2 x}{d\tau^2} = \sin x (\mu \cos x - 1). \qquad (3.37)$$

This equation may be written as a first-order system. If we put $x_1 = x$ and $x_2 = \dot{x}$, where now the dot represents derivation with respect to reduced time τ, Equation (3.37) is replaced by the system

$$\dot{x}_1 = x_2, \quad \dot{x}_2 = \sin x_1 (\mu \cos x_1 - 1). \tag{3.38}$$

The state of the system represented by $\mathbf{x} = (x_1, x_2)$ belongs to the cylinder $[0, 2\pi[\times \mathbb{R}$. In this phase space, there exist four equilibrium points

$$(0,0), \quad (\pi, 0), \quad (\pm \arccos \mu^{-1}, 0).$$

At these equilibrium points, the Jacobian of the system $J(x_1^*, x_2^*)$ is given by

$$J(0,0) = \begin{bmatrix} 0 & 1 \\ \mu - 1 & 0 \end{bmatrix}, \quad J(\pi, 0) = \begin{bmatrix} 0 & 1 \\ \mu + 1 & 0 \end{bmatrix},$$

$$J(\pm \arccos \mu^{-1}, 0) = \begin{bmatrix} 0 & 1 \\ \dfrac{1 - \mu^2}{\mu} & 0 \end{bmatrix}.$$

If $0 < \mu < 1$, $(0,0)$ is Lyapunov stable but not asymptotically stable, $(\pi, 0)$ is unstable, and $(\pm \arccos \mu^{-1}, 0)$ do not exist. For μ slightly greater than 1, $(0, 0)$ is unstable, and the two equilibrium points $(\pm \arccos \mu^{-1}, 0)$ are stable. The nonhyperbolic equilibrium point $(x, \mu) = (0, 1)$ is a pitchfork bifurcation point. For $0 < \mu < 1$, the bead oscillates around the point $x = 0$, while for μ slightly greater than 1, it oscillates around either $x = \arccos \mu^{-1}$ or $x = -\arccos \mu^{-1}$. Since the Lagrangian is invariant under the transformation $x \to -x$, this system exhibits a symmetry-breaking bifurcation similar to those characterizing second-order phase transitions in some simple magnetic systems.

3.5.2 Equivalent families of vector fields

In a one-dimensional system, Conditions (3.32) are necessary but not sufficient for the system to exhibit a bifurcation. For instance, the equation

$$\dot{x} = \mu - x^3$$

does not exhibit a bifurcation at the equilibrium point $(x^*, \mu^*) = (0, 0)$. A simple analysis shows that, for all real μ, the equilibrium point $x^* = (\mu)^{1/3}$ is always asymptotically stable. In such a case, it is said that the family of vector fields $X(x, \mu)$ does *unfold* the singularity in $X(x, 0)$. More precisely, any local family, $\mathbf{X}(\mathbf{x}, \boldsymbol{\mu})$, at $(\mathbf{x}^*, \boldsymbol{\mu}^*)$ is said to be an *unfolding* of the vector field $\mathbf{X}(\mathbf{x}, \boldsymbol{\mu}^*)$. When $\mathbf{X}(\mathbf{x}, \boldsymbol{\mu}^*)$ has a singularity at $\mathbf{x} = \mathbf{x}^*$, $\mathbf{X}(\mathbf{x}, \boldsymbol{\mu})$ is referred to as an *unfolding of the singularity*.

Definition 12. *Two local families* $\mathbf{X}(\mathbf{x}, \boldsymbol{\mu})$ *and* $\mathbf{Y}(\mathbf{y}, \boldsymbol{\nu})$ *are said to be equivalent if there exists a continuous mapping* $(\mathbf{x}, \boldsymbol{\mu}) \mapsto \mathbf{h}(\mathbf{x}, \boldsymbol{\mu})$, *defined in a neighborhood of* $(\mathbf{x}^*, \boldsymbol{\mu}^*)$, *satisfying* $\mathbf{h}(\mathbf{x}^*, \boldsymbol{\mu}^*) = \mathbf{y}^*$, *such that for each* $\boldsymbol{\mu}$, $\mathbf{x} \mapsto \mathbf{h}(\mathbf{x}, \boldsymbol{\mu})$ *is a homeomorphism that exhibits the topological equivalence*[45] *of the flows generated by* $\mathbf{X}(\mathbf{x}, \boldsymbol{\mu})$ *and* $\mathbf{Y}(\mathbf{y}, \boldsymbol{\nu})$.

---

[45] On topological equivalence, see Remark 2.

Example 20. The family of vector fields $X(x, \mu) = \mu x - x^2$ can be written

$$X(x, \mu) = \frac{\mu^2}{4} - \left(x - \frac{\mu}{2}\right)^2.$$

If $y = h(x, \mu) = x - \frac{1}{2}\mu$, then, for each μ,

$$\dot{y} = \dot{x} = \frac{\mu^2}{4} - \left(x - \frac{\mu}{2}\right)^2 = \frac{\mu^2}{4} - y^2 = Y(y, \nu),$$

where $\nu = \frac{1}{4}\mu^2$. Hence, the two families $X(x, \mu) = \mu x - x^2$ and $Y(y, \nu) = \nu - y^2$ are topologically equivalent.

3.5.3 Hopf bifurcation

If the differential equation $\dot{x} = X(x, \mu)$ exhibits a bifurcation of one of the types described above, clearly the two-dimensional system

$$\dot{x}_1 = X(x_1, \mu), \quad \dot{x}_2 = -x_2$$

also exhibits the same type of bifurcation. We will not insist. In this section we present a new type of bifurcation that does not exist in a one-dimensional system: the *Hopf bifurcation*. This type of bifurcation appears at a nonhyperbolic equilibrium point of a two-dimensional system whose eigenvalues of its linear part are pure imaginary.

We have already found Hopf bifurcations in two models: the prey-predator model used by Harrison, which predicts the outcome of Luckinbill's experiment with *Didinium* and *Paramecium* qualitatively (Section 2.3), and the van der Pol oscillator (Example 16).

The following theorem indicates under which conditions a two-dimensional system undergoes a Hopf bifurcation.[46]

Theorem 4. *Let $\dot{\mathbf{x}} = \mathbf{X}(\mathbf{x}, \mu)$ be a two-dimensional one-parameter family of vector fields ($\mathbf{x} \in \mathbb{R}^2$, $\mu \in \mathbb{R}$) such that*

1. $\mathbf{X}(\mathbf{0}, \mu) = \mathbf{0}$,
2. $\mathbf{X}$ *is an analytic function of* $\mathbf{x}$ *and* μ, *and*
3. $D_{\mathbf{x}}\mathbf{X}(\mathbf{0}, \mu)$ *has two complex conjugate eigenvalues* $\alpha(\mu) \pm i\omega(\mu)$ *with* $\alpha(0) = 0$, $\omega(0) \neq 0$, *and* $d\alpha/d\mu|_{\mu=0} \neq 0$;

then, in any neighborhood $U \subset \mathbb{R}^2$ of the origin and any given $\mu_0 > 0$ there is a $\mu < \mu_0$ such that the equation $\dot{\mathbf{x}} = \mathbf{X}(\mathbf{x}, \mu)$ has a nontrivial periodic orbit in U.

---

[46] For a proof, see Hassard, Kazarinoff, and Wan [166], who present various proofs and many applications; see also Hale and Koçak [164].

Example 21. The van der Pol system revisited. In the case of the van der Pol system (3.26), we have

$$D_{\mathbf{x}}\mathbf{X}(\mathbf{0}, \lambda) = \begin{bmatrix} 0 & 1 \\ -1 & \lambda \end{bmatrix}.$$

The eigenvalues of the linear part of the van der Pol system are $\frac{1}{2}(\lambda \pm i\sqrt{4 - \lambda^2})$. If $\lambda < 0$, the origin is asymptotically stable, and if $\lambda > 0$ the origin is unstable. Since the real part of both eigenvalues is equal to $\frac{1}{2}\lambda$, all the conditions of Theorem 4 are verified.

Example 22. Section 2.3 model. Harrison found that the model described by the system

$$\dot{H} = r_H H \left(1 - \frac{H}{K} \right) - \frac{a_H PH}{b + H},$$

$$\dot{P} = \frac{a_P PH}{b + H} - cP,$$

exhibits a stable limit cycle and is in good qualitative agreement with Luckinbill's experiment. Using the reduced variables

$$h = \frac{H}{H^*}, \quad p = \frac{P}{P^*}, \quad \tau = r_H t, \quad k = \frac{K}{H^*}, \quad \beta = \frac{b}{H^*}, \quad \gamma = \frac{c}{r},$$

where (H^*, P^*) denotes the nontrivial fixed point, the model can be written

$$\frac{dh}{d\tau} = h \left(1 - \frac{h}{k} \right) - \frac{\alpha_h ph}{\beta + h},$$

$$\frac{dp}{d\tau} = \frac{\alpha_p ph}{\beta + h} - \gamma p,$$

where

$$\alpha_h = \left(1 - \frac{1}{k} \right) (\beta + 1) \quad \text{and} \quad \alpha_p = \gamma(\beta + 1).$$

The Jacobian at the equilibrium point $(1, 1)$ is

$$\begin{bmatrix} \dfrac{k - 2 - \beta}{k(1 + \beta)} & -1 + \dfrac{1}{k} \\ \dfrac{\beta\gamma}{1 + \beta} & 0 \end{bmatrix}.$$

Its eigenvalues are

$$\frac{k - 2 - \beta \pm i\sqrt{4(k - 1)k^2\beta\gamma - (k - 2 - \beta)^2}}{2k(1 + \beta)}.$$

The equilibrium point is asymptotically stable if $k < 2+\beta$. Since the derivative of the real part of the eigenvalues with respect to the parameter k is different from zero at the bifurcation point, all the conditions of Theorem 4 are verified: there exists a limit cycle for $k > 2 + \beta$.

It is often desirable to prove that a two-dimensional system does not possess a limit cycle. The following theorem, known as the *Dulac criterion* [111], is often helpful to prove the nonexistence of a limit cycle.

Theorem 5. *Let Ω be a simply connected subset of $\mathbb{R}^2$ and $D : \Omega \to \mathbb{R}$ a C^1 function. If the function*

$$\mathbf{x} \mapsto \nabla D\mathbf{X}(\mathbf{x}) = \frac{\partial DX_1}{\partial x_1} + \frac{\partial DX_2}{\partial x_2}$$

has a constant sign and is not identically zero in Ω, then the two-dimensional system $\dot{\mathbf{x}} = \mathbf{X}(\mathbf{x})$ has no periodic orbit lying entirely in Ω.

If there exists a closed orbit γ in Ω, Green's theorem implies

$$\oint_\gamma D(\mathbf{x})X_1(\mathbf{x})\,dx_2 - D(\mathbf{x})X_2(\mathbf{x})\,dx_1 = \int_\Gamma \left(\frac{\partial DX_1}{\partial x_1} + \frac{\partial DX_2}{\partial x_2} \right)\,dx_1\,dx_2 \neq 0,$$

where Γ is the bounded *interior* of γ ($\partial\Gamma = \gamma$).[47] But if γ is an orbit, we also have

$$\oint_\gamma D(\mathbf{x})\big(X_1(\mathbf{x})\,\dot{x}_2 - X_2(\mathbf{x})\,\dot{x}_1\big)\,dt = \oint_\gamma D(\mathbf{x})\big(X_1(\mathbf{x})dx_2 - X_2(\mathbf{x})dx_1\big) = 0,$$

where we have taken into account that $\dot{x}_i = X_i(\mathbf{x})$ for $i = 1, 2$. This contradiction proves the theorem.

The function D is a *Dulac function*. There is no general method for finding an appropriate Dulac function. If $D(x_1, x_2) = 1$, the theorem is referred to as the *Bendixson criterion* [40].

3.5.4 Catastrophes

Catastrophe theory studies abrupt changes associated with smooth modifications of control parameters.[48] In this section, we describe one of the simplest types found in many models: the cusp catastrophe.

---

[47] Since a closed orbit in $\mathbb{R}^2$ does not intersect itself, it is a closed Jordan's curve and, according to Jordan's theorem, it separates the plane into two disjoint connected components such that only one of these two components is bounded. The closed Jordan's curve is the common boundary of the two components.

[48] On catastrophe theory, one should consult Thom [328]. On the mathematical presentation of the theory, refer to the recent book of Castrigiano and Hayes [82] with a preface by René Thom, the founder of the theory. On early applications of the theory, see Zeeman [354]. Many applications of catastrophe theory have been heavily criticized. For example, here is a quotation from the English translation by Wassermann and Thomas of Arnol'd [10], p. 9:

> I remark that articles on catastrophe theory are distinguished by a sharp and catastrophic lowering of the level of demands of rigour and also of novelty of published results.

In [329] Thom expounds his philosophical standpoint and shows that a qualitative approach may offer a subtler explanation than a purely quantitative description.

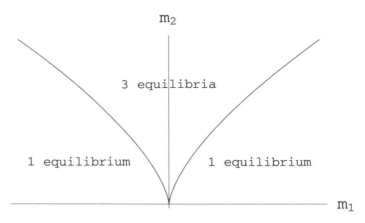

Fig. 3.19. *The bifurcation diagram of the differential equation $\dot{x} = m_1 + m_2 x - x^3$.*

Consider the two-parameter family of maps on $\mathbb{R}$

$$(x, m_1, m_2) \mapsto X(x, m_1, m_2) = m_1 + m_2 x - x^3.$$

Depending upon the values of the parameters, the differential equation $\dot{x} = X(x, m_1, m_2)$ has either one or three equilibrium points. Since at bifurcation points the differential equation must have multiple equilibrium points, bifurcation points are the solutions of the system

$$X(x, m_1, m_2) = 0, \quad \frac{\partial X}{\partial x}(x, m_1, m_2) = 0,$$

that is,

$$m_1 + m_2 x - x^3 = 0, \quad m_2 - 3x^2 = 0.$$

Eliminating x, we find

$$27 m_1^2 - 4 m_2^3 = 0,$$

which is the equation of the boundary between the domains in the parameter space in which the differential equation has either one or three equilibrium points. It is the equation of a cusp (Figure 3.19).

The bifurcation diagram represented in Figure 3.20 will help us understand the nature of the cusp catastrophe. In this figure, the solid line corresponds to asymptotically stable equilibrium points and the broken line to unstable points. Suppose that the parameter m_2 has a fixed value, say 1. If $m_1 > 2/3\sqrt{3}$, there exists only one equilibrium point, represented by point A. This equilibrium is asymptotically stable. As m_1 decreases, the point representing the equilibrium moves along the curve in the direction of the arrow. It passes point E, where nothing special occurs, to finally reach point B. If we try to further decrease m_1, the state of the system, represented by the x-coordinate of B, jumps to the value of the x-coordinate of C, and, for decreasing values of m_1, the asymptotically stable equilibrium points will be represented by the points on the solid line below C. If now we start increasing m_1, the point representing the asymptotically stable equilibrium will go back to C and proceed up to point D, where, if we try to further increase m_1, it will jump to E and move up along the solid line towards A as the parameter is varied.

The important fact is that the state of the system has experienced jumps at two different values of the control parameter m_1. The parameter values at which jumps take place depend upon the direction in which the parameter is varied. This phenomenon, well-known in physics, is referred to as *hysteresis*, and the closed path $BCDEB$ is called the *hysteresis loop*. It is important to note that the cusp catastrophe, described here, occurs in dynamical systems in which the vector field X is a *gradient field*; that is,

$$X(x, m_1, m_2) = -\nabla \Phi(x, m_1, m_2),$$

where

$$\Phi(x, m_1, m_2) = \tfrac{1}{4}\, x^4 - \tfrac{1}{2}\, m_2 x^2 - m_1 x.$$

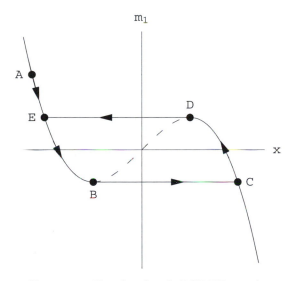

Fig. 3.20. *Hysteresis. The closed path BCDEB is a hysteresis loop.*

Example 23. Street gang control.[49] Street gangs have emerged as tremendously powerful institutions in many communities. In urban ghettos, they may very

<hr />

[49] I presented a first version of this model at a meeting of the Research Police Forum organized at Starved Rock (IL) by the Police Training Institute of the University of Illinois at Urbana-Champaign in the early 1990s and a more elaborate version a few weeks later at the Criminal Justice Authority (Chicago, IL). It was prepared

well be the most important institutions in the lives of a large proportion of adolescent and young adult males. To model the dynamics of a gang population, assume that the growth rate of the gang population size N is given by

$$\dot{N} = g(N) - p(N),$$

where g is the intrinsic growth function and p the police response function, which describes the amount of resources the police devote at each level of the population. It might be understandably objected that coercive methods are not sufficient; social programs and education also play an important role. The expression $p(N)$ should, therefore, represent the amount of resources devoted by society as a whole.

Using a slight variant of the logistic function, assume that

$$g(N) = r(N + N_0)\left(1 - \frac{N}{K}\right),$$

where r, K, and N_0 are positive constants. Note that the "initial condition" N_0 implies that a zero gang population is not an equilibrium.

For the response function p, assume that, as gang membership in a community grows, the society will devote more resources to the problem. p should, therefore, be monotonically increasing, but since resources are limited, the investment will ultimately approach some maximum level. This pattern can be modeled by

$$p(N) = \frac{aN^\xi}{b + N^\xi},$$

where a, b, and ξ are positive constants. The parameter a is the maximum response level. This maximum is approached faster for decreasing values of b and ξ.[50] The parameter ξ is a measure of how "tough" the society response is.

Introducing the dimensionless variables

$$\tau = rt, \quad n = \frac{N}{K}, \quad \alpha = \frac{a}{rK}, \quad \beta = \frac{b}{K^\xi},$$

the dynamics of the reduced gang population n is modeled by

$$\frac{dn}{d\tau} = (n + n_0)(1 - n) - \frac{\alpha n^\xi}{\beta + n^\xi}. \tag{3.39}$$

The equilibrium points are the solutions to the equation

---

in collaboration with Jonathan Crane from the Department of Sociology of the University of Illinois at Chicago and the Institute of Government and Public Affairs, Center for Prevention Research and Development.

[50] Note the similarity of this street gang control model with the Ludwig-Holling-Jones model of budworm outbreaks (Equation (1.6)).

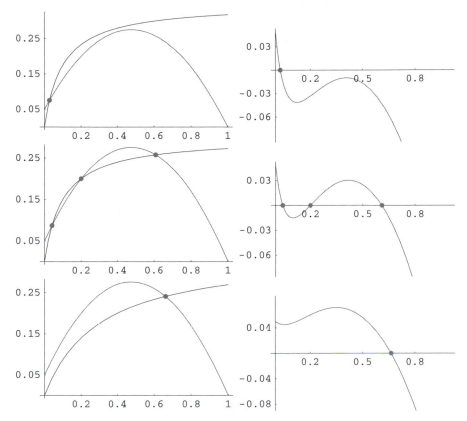

Fig. 3.21. *Equilibrium points of the street gang model. The graphs on the left show the intersections of the graphs of $N \mapsto g(N)$ and $N \mapsto p(N)$; on the right are represented the graphs of $N \mapsto g(n) - p(n)$. Top: one stable low-density equilibrium; middle: two stable equilibria, one at low density and one at high density, separated by one unstable equilibrium at an intermediate density; bottom: one stable equilibrium at high density.*

$$(n + n_0)(1 - n) = \frac{\alpha n^\xi}{\beta + n^\xi}. \tag{3.40}$$

As shown in Figure 3.21, depending upon the values of the parameters, there exist one or three equilibrium points. The equation of the boundary separating the domains in which there exist either one or three equilibrium points is determined by eliminating n between Equation (3.40) and

$$\frac{d}{dn}(n + n_0)(1 - n) = \frac{d}{dn}\frac{\alpha n^\xi}{\beta + n^\xi};$$

that is,

$$1 - n_0 - 2n = \frac{\xi \alpha \beta n^{\xi-1}}{(\beta + n^\xi)^2}. \tag{3.41}$$

Eliminating α between (3.40) and (3.41), we solve for β, and then, replacing the expression of β in one of the equations, we obtain α. The parametric equation of the boundary is then

$$\alpha = \frac{\xi(n+n_0)^2(1-n)^2}{(2-\xi)n^2 + (\xi-1)(1-n_0)n + \xi n_0},$$

$$\beta = \frac{n^{\xi+1}(1-n_0-2n)}{(2-\xi)n^2 + (\xi-1)(1-n_0)n + \xi n_0}.$$

It is represented in Figure 3.22. It shows the existence of a cusp catastrophe in the model.

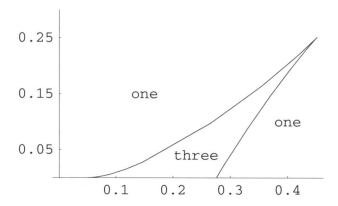

Fig. 3.22. *Boundary between the domains in which there are three equilibrium points and one equilibrium point in the $(\alpha - \beta)$ parameter space.*

When dramatic increases in gang populations have occurred, there are usually numerous attempts to try to reverse the rise. Youth employment programs, educational programs in schools, and increases in police resources are common interventions. Such interventions can prevent some individuals from joining the gang, but none of them seem to have had much effect on overall gang populations. This model suggests one possible reason for this.

Due to the existence of the hysteresis effect, small-scale interventions to reduce gang activity may have no permanent effect at all. An intervention may temporarily push gang membership to a slightly lower equilibrium, but unless the intervention is large enough to push the population all the way down to the unstable equilibrium, membership will move back up to the high equilibrium.

While this implication is pessimistic, the model does suggest a strategy that could succeed. If the intervention is large enough to push gang membership below the unstable equilibrium, the gang population will revert to a low equilibrium, and because of the stability of the low equilibrium, the intervention does not have to be continued once the low level is achieved. Thus a short-term but high-intensity intervention might succeed where a long-term, low-intensity strategy would fail.

3.6 Influence of diffusion

In all the models discussed so far, it has always been assumed that the various species were uniformly distributed in space. In other words, over the whole territory available to them, individuals were supposed to mix homogeneously. In a variety of problems, the spatial dimension of the environment cannot, however, be ignored.

In this section, we study the influence of diffusion on the evolution of various populations[51]; that is, we investigate the dynamics of populations assuming that the individuals move at random.[52] We will discover that diffusion can have a profound effect on the dynamics of populations.

3.6.1 Random walk and diffusion

Consider a *random walker* who, in a one-dimensional space, takes a step of length ξ during each time interval τ. Assuming that the steps are taken either in the positive or negative direction with equal probabilities, let $p(t, x)\, dx$ be the probability that the random walker is between x and $x + dx$ at time t. The random walker is between x and $x + dx$ at time t if he was either between $x + \xi$ and $x + \xi + dx$ or between $x - \xi$ and $x - \xi + dx$ at time $t - \tau$. Thus, the function p satisfies the following difference equation

$$p(x, t) = \tfrac{1}{2}(p(x + \xi, t - \tau) + p(x - \xi, t - \tau)). \qquad (3.42)$$

If p is continuously differentiable, and if we assume that τ and ξ are small, then

$$p(x \pm \xi, t - \tau) = p(x, t) - \frac{\partial p}{\partial t}(x, t)\tau + O(\tau^2) \pm \frac{\partial p}{\partial x}(x, t)\xi + \frac{1}{2}\frac{\partial^2 p}{\partial x^2}(x, t)\xi^2 + O(\xi^3),$$

and substituting in (3.42) yields

$$\frac{\partial p}{\partial t}(x, t) = \frac{\xi^2}{2\tau}\frac{\partial^2 p}{\partial x^2}(x, t) + \frac{1}{\tau}(O(\tau^2) + O(\xi^3)).$$

Hence, if there exists a positive constant D such that, when both τ and ξ tend to 0, $\xi^2/2\tau$ tends to a constant D, p satisfies the diffusion equation

$$\frac{\partial p}{\partial t} = D\frac{\partial^2 p}{\partial x^2}. \qquad (3.43)$$

---

[51] The standard text on diffusion in ecology is Okubo [272].

[52] The interest in random dispersal in populations was triggered by the paper of Skellam [314].

If, at $t = 0$, the random walker is at the origin, $p(t, x)$ satisfies the initial condition

$$\lim_{t \to 0} p(x, t) = \delta,$$

where δ is the Dirac distribution, and for $t > 0$, the solution of (3.43) is[53]

$$p(x, t) = \frac{1}{2\sqrt{\pi Dt}} \exp -\frac{x^2}{4Dt}. \qquad (3.44)$$

The random variable representing the distance of the random walker from the origin is, therefore, *normally distributed*. Its mean value is 0, and its variance $2Dt$ varies *linearly* with time. This result is important. It is indeed a direct consequence of the fact that p is a generalized homogeneous function of t and x.[54]

3.6.2 One-population dynamics with dispersal

As a simple example, consider the influence of random dispersal in a two-dimensional space on the evolution of a Malthusian population. Including a diffusion term in the equation for the population density $\dot{n} = an$, we obtain

---

[53] If

$$\widehat{p}(k, t) = \int_{-\infty}^{\infty} p(x, t) e^{ikx} \, dx$$

denotes the Fourier transform of $x \mapsto p(x, t)$, we have

$$\frac{d\widehat{p}}{dt} + Dk^2 \widehat{p} = 0,$$

and $\widehat{p}(k, 0) = 1$. Hence, $\widehat{p}(k, t) = \exp(-Dk^2 t)$. Therefore,

$$p(x, t) = \frac{1}{2\pi} \int_{-\infty}^{\infty} \exp(-Dk^2 t - ikx) \, dk = \frac{1}{\sqrt{2\pi Dt}} \exp -\frac{x^2}{4Dt}.$$

This particular solution is known as a *fundamental solution* of the diffusion equation. It can be shown (see Boccara [46]) that, if the initial condition is $p(x, 0) = f(x)$, then the solution of the diffusion equation is simply given by the convolution $g * f$, where g is a fundamental solution. Fundamental solutions are not unique. In physics, fundamental solutions are referred to as *Green's functions*.

[54] $f : \mathbb{R}^n \to \mathbb{R}$ is a *generalized homogeneous function* if, for all $\lambda \in \mathbb{R}$,

$$f(\lambda^{a_1} x_1, \lambda^{a_2} x_2, \ldots, \lambda^{a_n} x_n) \equiv \lambda^r f(x_1, x_2, \ldots, x_n),$$

where $a_1, a_2, \ldots, a_n$ and r are real constants. Since λ can be any real number, we can replace λ by $1/x_1^{1/a_1}$; f may then be written as a function of the $n - 1$ reduced variables $x_i / x_1^{a_i/a_1}$ ($i = 2, 3, \ldots, n$). It is said that x_i scales as $x_1^{a_i/a_1}$. Since the probability density p satisfies the relation $p(\lambda^2 t, \lambda x) \equiv \lambda p(t, x)$, x scales as $\sqrt{t}$.

$$\frac{\partial n}{\partial t} = D\left(\frac{\partial^2 n}{\partial x_1^2} + \frac{\partial^2 n}{\partial x_2^2}\right) + an. \tag{3.45}$$

The exponential growth of a spreading population is an acceptable approximation if, initially, the population consists of very few individuals that spread and reproduce in a habitat where natural enemies, such as competitors and predators, are lacking. If we assume that the dispersal is isotropic, the population dispersal is modeled by

$$\frac{\partial n}{\partial t} = \frac{D}{r}\frac{\partial}{\partial r}\left(r\frac{\partial n}{\partial r}\right) + an, \tag{3.46}$$

where $r = \sqrt{x_1^2 + x_2^2}$.

Introducing the dimensionless variables

$$\tau = at \quad \text{and} \quad \rho = r\sqrt{\frac{a}{D}}, \tag{3.47}$$

Equation (3.46) becomes

$$\frac{\partial n}{\partial \tau} = \frac{1}{\rho}\frac{\partial}{\partial \rho}\left(\rho\frac{\partial n}{\partial \rho}\right) + n. \tag{3.48}$$

Assuming that, at time $t = 0$, there are N_0 individuals concentrated at the origin, the solution of (3.48) is

$$n(\rho, \tau) = \frac{N_0}{4\pi\tau}\exp\left(\tau - \frac{\rho^2}{4\tau}\right). \tag{3.49}$$

The total number of individuals that, at time t, are at a distance greater than R from the origin is given by

$$N(R, t) = \int_R^\infty n(r\sqrt{a/D}, at)\, 2\pi r\, dr$$

$$= N_0\exp\left(at - \frac{R^2}{4Dt}\right).$$

For $R^2 = 4aDt^2$, the number of individuals that, at time t, are outside a circle of radius R, equal to N_0, is negligible compared to the total number of individuals $N_0 e^{at}$. Hence, the radius of an approximate boundary of the habitat occupied by the invading species is proportional to t. According to Skellam [314], this result is in agreement with data on the spread of the muskrat (*Ondatra zibethica*), an American rodent, introduced inadvertently into Central Europe in 1905. The fact that, for the random dispersal of a Malthusian population, R is proportional to t contrasts with simple diffusion, where space scales as the square root of time.

3.6.3 Critical patch size

Consider a *refuge* (*i.e.*, a patch of favorable environment surrounded by a region where survival is impossible). If the population is diffusing, individuals crossing the patch boundary will be lost. The problem is to find the *critical patch size* ℓ_c such that the population cannot sustain itself against losses from individuals crossing the patch boundary if the patch size is less than ℓ_c but can maintain itself indefinitely if the patch size is greater than ℓ_c. As a simplification, we discuss a one-dimensional problem (*i.e.*, we assume that in two dimensions, the patch Σ is an infinite strip of width ℓ):

$$\Sigma = \left\{ (x, y) \mid -\tfrac{1}{2}\ell < x < \tfrac{1}{2}\ell, -\infty < y < \infty \right\}.$$

Consider the scaled equation

$$\frac{\partial n}{\partial t} = \frac{\partial^2 n}{\partial x^2} + n, \tag{3.50}$$

where a, the growth rate of the population density, has been absorbed in t and $\sqrt{D/a}$, the square root of the diffusion coefficient divided by the growth rate, in x (see (3.47)). The condition that survival is impossible outside Σ implies

$$n(x, t) = 0 \quad \text{if} \quad x = \pm \tfrac{1}{2}\ell.$$

Assuming that the solution $n(x, t)$ of (3.50), as a function of x, can be represented as a convergent Fourier series, one finds that this solution can be written as

$$n(x, t) = \sum_{k=1}^{\infty} n_k \exp\left(\left(1 - \frac{k^2 \pi^2}{\ell^2}\right) t\right) \sin \frac{k\pi}{\ell} \left(x + \tfrac{1}{2}\ell\right). \tag{3.51}$$

If $\ell < \pi$, then, for all positive integers k, $1 - k^2/\pi^2\ell^2 < 0$, and $n(x, t)$ given by (3.51) goes to zero exponentially as t tends to infinity. Therefore, for $\ell < \pi$, the strip Σ is not a refuge. If $\ell > \pi$, then, for any initial population density, Σ is a refuge since the amplitude of the first Fourier component of $n(x, t)$ grows without limit when t tends to infinity. The reduced critical size is then $\ell_c = \pi$. In terms of the original space variable, the critical size is $\sqrt{D/a}\,\pi$.

Remark 5. If, for $-\tfrac{1}{2}\ell \le x \le \tfrac{1}{2}\ell$, the initial density $n(x, 0)$ is bounded by M, then, for all $t > 0$ and all $|x| \le \tfrac{1}{2}\ell$,

$$n(x, t) \le \frac{4M}{\pi} \sum_{k=0}^{\infty} \frac{1}{2k + 1} \exp\left(\left(1 - \frac{(2k+1)^2 \pi^2}{\ell^2}\right) t\right)$$

$$\times \sin \frac{(2k+1)\pi}{\ell} \left(x + \frac{1}{2}\ell\right). \tag{3.52}$$

If the growth of the population density obeys the reduced logistic equation $\dot{n} = n(1 - n)$, Ludwig, Aronson, and Weinberg [216] have shown that it grows less rapidly than if it were Malthusian and, therefore, goes to zero, as $t \to \infty$, if $\ell < \pi$. If $\ell > \pi$, then, as $t \to \infty$, $n(x,t)$ tends to the solution of $u'' + u(1-u) = 0$, satisfying the condition $u(\pm \ell/2) = 0$. In their paper, Ludwig, Aronson, and Weinberg apply their method to the Ludwig-Jones-Holling model of spruce budworm outbreak (Equation (1.4)). They show that, for this system, there exist two critical strip widths. The smaller one gives a lower bound for the strip width that can support a nonzero population. The larger one is the lower bound for the strip width that can support an outbreak.

3.6.4 Diffusion-induced instability

Since diffusion tends to mix the individuals, it seems reasonable to expect that a system eventually evolves to a homogeneous state. Thus, diffusion should be a stabilizing factor. This is not always the case, and the *Turing effect* [334], or *diffusion-induced instability*, is an important exception. Turing's paper, which shows that the reaction (interaction) and diffusion of chemicals can give rise to a spatial structure, suggests that this instability could be a key factor in the formation of biological patterns.[55] Twenty years later, Segel and Jackson [311] showed that diffusive instabilities could also appear in an ecological context.

Consider two populations N_1 and N_2 evolving according to the following one-dimensional diffusion equations[56]:

$$\frac{\partial N_1}{\partial t} = f_1(N_1, N_2) + D_1 \frac{\partial^2 N_1}{\partial x^2}, \tag{3.53}$$

$$\frac{\partial N_2}{\partial t} = f_2(N_1, N_2) + D_2 \frac{\partial^2 N_2}{\partial x^2}. \tag{3.54}$$

$f_1(N_1, N_2)$ and $f_2(N_1, N_2)$ denote the interaction terms, D_1 and D_2 are the diffusion coefficients, and x is the spatial coordinate. We assume the existence of an asymptotically stable steady state (N_1^*, N_2^*) in the absence of diffusion. To examine the stability of this uniform solution to perturbations, we write

$$N_1(t, x) = N_1^* + n_1(t, x) \quad \text{and} \quad N_2(t, x) = N_2^* + n_2(t, x). \tag{3.55}$$

If $n_1(t, x)$ and $n_2(t, x)$ are small, we can linearize the equations obtained upon substituting (3.55) in Equations (3.53) and (3.54). We obtain[57]

$$\frac{\partial n_1}{\partial t} = a_{11} n_1 + a_{12} n_2 + D_1 \frac{\partial^2 n_1}{\partial x^2},$$

$$\frac{\partial n_2}{\partial t} = a_{21} n_1 + a_{22} n_2 + D_2 \frac{\partial^2 n_2}{\partial x^2},$$

where the constants a_{ij}, for $i = 1, 2$ and $j = 1, 2$, are given by

---

[55] A rich variety of models is discussed in Murray [255].
[56] Extension to the more realistic two-dimensional case is straightforward.
[57] N_1^* and N_2^* are such that $f_1(N_1^*, N_2^*) = 0$ and $f_2(N_1^*, N_2^*) = 0$.

$$a_{ij} = \frac{\partial f_i}{\partial N_j}(N_1^*, N_2^*).$$

Solving the system of linear partial differential equations above is a standard application of Fourier transform theory.[58] Let

$$n_1(t, x) = \frac{1}{2\pi} \int_{-\infty}^{\infty} \widehat{n_1}(t, k) e^{-ikx} \, dk, \tag{3.56}$$

$$n_2(t, x) = \frac{1}{2\pi} \int_{-\infty}^{\infty} \widehat{n_2}(t, k) e^{-ikx} \, dx. \tag{3.57}$$

Replacing (3.56) and (3.57) in the system of linear partial differential equations yields the following system of ordinary linear differential equations

$$\frac{d\widehat{n_1}}{dt} = a_{11}\widehat{n_1} + a_{12}\widehat{n_2} - D_1 k^2 \widehat{n_1},$$

$$\frac{d\widehat{n_2}}{dt} = a_{21}\widehat{n_1} + a_{22}\widehat{n_2} - D_2 k^2 \widehat{n_2}.$$

This system has solutions of the form

$$\widehat{n_1}(t, k) = \widehat{n_{01}}(k) e^{\lambda t},$$

$$\widehat{n_2}(t, k) = \widehat{n_{02}}(k) e^{\lambda t},$$

where λ is an eigenvalue of the 2×2 matrix

$$\begin{bmatrix} a_{11} - D_1 k^2 & a_{12} \\ a_{21} & a_{22} - D_2 k^2 \end{bmatrix}.$$

For a diffusive instability to set in, at least one of the conditions

$$a_{11} - D_1 k^2 + a_{22} - D_2 k^2 < 0, \tag{3.58}$$

$$(a_{11} - D_1 k^2)(a_{22} - D_2 k^2) - a_{12} a_{21} > 0, \tag{3.59}$$

should be violated. Since (N_1^*, N_2^*) is asymptotically stable, the conditions

$$a_{11} + a_{22} < 0, \tag{3.60}$$

$$a_{11} a_{22} - a_{12} a_{21} > 0, \tag{3.61}$$

are satisfied. From Condition (3.60) it follows that (3.58) is always satisfied. Therefore, an instability can occur if, and only if, Condition (3.59) is violated. If $D_1 = D_2 = D$, the left-hand side of (3.59) becomes

$$a_{11} a_{22} - a_{12} a_{21} - Dk^2 (a_{11} + a_{22}) + D^2 k^4.$$

---

[58] For discussion of Fourier transform theory and its applications to differential equations, see Boccara [46], Chapter 2, Section 4.

In this case, (3.59) is always satisfied since it is the sum of three positive terms. Thus, *if the diffusion coefficients of the two species are equal, no diffusive instability can occur.*

If $D_1 \neq D_2$, Condition (3.59) may be written

$$H(k^2) = D_1 D_2 k^4 - (D_1 a_{22} + D_2 a_{11})k^2 + a_{11}a_{22} - a_{12}a_{21} > 0.$$

Since $D_1 D_2 > 0$, the minimum of $H(k^2)$ occurs at $k^2 = k_m^2$, where

$$k_m^2 = \frac{D_1 a_{22} + D_2 a_{11}}{2 D_1 D_2} > 0.$$

The condition $H(k_m^2) < 0$, which is equivalent to

$$a_{11}a_{22} - a_{12}a_{21} - \frac{(D_1 a_{22} + D_2 a_{11})^2}{4 D_1 D_2} < 0,$$

is a sufficient condition for instability. This criterion may also be written

$$\frac{D_1 a_{22} + D_2 a_{11}}{(D_1 D_2)^{1/2}} > 2(a_{11}a_{22} - a_{12}a_{21})^{1/2} > 0. \tag{3.62}$$

Since the first term on the left-hand side of (3.62) is a homogeneous function of D_1 and D_2, for given interactions between the two species, the occurrence of a diffusion-induced instability depends only on the ratio of the two diffusion coefficients D_1 and D_2.

From (3.60), a_{11} and a_{22} cannot both be positive. If they are both negative, Condition (3.62) is violated, and Condition (3.59) cannot be violated by increasing k; then, necessarily,

$$a_{11}a_{22} < 0 \tag{3.63}$$

and, from (3.61),

$$a_{12}a_{21} < 0. \tag{3.64}$$

If a_{11} (resp. a_{22}) is equal to zero, then, from (3.60), a_{22} (resp. a_{11}) must be negative, and here again Condition (3.59) cannot be violated by increasing k^2. Conditions (3.63)[59] and (3.64), which are strict inequalities, are useful. *An instability can be immediately ruled out if they are not verified.*

---

[59] This condition also follows from the fact that the expression of k_m^2 has to be positive.

Exercises

Exercise 3.1 *In a careful experimental study of the dynamics of populations of the metazoan* Daphnia magna, *Smith [316] found that his observations did not agree with the predictions of the logistic model. Using the mass M of the population as a measure of its size, he proposed the model*

$$\dot{M} = rM \left(\frac{K - M}{K + aM} \right),$$

where r, K, and a are positive constants. Find the equilibrium points and determine their stabilities.

Exercise 3.2 *The dimensionless Lotka-Volterra equations (2.4 and 2.5) are*

$$\frac{dh}{d\tau} = \rho h \left(1 - p \right),$$
$$\frac{dp}{d\tau} = -\frac{1}{\rho} p \left(1 - h \right),$$

where h and p denote, respectively, the scaled prey and predator populations, and ρ is a positive parameter.

(i) Show that there exists a function $(h, p) \mapsto f(h, p)$ that is constant on each trajectory in the (h, p) plane.

(ii) Use the result above to find a Lyapunov function defined in a neighborhood of the equilibrium point $(1, 1)$.

Exercise 3.3 *In order to develop a strategy for harvesting[60] a renewable resource, say fish, consider the equation*

$$\dot{N} = rN \left(1 - \frac{N}{K} \right) - H(N),$$

which is the usual logistic population model with an increase of mortality rate as a result of harvesting. $H(N)$ represents the harvesting yield per unit time.

(i) Assuming $H(N) = CN$, where C is the intrinsic catch rate, find the equilibrium population N^*, and determine the maximum yield.

(ii) If, as an alternative strategy, we consider harvesting with a constant yield $H(N) = H_0$, the model is

$$\dot{N} = rN \left(1 - \frac{N}{K} \right) - H_0.$$

Determine the stable equilibrium point, and show that when H_0 approaches $\frac{1}{4} rK$ from below, there is a risk for the harvested species to become extinct.

Exercise 3.4 *Assume that the system*

$$\dot{N}_1 = r_1 N_1 \left(1 - \frac{N_1}{K_1} \right) - \lambda_1 N_1 N_2 - C N_1, \quad \dot{N}_2 = r_2 N_2 \left(1 - \frac{N_2}{K_2} \right) - \lambda_2 N_1 N_2,$$

---

[60] For the economics of the sustainable use of biological resources, see Clark [89].

is an acceptable model of two competing fish species in which species 1 is subject to harvesting. For certain values of the parameters, we have seen in Example 18 that for $C = 0$ (no harvesting), this system has an unstable equilibrium point for nonzero values of both populations. In this particular case, what happens when the catching rate C increases from zero?

Exercise 3.5 The competition between two species for the same resource is described by the two-dimensional system

$$\dot{N}_1 = N_1 f_1(N_1, N_2), \quad \dot{N}_2 = N_2 f_2(N_1, N_2),$$

where f_1 and f_2 are differentiable functions.

(i) Show that the slope of the null clines is negative.

(ii) Assuming that there exists only one nontrivial equilibrium point (N_1^*, N_2^*), find the condition under which this equilibrium is asymptotically stable.

Exercise 3.6 In experiments performed on two species of fruit flies (Drosophila pseudoobsura and D. willistini), Ayala, Gilpin, and Ehrenfeld [20] tested 10 different models of interspecific competition, including the Lotka-Volterra model, presented in Example 18, as a special case. They found that the model that gave the best fit was the system

$$\dot{N}_1 = r_1 N_1 \left(1 - \left(\frac{N_1}{K_1}\right)^{\theta_1} - a_{12}\frac{N_2}{K_1}\right), \quad \dot{N}_2 = r_2 N_2 \left(1 - \left(\frac{N_2}{K_2}\right)^{\theta_2} - a_{21}\frac{N_1}{K_2}\right),$$

where r_1, r_2, K_1, K_2, θ_1, θ_2, a_{12}, and a_{21} are positive constants. Under which condition does this model exhibit an asymptotically stable nontrivial equilibrium point?

Exercise 3.7 Consider the two-dimensional system

$$\dot{x}_1 = x_2, \quad \dot{x}_2 = -x_1 + x_2(1 - 3x_1^2 - 2x_2^2),$$

which describes a perturbed harmonic oscillator. Use the Poincaré-Bendixson theorem to prove the existence of a limit cycle.
Hint: Using polar coordinates, show that there exists an invariant bounded set

$$\{(x_1, x_2) \in \mathbb{R}^2 \mid 0 < r_1 < x_1^2 + x_2^2 < r_2\},$$

that does not contain an equilibrium point.

Exercise 3.8 The second dimensionless Ludwig-Jones-Holling equation modeling budworm outbreaks (Equation (1.6)) reads

$$\frac{dx}{dt} = rx\left(1 - \frac{x}{k}\right) - \frac{x^2}{1 + x^2},$$

where x represents the scaled budworm density, and r and k are two positive parameters.

(i) Show that, according to the values of r and k, there exist either two or four equilibrium points and study their stabilities.

(ii) Determine analytically the domains in the (k, r)-space where this equation has either one or three positive equilibrium points. Show that the boundary between the two domains has a cusp point. Find its coordinates.

Solutions

Solution 3.1 *In terms of the dimensionless time and mass variables*

$$\tau = rt \quad and \quad m = \frac{M}{K},$$

the Smith model takes the form

$$\frac{dm}{d\tau} = m\left(\frac{1-m}{1+am}\right).$$

The parameter a is dimensionless.
   The equilibrium points are $m = 0$ and $m = 1$. Since

$$\frac{d}{dm}\left(\frac{m(1-m)}{1+am}\right) = \frac{1-2m-am^2}{(1+am)^2},$$

$m = 0$ is always unstable, and $m = 1$ is asymptotically stable if $a > 0$.

Solution 3.2 *(i) Eliminating $d\tau$ between the two equations yields*

$$\frac{dh}{dp} = -\rho^2 \frac{h(1-p)}{p(1-h)}$$

or

$$\frac{1-h}{h} dh = -\rho^2 \frac{1-p}{p} dp.$$

Integrating, we find that on any trajectory the function

$$f: (h, p) \mapsto \rho^2(p - \log p) + h - \log h$$

is constant. The constant depends upon the initial values $h(0)$ and $p(0)$.
(ii) Since $f(1, 1) = 1 + \rho^2$, define $V(h, p) = f(h, p) - (1 + \rho^2)$. It is straightforward to verify that, in the open set $]0, \infty[\times]0, \infty[$, the function V has the following properties

$$V(1, 1) = 0,$$
$$V(h, p) > 0 \quad for \ (h, p) \neq (1, 1),$$
$$\dot{V}(h, p) \equiv 0.$$

V is, therefore, a weak Lyapunov function, and the equilibrium point $(1, 1)$ is, consequently, Lyapunov stable.

Solution 3.3 *(i) It is simpler, as usual, to define dimensionless variables in order to reduce the number of parameters. If*

$$\tau = rt, \quad n = \frac{N}{K}, \quad c = \frac{C}{r},$$

the equation becomes

$$\frac{dn}{d\tau} = n(1-n) - cn.$$

The equilibrium point n^* is the solution of the equation $1 - n - c = 0$ (i.e., $n^* = 1 - c$). This result supposes that $c < 1$ or, in terms of the original parameters, $C < r$. That is, the intrinsic catch rate C must be less than the intrinsic growth rate r for the population not to become extinct. It is easy to check that n^* is asymptotically stable. The scaled yield is $cn^* = c(1 - c)$, and the maximum scaled yield, which corresponds to $c = \frac{1}{2}$, is equal to $\frac{1}{4}$, or, in terms of the original parameters, $\frac{1}{4} rK$.

(ii) If the harvesting goal is a constant yield H_0, then, with

$$\tau = rt, \quad n = \frac{N}{K}, \quad h_0 = \frac{H_0}{rK},$$

the equation becomes

$$\frac{dn}{d\tau} = n(1 - n) - h_0.$$

The equilibrium points are the solutions of $n(1-n) - h_0 = 0$ (i.e., $n_1^* = \frac{1}{2}(1 + \sqrt{1 - 4h_0})$ and $n_2^* = \frac{1}{2}(1 - \sqrt{1 - 4h_0})$). These solutions are positive numbers if the reduced constant yield h_0 is less than $\frac{1}{4}$. n_1^* and n_2^* are, respectively, asymptotically stable and unstable. When H_0 approaches $\frac{1}{4} rK$, the reduced variable h_0 approaches $\frac{1}{4}$ and the distance $|n_1^* - n_2^*| = \sqrt{1 - 4h_0}$ between the two equilibrium points becomes very small. Then, as a result of a small perturbation, $n(t)$ might become less than the unstable equilibrium value n_1^* and, in that case, will tend to zero. Therefore, if H_0 is close to $\frac{1}{4} rK$, there is a risk for the harvested species to become extinct.

Solution 3.4 *In terms of the dimensionless variables*

$$\tau = \sqrt{r_1 r_2} t, \quad n_1 = \frac{N_1}{K_1}, \quad n_2 = \frac{N_2}{K_2},$$

$$\rho = \sqrt{\frac{r_1}{r_2}}, \quad \alpha_1 = \frac{\lambda_1 K_2}{\sqrt{r_1 r_2}}, \quad \alpha_2 = \frac{\lambda_2 K_1}{\sqrt{r_1 r_2}}, \quad c = \frac{C}{\sqrt{r_1 r_2}},$$

the equations become

$$\frac{dn_1}{d\tau} = \rho n_1 (1 - n_1) - \alpha_1 n_1 n_2 - c n_1, \quad \frac{dn_2}{d\tau} = \frac{1}{\rho} n_2 (1 - n_2) - \alpha_2 n_1 n_2.$$

If $C = 0$ (i.e., $c = 0$), referring to Example 18, the nontrivial fixed point

$$\left(\frac{\alpha_1 - \rho}{\rho(\alpha_1 \alpha_2 - 1)}, \frac{\rho \alpha_2 - 1}{\alpha_1 \alpha_2 - 1} \right)$$

is a saddle if

$$\alpha_1 > \rho \quad \text{and} \quad \rho \alpha_2 > 1.$$

In this case, the equilibrium points $(0, 1)$ and $(1, 0)$ are both asymptotically stable. Depending upon environmental conditions, either species 1 or species 2 is capable of dominating the natural system. Since, in this model, species 1 is harvested we will assume that species 1 is dominant. As a result of a nonzero c value, the asymptotically stable equilibrium

$$\left(\frac{\rho - c}{\rho}, 0 \right),$$

which was equal to $(1, 0)$ for $c = 0$, moves along the n_1-axis, as indicated by the arrow on the left in Figure 3.23. For a c value such that

$$\frac{\rho - c}{\rho} = \frac{1}{\rho \alpha_2}, \quad that \ is, \quad c = \rho - \frac{1}{\alpha_2},$$

there is a bifurcation and, for c slightly greater than $\rho - 1/\alpha_2$, as on the right in Figure 3.23, species 1 does not exist anymore. This extinction is, however, not a direct consequence of harvesting (c is still less than ρ, that is, $C < r_1$) but the result of the competitive interaction. For $c < \rho - 1/\alpha_2$, the only stable equilibrium point is $(0,1)$, so species 2 should become dominant. But, we assumed that, before harvesting, the state of the system was $n_1^* = 1$ and $n_2^* = 0$. In nature this might not be entirely true, a small population n_2 could exist in a refuge and could grow according to the equations of the model once a sufficient increase of the catching rate had changed the values of the parameters.

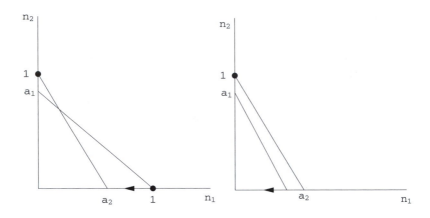

Fig. 3.23. *Modification of the null cline of species 1 as it is harvested.*

This model is a possible explanation of the elimination, in the late 1940s and early 1950s, of the Pacific sardines that have been replaced by an anchovy population.[61]

Solution 3.5 *(i) The equations of the null clines in the (N_1, N_2)-plane are*

$$f_1(N_1, N_2) = 0 \quad and \quad f_2(N_1, N_2) = 0,$$

and their slopes are given by

$$\frac{dN_2}{dN_1} = -\frac{\dfrac{\partial f_1}{\partial N_1}}{\dfrac{\partial f_1}{\partial N_2}} \quad for \ \dot{N}_1 = 0, \quad and \quad \frac{dN_2}{dN_1} = -\frac{\dfrac{\partial f_2}{\partial N_1}}{\dfrac{\partial f_2}{\partial N_2}} \quad for \ \dot{N}_1 = 0.$$

Since all the partial derivatives, which represent limiting effects of each species on itself or on its competitor, are negative in a competition model, both slopes are negative. That is, each population is a decreasing function of the other, as should be the case for a system of two competing species.

---

[61] See Clark [89].

(ii) The nontrivial equilibrium point (N_1^*, N_2^*), if it exists, is the unique solution, in the positive quadrant, of the system

$$f_1(N_1, N_2) = 0, \quad f_2(N_1, N_2) = 0.$$

This equilibrium point is asymptotically stable if the eigenvalues of the Jacobian matrix

$$J = \begin{bmatrix} N_1^* \dfrac{\partial f_1}{\partial N_1}(N_1^*, N_2^*) & N_1^* \dfrac{\partial f_1}{\partial N_2}(N_1^*, N_2^*) \\ N_2^* \dfrac{\partial f_2}{\partial N_1}(N_1^*, N_2^*) & N_2^* \dfrac{\partial f_2}{\partial N_2}(N_1^*, N_2^*) \end{bmatrix}$$

have negative real parts, that is, if

$$\operatorname{tr} J < 0 \quad \text{and} \quad \det J > 0.$$

All partial derivatives being negative, the condition on the trace is automatically satisfied. The only condition to be satisfied for the equilibrium point to be asymptotically stable is, therefore,

$$\frac{\partial f_1}{\partial N_1}(N_1^*, N_2^*)\frac{\partial f_2}{\partial N_2}(N_1^*, N_2^*) > \frac{\partial f_1}{\partial N_2}(N_1^*, N_2^*)\frac{\partial f_2}{\partial N_2}(N_1^*, N_2^*).$$

That is, a system of two competitive species exhibits a stable equilibrium if, and only if, the product of the intraspecific growth regulations is greater than the product of the interspecific growth regulations. This result has been given without proof by Gilpin and Justice [143]. It has been established by Maynard Smith [234] assuming, as suggested by Gilpin and Justice, that, at the point of intersection of the two null clines, the slope of the null cline of the species plotted along the x-axis is greater than the slope of the null cline of the species plotted along the y-axis. This property of the null clines is a consequence of the position of the stable equilibrium point found by Ayala [19] in his experimental study of two competing species of Drosophila.

Solution 3.6 *In terms of the dimensionless variables*

$$\tau = \sqrt{r_1 r_2}\, t, \quad \rho = \frac{r_1}{r_2}, \quad n_1 = \frac{N_1}{K_1}, \quad n_2 = \frac{N_2}{K_2}, \quad \alpha_{12} = a_{12}\frac{K_2}{K_1}, \quad \alpha_{21} = a_{21}\frac{K_1}{K_2},$$

the Ayala-Gilpin-Ehrenfeld model becomes

$$\frac{dn_1}{d\tau} = \rho\left(1 - n_1^{\theta_1} - \alpha_{12} n_2\right), \quad \frac{dn_1}{d\tau} = \frac{1}{\rho}\left(1 - n_2^{\theta_2} - \alpha_{21} n_1\right).$$

Assuming the existence of a nontrivial equilibrium point, from the general result derived in the preceding exercise, this point is asymptotically stable if the condition

$$\frac{\partial f_1}{\partial n_1}(n_1^*, n_2^*)\frac{\partial f_2}{\partial n_2}(n_1^*, n_2^*) > \frac{\partial f_1}{\partial n_2}(n_1^*, n_2^*)\frac{\partial f_2}{\partial n_2}(n_1^*, n_2^*),$$

where

$$f_1(n_1, n_2) = \rho\left(1 - n_1^{\theta_1} - \alpha_{12} n_2\right) \quad \text{and} \quad f_2(n_1, n_2) = \frac{1}{\rho}\left(1 - n_2^{\theta_2} - \alpha_{21} n_1\right),$$

is satisfied; that is, if

$$\theta_1 \theta_2 (n_1^*)^{\theta_1 - 1}(n_2^*)^{\theta_2 - 1} > \alpha_{12}\alpha_{21},$$

where n_1^* and n_2^* are the dimensionless coordinates of the equilibrium point.

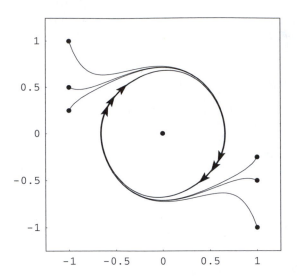

Fig. 3.24. *Phase portrait of the perturbed harmonic oscillator. Dots show initial values.*

Solution 3.7 *In polar coordinates, the equations become*

$$\dot{r} = r\sin^2\theta(1 - 2r^2 - r^2\cos^2\theta),$$
$$\dot{\theta} = -1 + \sin\theta\cos\theta(1 - 2r^2 - r^2\cos^2\theta).$$

For $r = \frac{1}{2}$, the first equation shows that

$$\dot{r} = \tfrac{1}{4}\sin^2\theta(1 - \tfrac{1}{2}\cos\theta) \geq 0$$

with equality only for $\theta = 0$ and $\theta = \pi$. The first equation also implies that

$$\dot{r} \leq r\sin^2(1 - 2r^2).$$

Hence, for $r = 1/\sqrt{2}$, $\dot{r} \leq 0$, with equality only for $\theta = 0$ and $\theta = \pi$. These two results prove that, for any point $\mathbf{x}$ in the annular domain $\left\{\mathbf{x} \in \mathbb{R}^2 \mid \frac{1}{2} < \|\mathbf{x}\| < \frac{1}{\sqrt{2}}\right\}$, the trajectory $\{\varphi_t(\mathbf{x}) \mid t > 0\}$ remains in D. Since the only equilibrium point of the planar system is the origin, which is not in D, D contains a limit cycle. The phase portrait of this system is represented in Figure 3.24.

Solution 3.8 *(i) Equilibrium points are the solutions of the equation*

$$f(x; r, k) = rx\left(1 - \frac{x}{k}\right) - \frac{x^2}{1 + x^2} = 0.$$

This is an algebraic equation of degree 4 that has either two or four real solutions. The graph of the function $f : x \mapsto rx(1 - x/k) - x^2/(1 + x^2)$ represented in Figure 3.25 shows that the equilibrium point $x_0 = 0$, which exists for all values of r and k, is always unstable (positive slope). Therefore, if there are only two equilibrium points, the nonzero

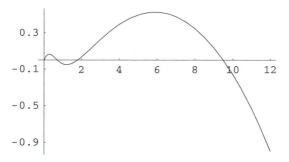

Fig. 3.25. *Typical graph of the function $x \mapsto rx(1 - x/k) - x^2/(1+x^2)$ when there exist four equilibrium points. For $k = 12$ and $r = 0.5$, the equilibrium points are located at $x_0 = 0$, $x_\ell = 0.704$, $x_u = 1.794$, and $x_h = 9.502$.*

equilibrium point is stable, and if there exist four equilibrium points (as in Figure 3.25) there exist two stable equilibrium points x_ℓ and x_h at, respectively, low and high density separated by an unstable point x_u.

(ii) The boundary between the domains in the (k, r)-space in which there exist either one or three positive equilibrium points is determined by expressing that the equation

$$r\left(1 - \frac{x}{k}\right) = \frac{x}{1 + x^2} \tag{3.65}$$

has a double root. This occurs when

$$r\frac{d}{dx}\left(1 - \frac{x}{k}\right) = \frac{d}{dx}\left(\frac{x}{1 + x^2}\right);$$

that is, if

$$-\frac{r}{k} = \frac{1 - x^2}{(1 + x^2)^2}. \tag{3.66}$$

Solving (3.65) and (3.66) for k and r, we obtain the parametric representation of the boundary of the domain in which Equation (3.65) has either one or three positive solutions (see Figure 3.26):

$$k = \frac{2x^3}{x^2 - 1},$$

$$r = \frac{2x^3}{(1 + x^2)^2}.$$

This model exhibits a cusp catastrophe. At the cusp point, the derivatives of k and r with respect to x both vanish. Its coordinates are $k = 3^{3/2}$ and $r = 3^{3/2}/8$ (i.e., $k = 5.196\ldots$ and $r = 0.650\ldots$). Note that

$$\lim_{x \to \infty} k(x) = \infty, \quad \lim_{x \to \infty} r(x) = 0,$$

and

$$\lim_{x \to 1} k(x) = \infty, \quad \lim_{x \to 1} r(x) = \frac{1}{2}.$$

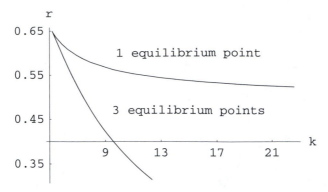

Fig. 3.26. *Boundary between domains in (k, r)-space in which there exist either one or three positive equilibrium points.*

4

Recurrence Equations

In this chapter, we study dynamical models described by recurrence equations of the form

$$\mathbf{x}_{t+1} = \mathbf{f}(\mathbf{x}_t), \tag{4.1}$$

where $\mathbf{x}$, representing the state of the system, belongs to a subset J of $\mathbb{R}^n$, $\mathbf{f} : J \to J$ is a map, and $t \in \mathbb{N}_0$. Since many notions are common to differential equations and recurrence equations, this chapter will be somewhat shorter than the preceding one.

As for differential equations, models formulated in terms of recurrence equations such as (4.1) ignore the short-range character of the interactions between the elements of a complex system.

4.1 Iteration of maps

Let $\mathbf{f}$ be a map defined on a subset J of $\mathbb{R}^n$; in what follows, for the iterates of $\mathbf{f}$, we shall adopt the notation

$$\mathbf{f}^{t+1} = \mathbf{f} \circ \mathbf{f}^t, \quad \text{where} \quad \mathbf{f}^1 = \mathbf{f}$$

and $t \in \mathbb{N}_0$ ($\mathbf{f}^0$ is the identity). If $\mathbf{f}$ is a diffeomorphism, then $\mathbf{f}^{-1}$ is defined and the definition of $\mathbf{f}^t$ above remains valid for all $t \in \mathbb{Z}$. In applications, the map $\mathbf{f}$ is seldom invertible.[1]

In the previous chapter on differential equations, a flow has been defined as a one-parameter group on the phase space. That is, if the phase space is an open set U of $\mathbb{R}^n$, for all $t \in \mathbb{R}$,[2] the flow is a mapping $\varphi_t : U \to U$

---

[1] In the mathematical literature, diffeomorphisms are favored. Among the different motivations, Smale [315] mentions "its natural beauty" and the fact that "problems in the qualitative theory of differential equations are present in their simplest form in the theory of diffeomorphisms."

[2] Or sometimes only for $t \in \mathbb{R}_+$. In this case, the flow is a one-parameter semigroup.

such that the group property holds: φ_0 is the identity and $\varphi_{t+s} = \varphi_t \circ \varphi_s$ (Relation (3.3)).

In the case of maps, the analogue of the flow is the mapping $\mathbf{f}^t : J \to J$, defined only for $t \in \mathbb{Z}$ or $t \in \mathbb{N}_0$.[3]

The *forward orbit* of $\mathbf{f}$ through $\mathbf{x}$ is the set $\{\mathbf{f}^t(\mathbf{x}) \mid t \in \mathbb{N}_0\}$ oriented in the sense of increasing t. If $\mathbf{f}$ is invertible, the *backward orbit* is the set $\{\mathbf{f}^{-t}(\mathbf{x}) \mid t \in \mathbb{N}_0\}$. In this case the *(whole) orbit* is the set $\{\mathbf{f}^t(\mathbf{x}) \mid t \in \mathbb{Z}\}$.

A point $\mathbf{x}^* \in J$ is an *equilibrium point* or a *fixed point* if $\mathbf{f}(\mathbf{x}^*) = \mathbf{x}^*$. The orbit of an equilibrium point is the equilibrium point itself. A point $\mathbf{x}^* \in J$ is a *periodic point* if $\mathbf{f}^\tau(\mathbf{x}^*) = \mathbf{x}^*$ for some nonzero τ. The smallest positive value of τ is the *period* of the point and is usually denoted by T. A periodic point of period T is a fixed point of $\mathbf{f}^T$. The set $\{\mathbf{f}^t(\mathbf{x}^*) \mid t = 0, 1, \ldots, T-1\}$ is a *periodic orbit* or a *T-point cycle*. A periodic orbit consists of a finite number of points.

Example 24. Graphical analysis. In the case of one-dimensional maps, there exists a simple graphical method to follow the successive iterates of an initial point x_0. First plot the graphs of $x \mapsto f(x)$ and $x \mapsto x$. Since the sequence of iterates is generated by the equation $x_{t+1} = f(x_t)$, the iterate of the initial value x_0 is on the graph of f at $(x_0, f(x_0))$, that is, (x_0, x_1). The horizontal line from this point intersects the diagonal at (x_1, x_1). The vertical line from this point intersects the graph of f at $(x_1, f(x_1))$; that is, (x_1, x_2). Repeating this process generates the sequence

$$(x_0, x_1), (x_1, x_1), (x_1, x_2), (x_2, x_2), (x_2, x_3), \ldots$$

The equilibrium point x^* is located at the intersection of the graphs of the two functions. The diagram (Figure 4.1) that consists of the graphs of the functions $x \mapsto f(x)$ and $x \mapsto x$ and the line joining the points of the sequence above is called a *cobweb*. It clearly shows if the sequence of iterates of the initial point x_0 does or does not converge to the equilibrium point.

Definition 13. *Two maps $\mathbf{f}$ and $\mathbf{g}$ defined on $J \subseteq \mathbb{R}^n$ are said to be* conjugate *if there exists a homeomorphism $\mathbf{h}$ such that*

$$\mathbf{h} \circ \mathbf{f} = \mathbf{g} \circ \mathbf{h}$$

or

$$\mathbf{h} \circ \mathbf{f} \circ \mathbf{h}^{-1} = \mathbf{g}.$$

This relation implies that, for all $t \in \mathbf{N}$,

$$\mathbf{h} \circ \mathbf{f}^t \circ \mathbf{h}^{-1} = \mathbf{g}^t.$$

---

[3] This mapping is sometimes called a *cascade* in the Russian literature on dynamical systems; see, for instance, Anosov and Arnold [7].

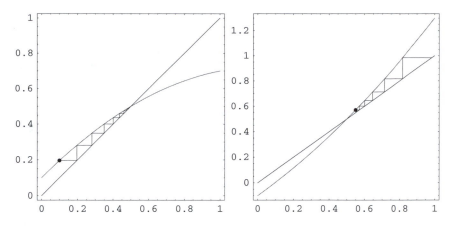

Fig. 4.1. *Cobwebs in the vicinity of equilibrium points. The equilibrium point in the left diagram is asymptotically stable ($|f(x^*)| < 1$), while the equilibrium point in the right diagram is unstable ($|f(x^*)| > 1$). Dots represent initial values.*

As for flows, $\mathbf{h}$ takes the orbits $\mathbf{f}$ into the orbits of $\mathbf{g}$. In particular, if $\mathbf{x}^*$ is a periodic point of $\mathbf{f}$ of period T, then $\mathbf{h}(\mathbf{x}^*)$ is a periodic point of period T of $\mathbf{g}$. This follows from

$$\mathbf{g}^T(\mathbf{h}(\mathbf{x}^*)) = \mathbf{h}(\mathbf{f}^T(\mathbf{x}^*)) = \mathbf{h}(\mathbf{x}^*),$$

which proves that $\mathbf{h}(\mathbf{x}^*)$ is a periodic point of $\mathbf{g}$. To verify that T is the (smallest) period, assume that there exists $T_0 < T$ such that $\mathbf{g}^{T_0}(\mathbf{h}(\mathbf{x}^*)) = \mathbf{h}(\mathbf{x}^*)$. Since $\mathbf{g}^{T_0}(\mathbf{h}(\mathbf{x}^*)) = \mathbf{h}(\mathbf{f}^{T_0}(\mathbf{x}^*))$, this would imply $\mathbf{f}^{T_0}(\mathbf{x}^*) = \mathbf{x}^*$, which is not true.

Example 25. The logistic map f_4 and the binary tent map T_2, defined on the interval $[0, 1]$ by

$$f_4(x) = 4x(1 - x) \quad \text{and} \quad T_2(x) = \begin{cases} 2x, & \text{if } 0 \le x < \frac{1}{2}, \\ 2 - 2x, & \text{if } \frac{1}{2} \le x \le 1, \end{cases}$$

are conjugate. To prove this result, it suffices to show that the function $h : x \mapsto \sin^2(\frac{\pi}{2} x)$ is a conjugacy. This is indeed the case since

1. $h : [0, 1] \to [0, 1]$ is continuous,
2. $h(x_1) = h(x_2) \Rightarrow x_1 = x_2$,
3. $h([0, 1]) = [0, 1]$,
4. h' exists and is positive, so $h^{-1} : [0, 1] \to [0, 1]$ exists and is continuous, and
5. $(f_4 \circ h)(x) = 4\sin^2(\frac{\pi}{2} x)(1 - \sin^2(\frac{\pi}{2} x)) = \sin^2(\pi x) = (h \circ T_2)(x)$.

4.2 Stability

Definition 14. *A fixed point $\mathbf{x}^*$ of the recurrence equation $\mathbf{x}_{t+1} = \mathbf{f}(\mathbf{x}_t)$ is said to be* Lyapunov stable *if, for any given positive ε, there exists a positive δ (which depends on ε only) such that $\|\mathbf{f}^t(\mathbf{x}) - \mathbf{x}^*\| < \varepsilon$ for all $\mathbf{x}$ such that $\|\mathbf{x} - \mathbf{x}^*\| < \delta$ and all $t > 0$. A fixed point $\mathbf{x}^*$ is said to be* asymptotically stable *if it is Lyapunov stable and if there exists a positive δ such that, for all $\mathbf{x}$ such that $\|\mathbf{x} - \mathbf{x}^*\| < \delta$, $\lim_{t\to\infty} \mathbf{f}^t(\mathbf{x}) = \mathbf{x}^*$. Fixed points that are not stable are* unstable.

That is, the iterates of points close to a Lyapunov stable fixed point remain close to it, while the iterates of points close to an asymptotically stable fixed point move towards it as t increases.

Since a T-periodic point $\mathbf{x}^*$ is a fixed point of $\mathbf{f}^T$, its stability is determined by the behavior of the sequence of iterates: $\mathbf{f}^T(\mathbf{x}), \mathbf{f}^{2T}(\mathbf{x}), \mathbf{f}^{3T}(\mathbf{x}), \ldots$, where $\mathbf{x}$ is close to $\mathbf{x}^*$.

Definition 15. *If a metric d is defined on the phase space $\mathcal{S}$, a map $\mathbf{f} : \mathcal{S} \to \mathcal{S}$ is said to be* contracting *if there exists a positive real $\lambda < 1$ such that, for any pair $(\mathbf{x}, \mathbf{y}) \in \mathcal{S} \times \mathcal{S}$,*

$$d\big(\mathbf{f}(\mathbf{x}), \mathbf{f}(\mathbf{y})\big) \leq \lambda d(\mathbf{x}, \mathbf{y}). \tag{4.2}$$

Iterating inequality (4.2) yields

$$d(\mathbf{f}^t(\mathbf{x}), \mathbf{f}^t(\mathbf{y})) \leq \lambda d(\mathbf{f}^{t-1}(\mathbf{x}), \mathbf{f}^{t-1}(\mathbf{y})) \leq \cdots \leq \lambda^t d(\mathbf{x}, \mathbf{y}), \tag{4.3}$$

so

$$\lim_{t\to\infty} d(\mathbf{f}^t(\mathbf{x}), \mathbf{f}^t(\mathbf{y})) = 0.$$

This result means that the asymptotic behavior of all points is the same. On the other hand, as a consequence of the triangular inequality, for $s > t$,

$$
\begin{aligned}
d(\mathbf{f}^s(\mathbf{x}), \mathbf{f}^t(\mathbf{x})) &\leq \big(d(\mathbf{f}^s(\mathbf{x}), \mathbf{f}^{s-1}(\mathbf{x})) + d(\mathbf{f}^{s-1}(\mathbf{x}), \mathbf{f}^{s-2}(\mathbf{x})) + \cdots \\
&\quad + d(\mathbf{f}^{t+1}(\mathbf{x}), \mathbf{f}^t(\mathbf{x}))\big) \\
&\leq \big(\lambda^{s-1} d(\mathbf{f}(\mathbf{x}), \mathbf{x}) + \lambda^{s-2} d(\mathbf{f}(\mathbf{x}), \mathbf{x}) + \cdots + \lambda^t d(\mathbf{f}(\mathbf{x}), \mathbf{x})\big) \\
&\leq \frac{\lambda^t}{1 - \lambda} d(\mathbf{f}(\mathbf{x}), \mathbf{x}),
\end{aligned}
\tag{4.4}
$$

that is,

$$\lim_{t\to\infty} d(\mathbf{f}^s(\mathbf{x}), \mathbf{f}^t(\mathbf{x})) = 0.$$

For a given $\mathbf{x} \in \mathcal{S}$, the sequence of iterates $\big(\mathbf{f}^t(\mathbf{x})\big)_{t\in\mathbb{N}}$ is a Cauchy sequence.[4] Therefore, if the space $\mathcal{S}$ is complete, the sequence $\mathbf{f}^t(\mathbf{x})$ is convergent, and

---

[4] A sequence (x_n) is a *Cauchy sequence* if, for any positive $\varepsilon > 0$, there exists a positive integer $n_0(\varepsilon)$ such that $d(x_n, x_{n+k}) < \varepsilon$ for all $n > n_0(\varepsilon)$ and $k > 0$. A metric space is said to be *complete* if every Cauchy sequence is convergent.

from (4.3) the limit of the sequence is the same for all $\mathbf{x} \in \mathcal{S}$. Denote this limit by $\mathbf{x}^*$. For any $\mathbf{x}$ and any positive integer t, using again the triangular inequality, we obtain

$$d(\mathbf{f}(\mathbf{x}^*), \mathbf{x}^*) \le d(\mathbf{f}(\mathbf{x}^*), \mathbf{f}^{t+1}(\mathbf{x})) + d(\mathbf{f}^{t+1}(\mathbf{x}), \mathbf{f}^t(\mathbf{x})) + d(\mathbf{f}^t(\mathbf{x}), \mathbf{x}^*)$$
$$\le (1+\lambda)d(\mathbf{f}^t(\mathbf{x}), \mathbf{x}^*) + \lambda^t d(\mathbf{f}(\mathbf{x}), \mathbf{x}).$$

Since $\lim_{t \to \infty} d(\mathbf{f}^t(\mathbf{x}), \mathbf{x}^*) = 0$ and $\lim_{t \to \infty} \lambda^t = 0$, we find that

$$\mathbf{f}(\mathbf{x}^*) = \mathbf{x}^*,$$

which shows that the limit $\mathbf{x}^*$ of the sequence $(\mathbf{f}^t(\mathbf{x}))$ is a fixed point of the recurrence equation (4.1). If, in the inequality (4.4) we take the limit $s \to \infty$, we find that

$$d(\mathbf{f}^t(\mathbf{x}), \mathbf{x}^*) \le \frac{\lambda^t}{1-\lambda} d(\mathbf{f}(\mathbf{x}), \mathbf{x}),$$

that is the iterates of all points $\mathbf{x} \in \mathcal{S}$ converge *exponentially* to the fixed point $\mathbf{x}^*$.

The considerations above prove the following fundamental result:

Theorem 6. *Let* $\mathbf{f} : \mathcal{S} \to \mathcal{S}$ *be a contracting map defined on a complete metric space. Then the sequence of iterates* $(\mathbf{f}^t(\mathbf{x}))_{t \in \mathbb{N}}$ *converges exponentially to the unique fixed point* $\mathbf{x}^*$ *of* $\mathbf{f}$.

Example 26. Newton's method. Any solution of the equation $f(x) = 0$ is a solution of the equation

$$x = x - \frac{f(x)}{g(x)}$$

for any function g that is not equal to zero in the vicinity of the solution. In order to optimize the convergence of the iterates of the function defined by the right-hand side of the equation above towards the root x^* of $f(x) = 0$, the absolute value of the derivative of this function at $x = x^*$ should be as small as possible. From

$$\frac{d}{dx}\left(x - \frac{f(x)}{g(x)}\right)\bigg|_{x=x^*} = 1 - \frac{f'(x^*)}{g(x^*)} + \frac{f(x^*)g'(x^*)}{(g(x^*))^2},$$

it follows that, if we take $g(x) = f'(x)$,

$$\frac{d}{dx}\left(x - \frac{f(x)}{g(x)}\right)\bigg|_{x=x^*} = 0.$$

The iteration

$$x_{t+1} = x_t - \frac{f(x_t)}{f'(x_t)}$$

is known as Newton's method for solving the equation $f(x) = 0$. For instance, the numerical values of the solutions of the equation $x^2 - 2 = 0$ (which are $\pm\sqrt{2}$) are the asymptotically stable equilibrium points of the map $x \mapsto \dfrac{x_t}{2} - \dfrac{1}{x_t}$.[5] Starting from the initial value $x = 3$, the first iterates are

$$1.833333333333, \quad 1.46212121212194, \quad 1.4149984298,$$
$$1.414213780047, \quad 1.414213562373, \ldots .$$

The convergence towards the equilibrium point is very fast. After only five iterations, we obtain the value of $\sqrt{2}$ with 12 exact digits after the decimal point.

Theorem 6 has many important consequences.[6] For instance, in the case of linear maps, if all the eigenvalues of a linear map $A : \mathbb{R}^n \to \mathbb{R}^n$ have absolute values less than one, the map is contracting and the iterates of every point converge to the origin exponentially.

In the case of a nonlinear recurrence equation $\mathbf{x}_{t+1} = \mathbf{f}(\mathbf{x}_t)$, as for nonlinear differential equations, under certain conditions, the stability of an equilibrium point $\mathbf{x}^*$ can be determined by the eigenvalues of the linear operator $D\mathbf{f}(\mathbf{x}^*)$.

Definition 16. *Let* $\mathbf{x}^* \in \mathcal{S}$ *be an equilibrium point of the recurrence equation* $\mathbf{x}_{t+1} = \mathbf{f}(\mathbf{x}_t)$; $\mathbf{x}^*$ *is said to be* hyperbolic *if none of the eigenvalues of the Jacobian matrix* $D\mathbf{f}(\mathbf{x}^*)$ *has modulus equal to one. The linear function*

$$\mathbf{x} \mapsto D\mathbf{f}(\mathbf{x}^*)\mathbf{x}$$

is called the linear part *of* $\mathbf{f}$ *at* $\mathbf{x}^*$.

The following result is the Hartman-Grobman theorem for maps. It is the analogue of Theorem 1 for flows.

Theorem 7. *If* $\mathbf{x}^*$ *is a hyperbolic equilibrium point of the recurrence equation* $\mathbf{x}_{t+1} = \mathbf{f}(\mathbf{x}_t)$, *the mapping* $\mathbf{f}$ *in the neighborhood of* $\mathbf{x}^*$ *is* C^0 *conjugate to the linear mapping* $D\mathbf{f}(\mathbf{x}^*)$ *in the neighborhood of the origin.*

Therefore, if $D\mathbf{f}(\mathbf{x}^*)$ has no eigenvalues with modulus equal to 1, the stability of the equilibrium point $\mathbf{x}^*$ of the nonlinear recurrence equation $\mathbf{x}_{t+1} = \mathbf{f}(\mathbf{x}_t)$ can be determined from the study of the linear recurrence equation $\mathbf{y}_{t+1} = D\mathbf{f}(\mathbf{x}^*)\mathbf{y}_t$, where $\mathbf{y} = \mathbf{x} - \mathbf{x}^*$. If $D\mathbf{f}(\mathbf{x}^*)$ has eigenvalues with modulus equal to 1, this is not the case. The following examples illustrate the various possibilities.

---

[5] The restrictions of the map to the open intervals $]\sqrt{2}, \infty[$ and $] - \infty, -\sqrt{2}[$ are obviously contracting. If the initial point x_0 does not lie in these intervals, then its first iterate $f(x_0) = x_1$ is either in $]\sqrt{2}, \infty[$, for $0 < x_0 < \sqrt{2}$, or in $]-\infty, -\sqrt{2}[$, for $-\infty < x_0 < -\sqrt{2}$.

[6] For more details, refer to Katok and Hasselblatt [181].

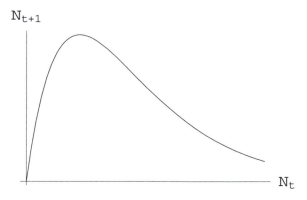

N_{t+1}

N_t

Fig. 4.2. *Typical graph of a function f in a one-population model of the form $N_{t+1} = f(N_t)$.*

Example 27. One-population models. If a species breeds only at a particular time of the year, whether adults do or do not survive to breed in the next season has an important effect on their population dynamics. Due to the existence of breeding seasons, the population growth may be described by a recurrence equation of the form $N_{t+1} = f(N_t)$.[7] A reasonable function f should satisfy the following two conditions:

- the image $f(N)$ of any positive N should be positive, and
- f should be increasing for small N and decreasing for large N.

The graph of a function f satisfying these two conditions is represented in Figure 4.2. A cobweb analysis of single-population models is presented in Figure 4.3.

As a first model, consider

$$N_{t+1} = N_t \exp\left(r\left(1 - \frac{N_t}{K}\right)\right), \tag{4.5}$$

where r and K are positive constants. If n_t denotes the dimensionless population N_t/K, we have

$$n_{t+1} = n_t \exp\left(r(1 - n_t)\right). \tag{4.6}$$

There are two equilibrium points 0 and 1. Since the values of the derivative of the function $n \mapsto n \exp(r(1 - n))$ at $n = 0$ and $n = 1$ are, respectively, e^r and $1 - r$, 0 is always unstable and 1 is asymptotically stable if $0 < r < 2$.

Another example is that of Hassel [167]:

$$N_{t+1} = \frac{rN_t}{\left(1 + \dfrac{N_t}{K}\right)^a}, \tag{4.7}$$

---

[7] Various examples are given in May and Oster [231].

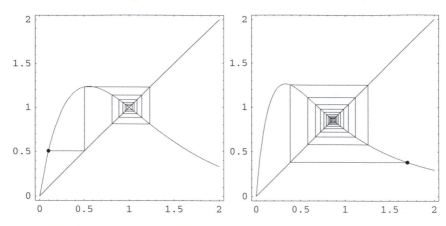

Fig. 4.3. *Cobweb analysis of single-population models. Left figure: $n_{t+1} = n_t \exp(r(1 - n_t))$ for $r = 1.8$ and $n_0 = 0.1$; the equilibrium point $n^* = 1$ is asymptotically stable. Right figure: $n_{t+1} = rn_t/(1 - n_t)^a$ for $r = 12$, $a = 4$, and $n_0 = 1.5$; the equilibrium point $n^* = r^{1/a} - 1$ is asymptotically stable.*

where r, K, and a are constants greater than 1. If n_t denotes the dimensionless population N_t/K, we have

$$n_{t+1} = \frac{rn_t}{(1 + n_t)^a}. \tag{4.8}$$

There are two equilibrium points, 0 and $r^{1/a} - 1$. Since the values of the derivative of the function $n \mapsto rn/(1 + n)^a$ at $n = 0$ and $n = r^{1/a} - 1$ are, respectively, r and $1 - a(1 - r^{-1/a})$, 0 is always unstable and $r^{1/a} - 1$ is asymptotically stable if $0 < a(1 - r^{-1/a}) < 2$.

Example 28. Host-parasitoid models. A parasitoid is an insect having a lifestyle intermediate between a parasite and a usual predator. Parasitoid larvae live inside their hosts, feeding on the host tissues and generally consuming them almost completely. If we assume that the host has discrete nonoverlapping generations and is attacked by a parasitoid during some interval of its life cycle, a simple model is

$$H_{t+1} = rH_t e^{-aP_t}, \tag{4.9}$$

$$P_{t+1} = cH_t(1 - e^{-aP_t}), \tag{4.10}$$

where H_t and P_t are, respectively, the host and parasitoid populations at time t, and r, a, and c are positive constants. This model, whose interest is essentially historical, was proposed by Nicholson and Bailey [268] in 1935. The term e^{aP_t} represents, at time step t, the fraction of hosts that escape detection. They are the only ones that can live and reproduce. For the parasitoid, the factor $1 - e^{aP_t}$ is the probability that, at time t, a host is discovered. Each discovered host gives rise to c new parasitoids at the next time step.

The model has two equilibrium points, $(0,0)$ and (H^*, P^*), where

$$H^* = \frac{r \log r}{ac(r-1)}, \qquad P^* = \frac{\log r}{a}.$$

The nontrivial fixed point exists only for $r > 1$. The eigenvalues of the Jacobian matrix

$$\begin{bmatrix} re^{-aP} & -arHe^{-aP} \\ c(1-e^{-aP}) & acHe^{-aP} \end{bmatrix}$$

at $(0,0)$ are 0 and r. For (H^*, P^*) to exist, r must be greater than 1, and $(0,0)$ is unstable. At the equilibrium point (H^*, P^*) the eigenvalues are

$$\frac{1}{2(r-1)} \left(r - 1 + \log r \pm i\sqrt{4(r-1)r \log r - (r-1+\log r)^2} \right).$$

The modulus of these two complex conjugate eigenvalues is equal to

$$\sqrt{\frac{r \log r}{r-1}}.$$

For $r > 1$, it is greater than 1, and (H^*, P^*) is unstable (see the left panel of Figure 4.4). Nicholson and Bailey showed numerically that the oscillations of the populations of the two species seemed to increase without limit. They mentioned that such a behavior "is certainly not compatible with what we observe to happen." To eliminate the unrealistic behavior of the Nicholson-Bailey model, Beddington, Free, and Lawton [38] introduced a density-dependent self-regulation for the host. Their model is

$$H_{t+1} = H_t \exp\left(r(1 - H_t/K) - aP_t \right), \qquad (4.11)$$

$$P_{t+1} = cH_t\left(1 - \exp(-aP_t) \right). \qquad (4.12)$$

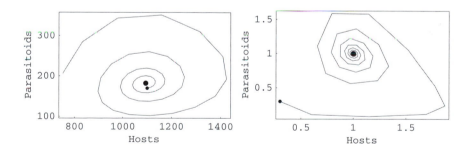

Fig. 4.4. *Host-parasitoid models. Small dots • represent initial states and bigger dots • equilibrium points. For the Nicholson-Bailey model (left figure), the nontrivial equilibrium point is unstable for all values of the parameters (here $r = 1.2$, $a = 0.001$, and $c = 1$). For the reduced Beddington-Free-Lawton model (right figure), there exists in the (r, q) parameter space a domain in which the nontrivial fixed point is stable (here $r = 1$ and $q = 0.5$).*

In the absence of parasitoids, the host population evolves according to the recurrence equation (4.5), which has a stable equilibrium for $0 < r < 2$.

To simplify the discussion of this model, we introduce reduced variables. If (H^*, P^*) is the nontrivial equilibrium point, let

$$h = \frac{H}{H^*}, \quad p = \frac{P}{P^*}, \quad q = \frac{H^*}{K}.$$

If, in the presence of parasitoids, the host population in the steady state is nonzero, it is expected to be less than the carrying capacity K. The scaled parameter q is a measure of the extent to which the host population at equilibrium is reduced with respect to K. The evolution of the scaled populations h and p is the solutions of the equations

$$h_{t+1} = h_t \exp(r(1 - qh_t) - \alpha p_t), \tag{4.13}$$

$$p_{t+1} = \gamma h_t (1 - \exp(-\alpha p_t)), \tag{4.14}$$

where the scaled constants α and γ depend on the two independent scaled parameters r and q. Their expressions satisfy the relations

$$\alpha = r(1 - q), \quad \gamma = \frac{1}{1 - e^{-\alpha}}.$$

The Jacobian matrix at the nontrivial equilibrium point $(h^*, p^*) = (1, 1)$ is

$$J(1, 1) = \begin{bmatrix} 1 - qr & -r(1 - q) \\ 1 & \dfrac{r(1 - q)}{e^{r(1-q)} - 1} \end{bmatrix}.$$

Its eigenvalues are the solutions to the quadratic equation

$$\lambda^2 - \operatorname{tr} J(1, 1)\, \lambda + \det J(1, 1) = 0. \tag{4.15}$$

The equilibrium point $(1, 1)$ is asymptotically stable if the solutions of this equation

$$\lambda_{1,2} = \frac{1}{2}\left(T \pm \sqrt{T^2 - 4D}\right),$$

where

$$T = \operatorname{tr} J(1, 1), \quad D = \det J(1, 1),$$

are such that

$$\left| T \pm \sqrt{T^2 - 4D} \right| < 2. \tag{4.16}$$

To find the conditions under which (4.16) is satisfied, we have to consider two cases:

1. If $T^2 - 4D < 0$, the eigenvalues are complex conjugate and their modulus is equal to $\sqrt{D}$. The stability conditions, in this case, are $D < 1$ and $|T| < 2$.

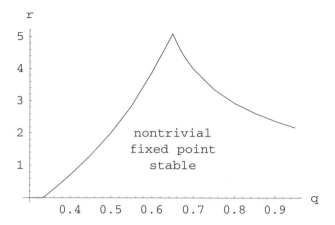

Fig. 4.5. *Domain in which* $(1,1)$ *is asymptotically stable determined using the conditions (4.17).*

2. If $T^2 - 4D > 0$, the eigenvalues are real and the stability conditions (4.16) may be written $(T \pm \sqrt{T^2 - 4D})^2 < 4$. It is easy to verify that, solving for T the equations $(T \pm \sqrt{T^2 - 4D})^2 = 4$, we find $T = \pm(1 + D)$. Hence, if $|T| < 1 + D$, (4.16) is satisfied.

The trace and the determinant of the Jacobian $J(1,1)$ should, therefore, satisfy the conditions

$$|\operatorname{tr} J(1,1)| < 1 + \det J(1,1) < 2 \qquad (4.17)$$

for the eigenvalues, either real or complex, to have a modulus less than 1 (see the right panel of Figure 4.4). Figure 4.5 shows the domain in which the nontrivial fixed point $(1,1)$ is asymptotically stable.

As for models described in terms of differential equations, models represented by maps should not change significantly when the model is slightly modified. A result similar to Theorem 3 can be stated for maps. First we define, as we did for flows, when one can say that two maps are close. In the space $\mathcal{M}(U)$ of all C^1 maps defined on an open set $U \subseteq \mathbf{R}^n$, an ε-*neighborhood* of $\mathbf{f} \in \mathcal{M}(U)$ is defined by

$$N_\varepsilon(\mathbf{f}) = \{\mathbf{g} \in \mathcal{M}(U) \mid \|\mathbf{f} - \mathbf{g}\|_1 < \varepsilon\}.$$

A map $\mathbf{g}$ that belongs to an ε-*neighborhood* of $\mathbf{f}$ is said to be ε-C^1-*close* to $\mathbf{f}$ or an ε-C^1-*perturbation* of $\mathbf{f}$. In this case, the components $(f_1(\mathbf{x}), \ldots, f_n(\mathbf{x}))$ and $(g_1(\mathbf{x}), \ldots, g_n(\mathbf{x}))$ of, respectively, $\mathbf{f}$ and $\mathbf{g}$ and their first derivatives are close throughout U.

Theorem 8. *Let* $\mathbf{x}^*$ *be a hyperbolic equilibrium point of the* C^1 *map* $\mathbf{f} \in \mathcal{M}(U)$. *Then, there exists a neighborhood* $V(\mathbf{x}^*)$ *of* $\mathbf{x}^*$ *and a neighborhood*

$N(\mathbf{f})$ *of* $\mathbf{f}$ *such that each* $\mathbf{g} \in N(\mathbf{f})$ *has a unique hyperbolic equilibrium point* $\mathbf{y}^* \in V(\mathbf{x}^*)$ *of the same type as* $\mathbf{x}^*$. *In this case, the map* $\mathbf{f}$ *is said to be* locally structurally stable *at* $\mathbf{x}^*$.

4.3 Poincaré maps

Definition 17. *Let* φ_t *be a flow on the phase space* $\mathcal{S}$, *generated by the vector field* $\mathbf{X}$, *and* Σ *a subset of* $\mathcal{S}$ *of codimension* 1^8 *such that:*

1. For all values of t, *every orbit of* φ_t *meets* Σ.
2. If $\mathbf{x} \in \Sigma$, $\mathbf{X}(\mathbf{x})$ *is not tangent to* Σ.

Then, Σ *is said to be a* global cross-section *of the flow.*
Let $\mathbf{y} \in \Sigma$ *and* $\tau(\mathbf{y})$ *be the least positive time for which* $\varphi_{\tau(\mathbf{y})}(\mathbf{y}) \in \Sigma$; *then the map* $\mathbf{P} : \Sigma \to \Sigma$ *defined by*

$$\mathbf{P}(\mathbf{y}) = \varphi_{\tau(\mathbf{y})}(\mathbf{y})$$

is the Poincaré map[9] *for* Σ *of the flow* φ_t.

Example 29. A trivial example. In Example 17, it was shown that, according to the Poincaré-Bendixson theorem, the two-dimensional system

$$\dot{x}_1 = x_1 - x_2 - x_1(x_1^2 + x_2^2), \qquad \dot{x}_2 = x_1 + x_2 - x_2(x_1^2 + x_2^2),$$

had a limit cycle. Using polar coordinates, this system becomes

$$\dot{r} = r - r^3, \qquad \dot{\theta} = 1.$$

Under this particularly simple form, we can integrate these two decoupled differential equations. We find

$$r(t) = \left(1 + \left(\frac{1}{r_0^2 - 1}\right) e^{-2t}\right)^{-1/2}, \qquad \theta(t) \qquad = \theta_0 + t,$$

where $r_0 = r(0)$ and $\theta_0 = \theta(0)$. If r_0 is close to 1, it is clear that the orbit tends to the limit cycle $r(t) = 1$, which is stable. This example is so simple that we can determine a Poincaré map explicitly. If Σ is the ray $\theta = \theta_0$ through the origin, then Σ is orthogonal to the limit cycle $r(t) = 1$, and the orbit passing through the point $(r_0, \theta_0) \in \Sigma$ at $t = 0$ intersects Σ again at $t = 2\pi$. Thus, the map defined on Σ by

$$P(r_0) = \left(1 + \left(\frac{1}{r_0^2} - 1\right) e^{-4\pi}\right)^{-1/2}$$

is a Poincaré map. $r_0 = 1$ is a fixed point for this map, and it is stable since

$$P'(1) = e^{-4\pi} < 1.$$

---

[8] The codimension of Σ is $\dim(\mathcal{S}) - \dim(\Sigma)$.
[9] Also called the *first return map.*

If φ_t is the flow generated by the vector field $\mathbf{X}$, taking into account (3.4), we have

$$\mathbf{X}(\varphi_t(\mathbf{x})) = \frac{d}{d\tau}\varphi_\tau(\mathbf{x})\Big|_{\tau=t}$$

$$= \frac{d}{d\tau}\varphi_t(\varphi_\tau(\mathbf{x}))\Big|_{\tau=0}$$

$$= D_{\mathbf{x}}\varphi_t \frac{d}{d\tau}\varphi_\tau(\mathbf{x})\Big|_{\tau=0}$$

$$= D_{\mathbf{x}}\varphi_t\big(\mathbf{X}(\mathbf{x})\big). \qquad (4.18)$$

If x_0 is a point on a periodic orbit of period T, then $\varphi_T(\mathbf{x}_0) = \mathbf{x}_0$, and from (4.18) it follows that

$$\mathbf{X}(\mathbf{x}_0) = \mathbf{X}(\varphi_T(\mathbf{x}_0)) = D_{\mathbf{x}_0}\varphi_T\big(\mathbf{X}(\mathbf{x}_0)\big).$$

That is, $\mathbf{X}(\mathbf{x}_0)$ is an eigenvector of the linear operator $D_{\mathbf{x}_0}\varphi_T$ associated with the eigenvalue 1.

If $\mathbf{x}_0$ and $\mathbf{x}_1$ are two points on a periodic orbit of period T, there exists s such that $\mathbf{x}_1 = \varphi_s(\mathbf{x}_0)$. But, we always have

$$\varphi_T \circ \varphi_s(\mathbf{x}) = \varphi_s \circ \varphi_T(\mathbf{x}).$$

Hence, taking the derivative of both sides of the relation above at $\mathbf{x}_0$, we obtain

$$D_{\mathbf{x}_1}\varphi_T D_{\mathbf{x}_0}\varphi_s = D_{\mathbf{x}_0}\varphi_s D_{\mathbf{x}_0}\varphi_T,$$

which shows that the linear operators $D_{\mathbf{x}_1}\varphi_T$ and $D_{\mathbf{x}_0}\varphi_T$ are (linearly) conjugate by $D_{\mathbf{x}_0}\varphi_s$.

From the considerations above, it follows that if γ is a periodic orbit of the flow φ_t in $\mathbb{R}^n$ and $\mathbf{x}_0 \in \gamma$, the n eigenvalues of the linear operator $D_{\mathbf{x}_0}\varphi_T$ are $1, \lambda_1, \lambda_2, \ldots, \lambda_{n-1}$. The $n-1$ eigenvalues $\lambda_1, \lambda_2, \ldots, \lambda_{n-1}$ are called the *characteristic multipliers* of the periodic orbit. Since $D_{\mathbf{x}_1}\varphi_T$ and $D_{\mathbf{x}_0}\varphi_T$ are conjugate, these eigenvalues do not depend upon the point $\mathbf{x}_0$ on the orbit. If, for all $i = 1, 2, \ldots, n-1$, $|\lambda_i| \neq 1$, the orbit is said to be hyperbolic. If the absolute values of all the characteristic multipliers are less than 1, the orbit is a *periodic attractor*, if the absolute values of all the characteristic multipliers are greater than 1, the orbit is a *periodic repeller*, and if the hyperbolic orbit is neither a periodic attractor nor a periodic repeller, it is a *periodic saddle*.

It can be shown that the eigenvalues of the derivative of the Poincaré map at a point $\mathbf{x}_0$ on a periodic orbit γ are the characteristic multipliers of the periodic orbit γ.[10] This result reduces the problem of the stability of a periodic orbit of a flow to the problem of the stability of a fixed point of a map.

---

[10] See Robinson [299].

4.4 Local bifurcations of maps

The bifurcation theory of equilibrium points of maps is similar to the theory for vector fields. We shall, therefore, only highlight the differences when they occur.

Let $(\mathbf{x}, \boldsymbol{\mu}) \mapsto \mathbf{f}(\mathbf{x}, \boldsymbol{\mu})$ be a $C^{k\ 11}$ r-parameter family of maps on $\mathbb{R}^n$. If $(\mathbf{x}^*, \boldsymbol{\mu}^*)$ is an equilibrium point (*i.e.*, $\mathbf{x}^* = \mathbf{f}(\mathbf{x}^*, \boldsymbol{\mu}^*)$), is the stability of the equilibrium point affected as $\boldsymbol{\mu}$ is varied?

If the equilibrium point is hyperbolic (*i.e.*, if none of the eigenvalues of the Jacobian matrix $D\mathbf{f}(\mathbf{x}^*, \boldsymbol{\mu}^*)$ has unit modulus), then for values of $\boldsymbol{\mu}$ sufficiently close to $\boldsymbol{\mu}^*$, the stability of the equilibrium point is not affected. Moreover, in this case, from the implicit function theorem, it follows that there exists a unique C^k function $\mathbf{x} : \boldsymbol{\mu} \mapsto \mathbf{x}(\boldsymbol{\mu})$ such that, for $\boldsymbol{\mu}$ sufficiently close to $\boldsymbol{\mu}^*$,

$$\mathbf{f}(\mathbf{x}(\boldsymbol{\mu}), \boldsymbol{\mu}) = \mathbf{x}(\boldsymbol{\mu}) \quad \text{with} \quad \mathbf{x}(\boldsymbol{\mu}^*) = \mathbf{x}^*.$$

Since the eigenvalues of the Jacobian matrix $D\mathbf{f}(\mathbf{x}(\boldsymbol{\mu}), \boldsymbol{\mu})$ are continuous functions of $\boldsymbol{\mu}$, for $\boldsymbol{\mu}$ sufficiently close to $\boldsymbol{\mu}^*$, none of the eigenvalues of this Jacobian matrix have unit modulus. Hence, equilibrium points close to $(\mathbf{x}^*, \boldsymbol{\mu}^*)$ are hyperbolic and have the same type of stability as $(\mathbf{x}^*, \boldsymbol{\mu}^*)$.

If the equilibrium point $(\mathbf{x}^*, \boldsymbol{\mu}^*)$ is nonhyperbolic (*i.e.*, if some eigenvalues of $D\mathbf{f}(\mathbf{x}^*, \boldsymbol{\mu}^*)$ have a modulus equal to one), for values of $\boldsymbol{\mu}$ close to $\boldsymbol{\mu}^*$, a totally new dynamical behavior can occur.

We shall only study the simplest bifurcations that occur at nonhyperbolic equilibrium points; that is, we investigate one-parameter maps f on $\mathbb{R}$ such that

$$f(0,0) = 0 \quad \text{and} \quad \frac{\partial f}{\partial x}(0,0) = 1 \text{ or } -1,$$

and one-parameter maps $\mathbf{f}$ on $\mathbb{R}^2$ such that

$$\mathbf{f}(\mathbf{0},0) = \mathbf{0} \quad \text{and} \quad \text{spec}\left(D_{\mathbf{x}}(\mathbf{0},0)\right) = \{\alpha \pm i\omega \mid (\alpha, \omega) \in \mathbb{R}^2, \alpha^2 + \omega^2 = 1\},$$

where $\text{spec}(A)$ is the *spectrum* of the linear operator A; that is, the set of complex numbers λ such that $A - \lambda I$ is not an isomorphism.

4.4.1 Maps on $\mathbb{R}$

Saddle-node bifurcation

Consider the one-parameter family of maps

$$x \mapsto f(x, \mu) = x + \mu - x^2. \tag{4.19}$$

---

[11] The degree of differentiability has to be as high as needed in order to satisfy the conditions for the family of maps to exhibit a given type of bifurcation. See the necessary and sufficient conditions below.

We have

$$f(0,0) = 0 \quad \text{and} \quad \frac{\partial f}{\partial x}(0,0) = 1.$$

In the neighborhood of the nonhyperbolic equilibrium point $(0,0)$, equilibrium points are solutions to the equation

$$f(x,\mu) - x = \mu - x^2 = 0.$$

For $\mu < 0$ there are no equilibrium points, while for $\mu > 0$ there are two hyperbolic equilibrium points $x^* = \pm\sqrt{\mu}$. Since $f'(\pm\sqrt{\mu},\mu) = 1 \mp 2\sqrt{\mu}$, $\sqrt{\mu}$ is asymptotically stable, and $-\sqrt{\mu}$ is unstable. This type of bifurcation is called a *saddle-node bifurcation*. Phase portraits and the bifurcation diagram for the recurrence equation $x_{t+1} = f(x_t,\mu)$ are identical to those obtained for the differential equation (3.33). See Figures 3.14 and 3.17.

As for flows, we can derive the necessary and sufficient conditions for a one-parameter map on $\mathbb{R}$ to undergo a saddle-node bifurcation. The derivation of these conditions makes use of the implicit function theorem and is similar to what has been done for one-parameter flows on $\mathbb{R}$. A one-parameter family of C^k ($k \geq 2$) maps $(x,\mu) \mapsto f(x,\mu)$ exhibits a saddle-node bifurcation at $(x,\mu) = (0,0)$ if

$$f(0,0) = 0, \quad \frac{\partial f}{\partial x}(0,0) = 1,$$

and

$$\frac{\partial f}{\partial \mu}(0,0) \neq 0, \quad \frac{\partial^2 f}{\partial x^2}(0,0) \neq 0.$$

Transcritical bifurcation

Consider the one-parameter family of maps

$$x \mapsto f(x,\mu) = x + \mu x - x^2. \tag{4.20}$$

We have

$$f(0,0) = 0 \quad \text{and} \quad \frac{\partial f}{\partial x}(0,0) = 1.$$

In the neighborhood of the nonhyperbolic fixed point $(0,0)$, equilibrium points are solutions to the equation

$$f(x,\mu) - x = \mu x - x^2 = 0.$$

For all values of μ, there exist two fixed points: 0 and μ. If $\mu < 0$, 0 is stable and μ unstable, while for $\mu > 0$, 0 becomes unstable and μ stable. Hence, there exist two curves of fixed points that exchange their stabilities when passing through the bifurcation point. This type of bifurcation is called a *transcritical bifurcation*. Phase portraits and the bifurcation diagram for the recurrence equation $x_{t+1} = f(x_t, \mu)$ are identical to those obtained for the differential equation (3.34). See Figures 3.15 and 3.17.

Here again, using the implicit function theorem, it may be shown that a one-parameter family of C^k ($k \geq 2$) maps $(x, \mu) \mapsto f(x, \mu)$ on $\mathbb{R}$ undergoes a transcritical bifurcation at $(x, \mu) = (0, 0)$ if

$$f(0,0) = 0, \quad \frac{\partial f}{\partial x}(0,0) = 1,$$

and

$$\frac{\partial f}{\partial \mu}(0,0) = 0, \quad \frac{\partial^2 f}{\partial x^2}(0,0) \neq 0, \quad \frac{\partial^2 f}{\partial \mu \partial x}(0,0) \neq 0.$$

Pitchfork bifurcation

Consider the one-parameter family of maps

$$x \mapsto f(x, \mu) = x + \mu x - x^3. \tag{4.21}$$

We have

$$f(0,0) = 0 \quad \text{and} \quad \frac{\partial f}{\partial x}(0,0) = 1.$$

In the neighborhood of the nonhyperbolic fixed point $(0,0)$, equilibrium points are solutions to the equation

$$f(x, \mu) - x = \mu x - x^3 = 0.$$

For $\mu \leq 0$, 0 is the only fixed point and it is asymptotically stable. For $\mu > 0$, there are three fixed points, 0 is unstable, and $\pm\sqrt{\mu}$ are both asymptotically stable. The curve $x = 0$ exists on both sides of the bifurcation point $(0,0)$, and the fixed points on this curve change their stabilities on passing through this point. The curve $\mu = x^2$ exists only on one side of $(0,0)$ and is tangent at this point to the line $\mu = 0$. This type of bifurcation is called a *pitchfork bifurcation*. Phase portraits and the bifurcation diagram for the recurrence equation $x_{t+1} = f(x_t, \mu)$ are identical to those obtained for the differential equation (3.35). See Figures 3.16 and 3.17.

Using the implicit function theorem, it may be shown that a one-parameter family of C^k ($k \geq 3$) maps $(x, \mu) \mapsto f(x, \mu)$ on $\mathbb{R}$ undergoes a pitchfork bifurcation at $(x, \mu) = (0, 0)$ if

$$f(0,0) = 0, \quad \frac{\partial f}{\partial x}(0,0) = 1,$$

and

$$\frac{\partial f}{\partial \mu}(0,0) = 0, \quad \frac{\partial^2 f}{\partial x^2}(0,0) = 0, \quad \frac{\partial^2 f}{\partial \mu \partial x}(0,0) \neq 0, \quad \frac{\partial^3 f}{\partial x^3}(0,0) \neq 0.$$

Period-doubling bifurcation

Consider the one-parameter family of maps

$$x \mapsto f(x, \mu) = -(1 + \mu)x + x^3. \tag{4.22}$$

We have

$$f(0, 0) = 0 \quad \text{and} \quad \frac{\partial f}{\partial x}(0, 0) = -1.$$

The fixed point $(0, 0)$ is a nonhyperbolic fixed point. The other fixed points are the solutions to the equation

$$f(x, \mu) - x = x(x^2 - 2 - \mu).$$

For all values of μ, 0 is a fixed point; it is stable for $-2 < \mu < 0$ and unstable for either $\mu \leq -2$ or $\mu \geq 0$. For $\mu \geq -2$, there exist two other fixed points, $\pm\sqrt{2 + \mu}$ that are both unstable. Hence for $\mu > 0$, the map $(x, \mu) \mapsto f(x, \mu)$ has three fixed points all of them being unstable. This behavior is new; it could not exist for one-dimensional vector fields. The second iterate f^2 of f is

$$(x, \mu) \mapsto f^2(x, \mu) = (1 + \mu)^2 x - (2 + 4\mu + 3\mu^2 + \mu^3)x^3$$
$$+ 3(1 + m)^2 x^5 - 3(1 + m)x^7 + x^9.$$

It can be verified that

$$f^2(0, 0) = 0 \quad \text{and} \quad \frac{\partial f^2}{\partial x}(0, 0) = 1,$$

and

$$\frac{\partial f^2}{\partial \mu}(0, 0) = 0, \quad \frac{\partial^2 f^2}{\partial x^2}(0, 0) = 0, \quad \frac{\partial^2 f^2}{\partial \mu \partial x}(0, 0) = 2, \quad \frac{\partial^3 f^2}{\partial x^3}(0, 0) = -12.$$

Hence, the one-parameter family of maps $(x, \mu) \mapsto f^2(x, \mu)$ undergoes a pitchfork bifurcation at the point $(0, 0)$ (*i.e.*, the new fixed points of f^2 are period 2 points of f). This new type of bifurcation is called a *period-doubling bifurcation*.[12]

Using the implicit function theorem, it is possible to find the necessary and sufficient conditions for a one-parameter family of C^k ($k \geq 3$) maps $(x, \mu) \mapsto f(x, \mu)$ on $\mathbb{R}$ to exhibit a period-doubling bifurcation at $(x, \mu) = (0, 0)$. From the discussion above, it follows that $(0, 0)$ should be a nonhyperbolic fixed point of f such that

$$f(0, 0) = 0 \quad \text{and} \quad \frac{\partial f}{\partial x}(0, 0) = -1,$$

and, at $(0, 0)$, f^2 should exhibit a pitchfork bifurcation; *i.e.*,

---

[12] Also called a *flip bifurcation*.

$$\frac{\partial f^2}{\partial \mu}(0,0) = 0, \quad \frac{\partial^2 f^2}{\partial x^2}(0,0) = 0, \quad \frac{\partial^2 f^2}{\partial \mu \partial x}(0,0) \neq 0, \quad \frac{\partial^3 f^2}{\partial x^3}(0,0) \neq 0.$$

For a small positive value of the parameter μ, the map f^2 has two fixed points, which are $x_1^* = -\sqrt{\mu}$ and $x_2^* = \sqrt{\mu}$, such that

$$f(x_1^*, \mu) = x_2^* \quad \text{and} \quad f(x_2^*, \mu) = x_1^*.$$

Hence, x_1^* and x_2^* are period 2 points of the map f, and the periodic orbit $\{x_1^*, x_2^*\}$ is asymptotically stable if x_1^* and x_2^* are asymptotically stable fixed points of f^2. Since

$$\frac{\partial f^2}{\partial x}(x_1^*, \mu) = \frac{\partial f}{\partial x}\left(\frac{\partial f}{\partial x}(x_1^*, \mu)\right)\frac{\partial f}{\partial x}(x_1^*, \mu)$$

$$= \frac{\partial f}{\partial x}(x_2^*, \mu)\frac{\partial f}{\partial x}(x_1^*, \mu),$$

the stability condition is

$$\left|\frac{\partial f}{\partial x}(x_1^*, \mu)\frac{\partial f}{\partial x}(x_2^*, \mu)\right| < 1. \tag{4.23}$$

With $x_1^* = -\sqrt{\mu}$ and $x_2^* = \sqrt{\mu}$, we find

$$\frac{\partial f}{\partial x}(\sqrt{\mu}, \mu) = \frac{\partial f}{\partial x}(-\sqrt{\mu}, \mu) = -1 + 2\mu.$$

Hence, the 2-point cycle $\{\sqrt{\mu}, -\sqrt{-\mu}\}$ is asymptotically stable for μ sufficiently small.

Example 30. *Period-doubling bifurcation in a one-population model.* In Example 27, two different one-population models were presented. In the first model, the reduced population n evolves according to the recurrence equation

$$n_{t+1} = n_t \exp(r(1 - n_t)).$$

Does this model exhibit a period-doubling bifurcation? Let

$$f(n, r) = n \exp(r(1 - n)).$$

Since

$$f(1, 2) = 1 \quad \text{and} \quad \frac{\partial f}{\partial n}(1, 2) = -1,$$

the nonhyperbolic fixed point $(1, 2)$ may be a period-doubling bifurcation point. Let us check the remaining four other conditions. The second iterate of f is

$$(n, r) \mapsto f^2(n, r) = n e^{r(1-n)} \exp\left(r(1 - n e^{r(1-n)})\right),$$

and we have

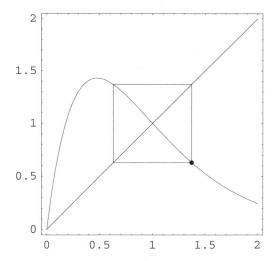

Fig. 4.6. *Stable 2-point cycle of the one-population model $n_{t+1} = n_t \exp(r(1 - n_t)$ obtained for $r = 2.1$. A transient of 40 iterations has been discarded.*

$$\frac{\partial f^2}{\partial r}(1, 2) = 0, \quad \frac{\partial^2 f^2}{\partial^2 n}(1, 2) = 0, \quad \frac{\partial^2 f^2}{\partial r \partial n}(1, 2) = 2, \quad \frac{\partial^3 f^2}{\partial^3 n}(1, 2) = -8.$$

The one-parameter family of maps $(n, r) \mapsto n \exp(r(1-n))$ exhibits, therefore, a period-doubling bifurcation at the point $(n, r) = (1, 2)$.

If r is slightly greater than 2, the map $f(n, r)$ has two period 2 points, n_1^* and n_2^*, such that $f(n_1^*, r) = n_2^*$ and $f(n_2^*, r) = n_1^*$. To check the stability of these periodic points, we have to verify that Condition (4.23) is satisfied. If, for example, $r = 2.1$, we find that $(n_1^*, n_2^*) = (0.629294 \ldots, 1.370706 \ldots)$ and

$$\frac{\partial f}{\partial n}(0.629294, 2.1) \, \frac{\partial f}{\partial 2}(1.370706, 2.1) = 0.603965.$$

The 2-point cycle $\{0.629294 \ldots, 1.370706 \ldots\}$, shown in Figure 4.6, is asymptotically stable.

4.4.2 The Hopf bifurcation

As in the case of families of maps on $\mathbb{R}$, nonhyperbolic points in families of maps on $\mathbb{R}^2$ are usually bifurcation points. The various types of bifurcations that may occur for one-dimensional systems may also occur for higher-dimensional systems. There exist, however, other possible types of bifurcations. In this section, we discuss a bifurcation that appears when the eigenvalues of the linear part of the two-dimensional part in the neighborhood of a nonhyperbolic equilibrium point cross the unit circle.

Example 31. Delayed logistic model. One-population models of the form $x_{t+1} = f(x_t)$ are not adequate for species whose individuals need a certain time to reach sexual maturity. By analogy with the continuous-time equation (1.12), consider the Maynard Smith delayed logistic recurrence equation [233]

$$n_{t+1} = rn_t(1 - n_{t-1}).$$

In order to write this second-order equation as a first-order two-dimensional system, let

$$x_t = n_{t-1} \quad \text{and} \quad y_t = n_t.$$

The delayed logistic equation then takes the form

$$x_{t+1} = y_t, \quad y_{t+1} = ry_t(1 - x_t).$$

The map $\mathbf{f} : (x, y) \mapsto (y, ry(1 - x))$ has two fixed points,

$$(0, 0) \quad \text{and} \quad \left(\frac{r-1}{r}, \frac{r-1}{r} \right).$$

The eigenvalues of the linear operator at the nontrivial fixed point

$$D\mathbf{f} \left(\left(\frac{r-1}{r}, \frac{r-1}{r} \right), r \right)$$

are $\frac{1}{2}(1 \pm \sqrt{5 - 4r})$. If $r > \frac{5}{4}$, the eigenvalues are complex conjugate. Their modulus is equal to 1 when

$$\frac{1}{4}1 + (4r - 5) = 1, \quad \text{that is, for } r = 2,$$

and, in this case, they are equal to $e^{\pm i\pi/3}$ (*i.e.*, to two of the sixth roots of unity).[13] For $r < 2$, the nontrivial fixed point is asymptotically stable, hence, for $r > 2$, we expect the existence of a stable limit cycle. This is the case, as shown in Figure 4.7. The nonhyperbolic point $((\frac{1}{2}, \frac{1}{2}), 2)$ is, therefore, a bifurcation point. Because of its connection with the Hopf bifurcation for flows, this bifurcation is also named after Hopf.

Theorem 9. *Let* $(\mathbf{x}, \mu) \mapsto \mathbf{f}(\mathbf{x}, \mu)$ *be a one-parameter* C^k ($k \geq 5$) *family of maps in* $\mathbb{R}^2$ *such that:*

1. *For* μ *near 0,* $\mathbf{f}(\mathbf{0}, \mu) = 0$.
2. *For* μ *near 0,* $D\mathbf{f}(\mathbf{0}, \mu)$ *has two complex conjugate eigenvalues* $\lambda(\mu)$ *and* $\overline{\lambda}(\mu)$, *with* $|\lambda(0)| = 1$.
3. $\lambda(0)$ *is not a qth root of unity for* $q = 1, 2, 3, 4$ *or* 5.
4. $\dfrac{d}{d\mu}|\lambda(\mu)| > 0$ *for* $\mu = 0$.

---

[13] See Theorem 9 for the importance of this remark.

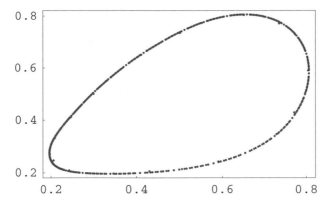

Fig. 4.7. *Stable cycle for the two-dimensional map* $(x, y) \mapsto (y, ry(1-x))$ *(delayed logistic model) for* $r = 2.1$. *Initial point:* $(0.3, 0.5)$, *number of iterations:* 400.

Then, there exists a smooth change of coordinates that brings **f** *in polar coordinates under the form*

$$\mathbf{g}(r, \theta, \mu) - (|\lambda(\mu)|r + a(\mu)r^3, \theta + b(\mu) + c(\mu)r^2),$$

where terms of degree 5 and higher in r *have been neglected.* a, b, *and* c *are smooth functions of* μ. *If* $a(0) < 0$ *(resp.* $a(\mu) > 0$*) then, for* $\mu < 0$ *(resp.* $\mu > 0$*), the origin is stable (resp. unstable) and, for* $\mu > 0$ *(resp.* $\mu < 0$*)), the origin is unstable (resp. stable) and surrounded by an attracting (resp. repelling) invariant circle.*

We could verify that in the case of the delayed logistic equation of Example 31, the conditions of the theorem are satisfied (for $r = 2$ the eigenvalues are the sixth root of unity). The proof of this theorem is rather laborious. See either Devaney [102] or Arrowsmith and Place [11].

4.5 Sequences of period-doubling bifurcations

When a one-parameter family of maps on $\mathbb{R}$, $(x, \mu) \mapsto f(x, \mu)$, exhibits a period-doubling bifurcation at a nonhyperbolic fixed point (x_1, μ_1), the family of second iterates $(x, \mu) \mapsto f^2(x, \mu)$ exhibits at (x_1, μ_1) a pitchfork bifurcation. Then, when μ is slightly modified, say increased, this bifurcation gives birth to a 2-point cycle, and both periodic points of this cycle are hyperbolic fixed points of f^2. As μ is further increased, it might happen that the family of second iterates undergoes a period-doubling bifurcation at a nonhyperbolic fixed point (x_2, μ_2), giving birth to a 2-point cycle for f^2 (*i.e.*, a 4-point cycle for f). This process may occur repeatedly, generating an infinite sequence of period-doubling bifurcations. The interesting fact is that these sequences possess some universal properties, which are discussed below.

4.5.1 Logistic model

The discrete logistic model, described by the recurrence equation

$$n_{t+1} = rn_t(1 - n_t), \tag{4.24}$$

has been extensively studied. It is the simplest nonlinear map, but nevertheless its properties are far from being trivial. In an earlier study of the iteration of quadratic polynomials, Myrberg [256] already mentioned the existence of considerable difficulties:

> *En nous limitant dans notre travail au cas le plus simple non linéaire, c'est-à-dire aux polynômes réels du second degré, nous observons que même dans ce cas spécial on rencontre des difficultés considérables, dont l'explication exigera des recherches ultérieures.*[14]

As a one-population model, Equation (4.24) represents the evolution of a reduced population $n = N/K$, where K is the carrying capacity. From a purely mathematical point of view, it has the advantage of being extremely simple, but, as a population model, it has the disadvantage of requiring n to belong to the closed interval $[0, 1]$ since, for $n \notin [0, 1]$, the sequence of iterates of n tends to $-\infty$.

The fixed points of (4.24) are the solutions of

$$n = f(n, r) = rn(1 - n);$$

i.e., $n = 0$ and $n = (r - 1)/r$. The derivative of f with respect to n being equal to $r(1 - 2n)$, if $r < 1$, $n = 0$ is asymptotically stable and $n = (r - 1)/r$, which does not belong to $[0, 1]$, is unstable. The nonhyperbolic fixed point $(n, r) = (0, 1)$ is a transcritical bifurcation point. If $1 < r < 3$, $n = 0$ is unstable and $n = (r - 1)/r$ is asymptotically stable.

For $r = 3$, the derivative of f at $n = (r - 1)/r = 2/3$ is equal to -1. Since

$$f^2(n, r) = r^2 n(1 - n)(1 - (rn(1 - n))),$$

we verify that

$$f^2(2/3, 3) = 2/3, \quad \frac{\partial f^2}{\partial r}(2/3, 3) = 1,$$

and

$$\frac{\partial f^2}{\partial r}(2/3, 3) = 0, \quad \frac{\partial^2 f^2}{\partial^2 n}(2/3, 3) = 0,$$

$$\frac{\partial^2 f^2}{\partial r \partial n}(2/3, 3) = 2, \quad \frac{\partial^3 f^2}{\partial^3 n}(2/3, 3) = -108.$$

---

[14] Limiting ourselves to the simplest nonlinear case; that is, to quadratic real polynomials, it is observed that even in this special case considerable difficulties are encountered, whose explanation will require more work in the future.

These results show that the nonhyperbolic fixed point $(2/3, 3)$ is a period-doubling bifurcation point. The map f^2 has four fixed points that are the solutions of the equation

$$f^2(n, r) = n.$$

Two solutions are already known, namely the two unstable fixed points, $n = 0$ and $n = (r-1)/r$, of f. The remaining two solutions are[15]

$$n = \frac{1}{2r}(r + 1 \pm \sqrt{r^2 - 2r - 3}).$$

They are the two components of the 2-point cycle. They are defined only for $r \geq 3$. The domain of stability of this cycle is determined by the condition

$$\left| \frac{\partial f}{\partial n}\left(\frac{1}{2r}(1 + r + \sqrt{r^2 - 2r - 3}), r \right) \frac{\partial f}{\partial n}\left(\frac{1}{2r}(1 + r - \sqrt{r^2 - 2r - 3}), r \right) \right| < 1.$$

That is, the 2-point cycle is asymptotically stable if $3 < r < 1 + \sqrt{6}$. For $r > 1 + \sqrt{6}$, the system undergoes an infinite sequence of period-doubling bifurcations.

A few iterations of the logistic map for increasing values of r are shown in Figure 4.8.

Let $(r_k)_{k \in \mathbb{N}}$ be the sequence of parameter values at which a period-doubling bifurcation occurs. This sequence is such that the 2^k-point cycle is stable for $r_k < r < r_{k+1}$. Then, if $\{n_{k,1}, n_{k,2}, \ldots, n_{k,2^k}\}$ denotes the 2^k-point cycle, for $i = 1, 2, \ldots, 2^k$ and $r_k < r < r_{k+1}$, we have

$$f^{2^k}(n_{k,i}, r) = n_{k,i}. \tag{4.25}$$

Moreover, for $i = 1, 2, \ldots, 2^k$,[16]

$$\frac{\partial f^{2^k}}{\partial n}(n_{k,i}, r_k) = +1, \qquad \frac{\partial f^{2^k}}{\partial n}(n_{k,i}, r_{k+1}) = -1. \tag{4.26}$$

Numerically, one can easily discover that (r_k) is an increasing bounded sequence of positive numbers. This sequence has, therefore, a limit r_∞. If we assume that the asymptotic behavior of r_k is of the form

---

[15] Let $n_{1,1}$ and $n_{1,2}$ be these solutions. They are such that

$$f(n_{1,1}, r) = n_{1,2}, \qquad f(n_{1,2}, r) = n_{1,1}.$$

Hence,

$$n_{1,1} + n_{1,2} = \frac{1 + r}{r}, \qquad n_{1,1} n_{1,2} = \frac{1 + r}{r^2},$$

which shows that $n_{1,1}$ and $n_{1,2}$ are the solutions of the quadratic equation

$$r^2 n^2 - r(1 + r)n + 1 + r = 0.$$

[16] The numerical values of the $n_{k,i}$ $(i = 1, 2, \ldots, 2^k)$ depend on r.

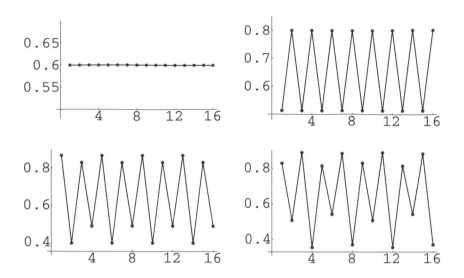

Fig. 4.8. *Iterations of the logistic map* $f : (n, r) \mapsto rn(1 - n)$ *for* $r = 2.5$ *(fixed point),* $r = 3.2$ *(2-point cycle),* $r = 3.48$ *(4-point cycle), and* $r = 3.55$ *(8-point cycle). Sixteen iterates are shown after 100 have been discarded.*

$$r_k \sim r_\infty - \frac{a}{\delta^k}, \tag{4.27}$$

where a and δ are two positive numbers, then, according to (4.27),

$$\lim_{k \to \infty} \frac{r_k - r_{k-1}}{r_{k+1} - r_k} = \delta; \tag{4.28}$$

δ is known as the *Feigenbaum number*. To determine the numerical values of $r_1, r_2, r_3, \ldots$ and δ, one proceeds as follows. We have seen that the exact values of r_1 and r_2 are, respectively, 3 and $1 + \sqrt{6}$. Once the approximate value of r_3 is found,

$$\delta_1 = \frac{r_2 - r_1}{r_3 - r_2}$$

gives an approximate value of δ, which is used to estimate

$$r_4 \approx r_3 + \frac{r_3 - r_2}{\delta_1}.$$

As k increases, solving Equation (4.25), we locate r_k and determine a better value of δ.[17] This is the method followed by Feigenbaum [121, 122] who found $\delta = 4.6692016091029\ldots$.

---

[17] The first period-doubling bifurcations occur for the following parameter values:

4.5.2 Universality

The interesting fact, found by Feigenbaum [121], is that the rate of convergence δ of the sequence (r_k) is *universal* in the sense that it is the same for all recurrence equations of the form $x_{t+1} = f(x_t, r)$ that exhibit an infinite sequence of period-doubling bifurcations, if f is continuous and has a unique maximum x_c (such a map is said to be *unimodal*) with $f(x_c) - f(x) \sim (x_c - x)^2$. If the order of the maximum is changed, the ratio will also change. Consider the map

$$g(x, a) = 1 - ax^z, \quad (z > 1).$$

It is easy to verify that, for $z = 2$, f and g are topologically equivalent; *i.e.*, there exists a map ℓ such that

$$\ell \circ g = f \circ \ell,$$

where

$$\ell(x) = \left(\tfrac{1}{4} r - \tfrac{1}{2} \right) x + \tfrac{1}{2}.$$

The parameters a and r are such that $r^2 - 2r - 4a = 0$.

The following table gives the value of δ for different values of the exponent z.

z	2	4	6	8
δ	4.669	7.284	9.296	10.948

The existence of an infinite sequence of period-doubling bifurcations for $r \geq 3$ can be made plausible by a simple geometric argument.

Consider the graphs of f^2 for two parameter values, one slightly less than 3 and the other one slightly greater than 3 (Figure 4.9). They show how the asymptotically stable point for $r < 3$ splits into an asymptotically stable 2-point orbit for $r > 3$.

Consider now the graph of f^2 for a value of r between r_1 and r_2. As shown in Figure 4.10, it has four fixed points. Two of them, 0 and $(r-1)/r$, are unstable and the other two correspond to the stable 2-point cycle. In the interval $[1/r, (r-1)/r]$, the graph of f^2, after a symmetry about a horizontal axis, is similar to the graph of f for a certain parameter value s. More precisely, the map F, defined by

$$F = L \circ f^2 \circ L^{-1}, \tag{4.29}$$

where L is the linear map

$$L : x \mapsto \frac{x - (r-1)/r}{(2-r)/r}, \tag{4.30}$$

$r_1 = 3.0$, $r_2 = 3.449499...$, $r_3 = 3.544090...$, $r_4 = 3.564407...$,

$r_5 = 3.568759...$, $r_6 = 3.569692...$, $r_7 = 3.569891...$, $r_8 = 3.569934...$.

The limit of the sequence (r_k) is $r_\infty = 3.5699456...$, and the value of a in Relation (4.27) is $2.6327...$.

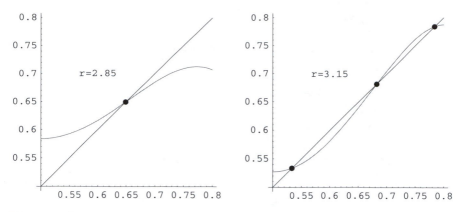

Fig. 4.9. *Graphs of the second iterates of the logistic map for $r = 2.85$ and $r = 3.15$ showing how the period-doubling bifurcation at the point $(2/3, 3)$ gives birth to a stable 2-point orbit.*

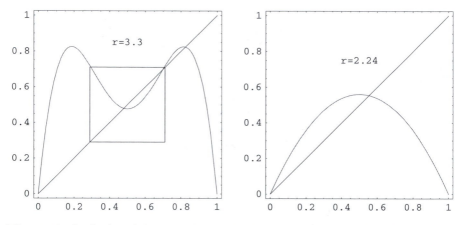

Fig. 4.10. *In the boxed domain, the graph of the second iterate of the logistic map for $r = 3.3$ is similar to the graph of the logistic map for $r = 2.24$.*

is such that $F(0) = F(1) = 0$, and if we choose a parameter value s such that $f(\frac{1}{2}, s) = F(\frac{1}{2}, r)$, the graphs of $(x, s) \mapsto f(x, s)$ and $(x, r) \mapsto F(x, r)$ are very close, as shown in Figure 4.11. For r close to r_2, in the boxed domain, the scenario represented in Figure 4.9 will this time occur for f^4; that is, the stable fixed point of f^2 in the boxed domain, the asymptotically stable point for $r < r_2$, splits into an asymptotically stable 2-point cycle for $r > r_2$. This 2-point cycle for f^2 is a 4-point cycle for f, which is asymptotically stable for $r_2 < r < r_3$, and so on ... .

This argument can be made more rigorous using renormalization-group methods.[18]

---

[18] See Coullet and Tresser [95] and Feigenbaum [121].

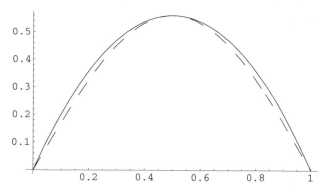

Fig. 4.11. *Graph of the transform F of f^2 (see text) for $r = 3.3$ (broken line) compared to the graph of the logistic curve for $r = 2.24$ (full line).*

A fixed point x^* of a map f is said to be *superstable* if $f'(x^*) = 0$.[19] In the case of the logistic map $f : (n, r) \mapsto rn(1 - n)$, $(\frac{1}{2}, 2)$ is a superstable fixed point. Superstable cycles are defined in a similar way. In the case of the logistic map, a 2^k-point cycle that includes the point $\frac{1}{2}$ is superstable if, for the parameter value $r = R_k$, $(f^{2^k})'(\frac{1}{2}, R_k) - 0$. Since $r_k < R_k < r_{k+1}$, we have

$$\lim_{k \to \infty} \frac{R_k - R_{k-1}}{R_{k+1} - R_k} = \delta.$$

It is actually easier to calculate δ accurately from the relation above than from (4.28) because, when r is close to r_k from above, the 2^k-cycle is weakly stable, and the method to determine r_k converges slowly.

The sequence (d_k), where

$$d_k = f^{2^{k-1}}(\tfrac{1}{2}, R_k) - \tfrac{1}{2}$$

represents the distance between the point $\frac{1}{2}$ and its nearest point on the 2^k-point cycle, is such that the limit[20]

$$-\lim_{k \to \infty} \frac{d_k}{d_{k+1}} = \alpha$$

is also a universal constant equal to $2.5029078751\ldots$ [121].

While Feigenbaum's universality is quantitative, in 1973 Metropolis, Stein, and Stein [241] had already discovered a *qualitative universality* of finite limit sets of families of unimodal maps f_r defined on the unit interval $[0, 1]$ such that $f_r(0) = f_r(1) = 0$.[21]

---

[19] Such a point is said to be *critical*.

[20] The minus sign in the definition of α is related to the fact that the nearest point to $\frac{1}{2}$ on the 2^k-point cycle is alternatively below and above $\frac{1}{2}$ as k increases.

[21] In their paper, Metropolis, Stein, and Stein enumerate a list of conditions the maps should satisfy. In particular, they allow maps such as, for example,

In order to illustrate this qualitative universality, consider the logistic map $(x, r) \mapsto f_r(x) = rx(1 - x)$. $f_r(x)$ is maximum for $x = \frac{1}{2}$, and $f_r'(\frac{1}{2}) = 0$. For a given integer $k > 1$, solving for r the equation

$$f_r^k(\tfrac{1}{2}) = \tfrac{1}{2}$$

gives the values of the parameter r for which there exists a superstable k-point cycle. For example, if $k = 5$, there are three different solutions: $3.7389149\ldots$, $3.9057065\ldots$, and $3.9902670\ldots$. In each case, the successive iterates of $x = \frac{1}{2}$ are either greater than or less than $\frac{1}{2}$. Following Metropolis, Stein, and Stein, according to whether an iterate is greater than or less than $\frac{1}{2}$, it is said to be of type R or of type L. Using this convention, the itinerary along a superstable 5-point cycle is characterized by a pattern of four letters:

$$\tfrac{1}{2} \to R \to L \to R \to R \to \tfrac{1}{2} \quad \text{for } r = 3.7389149\ldots,$$
$$\tfrac{1}{2} \to R \to L \to L \to R \to \tfrac{1}{2} \quad \text{for } r = 3.9057065\ldots,$$
$$\tfrac{1}{2} \to R \to L \to L \to L \to \tfrac{1}{2} \quad \text{for } r = 3.9902670\ldots.$$

Omitting the initial and final points $\frac{1}{2}$, the patterns are written in the simplified form

$$RLR^2, \quad RL^2R, \quad RL^3.$$

These sequences of R and L symbols are called *U-sequences*, where U stands for *universal*. They are universal in the sense that they appear for all unimodal maps in the same order for increasing values of the parameter r.

Note that any U-sequence necessarily starts with R, followed by L if the sequence contains more than one symbol; that is, if the period of the superstable cycle is greater than 2.

Many one-population models, described by continuous maps with only one quadratic maximum, exhibit universal infinite sequences of period-doubling bifurcations. This is the case, for example, for the map $(n, r) \mapsto n \exp(r(1 - n))$ (see Figure 4.12), that undergoes period-doubling bifurcations for the following parameter values[22]:

$$r_1 = 2, \quad r_2 = 2.526\ldots, \quad r_3 = 2.656\ldots, \quad r_4 = 2.682\ldots, \quad r_\infty = 2.692\ldots.$$

Since $n \exp(r(1 - n))$ is maximum for $x = 1/r$, superstable cycles must include this point. The one-population model exhibits superstable 2^k-point cycles for the following parameter values:

$$f_r(x, a) = \begin{cases} \dfrac{r}{a}\, x, & \text{if } 0 \le x \le a, \\[2mm] r, & \text{if } a \le x \le 1 - a, \\[2mm] \dfrac{r}{a}\,(1 - x), & \text{if } 1 - a \le x \le 1, \end{cases}$$

where a and r are positive reals such that $1 - a < r < 1$.

[22] See also Exercise 4.6.

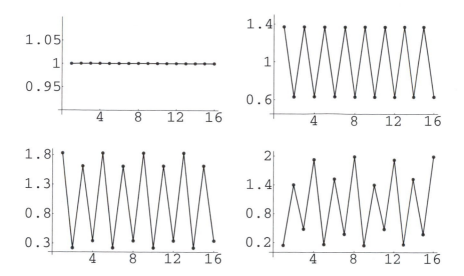

Fig. 4.12. *Iterations of the first one-population map of Example 27 $(n,r) \mapsto ne^{r(1-n)}$ for $r = 1.8$ (fixed point), $r = 2.1$ (2-point cycle), $r = 2.55$ (4-point cycle), and $r = 2.67$ (8-point cycle). Sixteen iterates are shown after 200 have been discarded.*

$$R_0 = 1, \quad R_1 = 2.25643\ldots, \quad R_2 = 2.59352\ldots, \quad R_3 = 2.671\ldots$$

It can be verified that the U-sequences corresponding to the 2-, 4-, and 8-point cycles are

$$R, \quad RLR, \quad RLR^3LR,$$

the same as for the superstable cycles of the logistic map.

Remark 6. The patterns in Figure 4.13 illustrate an interesting general feature mentioned by Metropolis, Stein, and Stein [241]. If P is a pattern for a given superstable cycle, then the pattern of its first harmonic is PXP, where $X = L$ if P contains an odd number of Rs, and $X = R$ otherwise.

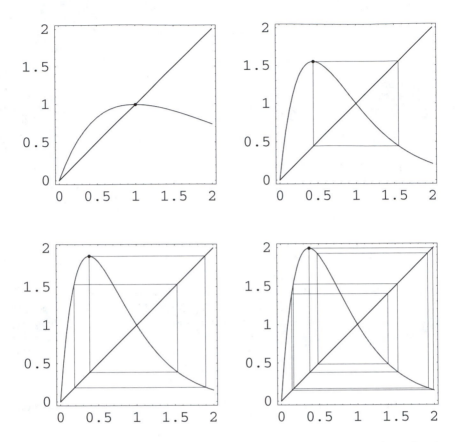

Fig. 4.13. *Superstable cycles of the one-population map of Example 27 $(n, r) \mapsto ne^{r(1-n)}$ for $r = 1$ (fixed point), $r = 2.25643$ (2-point cycle), $r = 2.59352$ (4-point cycle), and $r = 2.671$ (8-point cycle).*

Exercises

Exercise 4.1 *An acceptable discrete one-population model of sexually reproducing organisms should be such that, if the population density is small, the organisms are sparsely distributed in their habitat, resulting in a low mating rate and a density-dependent growth rate less than 1. This effect is named after Allee [6]. If the population density is large, intraspecific competition is strong, and it is reasonable to assume that, in this case, the density-dependent growth rate has to be less than 1. The model*

$$n_{t+1} = rn_t^x(1 - n_t),$$

where n_t is the population density at time t, and r and x are positive constants, is a generalization of the logistic model. Under which condition(s) does it exhibit the Allee effect? In this case, find its equilibrium points and determine their stabilities.

Exercise 4.2 *The recurrence equation*

$$n_{t+1} = rn_t^2(1 - n_t) - cn_t,$$

is a model of harvesting a species that exhibits the Allee effect. The constant c is the intrinsic catching rate. Show that, for a fixed value of r greater than a critical value r_c, the system undergoes a saddle-node bifurcation at a point (n^*, c^*). What does this bifurcation imply from an ecological point of view?

Exercise 4.3 *In an SIR model for the spread of an infectious disease, based on disease status, the individuals are divided into three disjoint groups:*

1. Susceptible *individuals are capable of contracting the disease and becoming infective.*
2. Infective *individuals are capable of transmitting the disease to susceptibles.*
3. Removed *individuals have had the disease and are dead, have recovered and are permanently immune, or are isolated until recovery and permanent immunity occur.*

Let S_t, I_t, and R_t denote the densities of individuals belonging to the three different groups defined above. If we assume that the interactions responsible for the spread of the disease have a very long range (i.e., each susceptible individual can be equally infected by any infective individual), it can be shown that the spread of the infectious disease can be modeled by the recurrence equations[23]

$$S_{t+1} + I_{t+1} + R_{t+1} = \rho,$$
$$I_{t+1} = I_t + S_t\big(1 - \exp(-iI_t)\big) - rI_t,$$
$$R_{t+1} = R_t + rI_t,$$

where i is a coefficient proportional to the probability for a susceptible individual to be infected, r is the probability for an infective individual to be removed, and ρ is the total initial density.

(i) Show that as time t goes to infinity, the limits S_∞, I_∞, and R_∞ exist. What is the value of I_∞?

(ii) Show that, as for the Kermack-McKendrick epidemic model (Example 8), there is an epidemic if, and only if, the initial density of susceptible individuals S_0 is greater than

---

[23] If each susceptible individual can be equally infected by any infective individual, the probability for a susceptible to be infected can be written i/N, where i is a positive real number and N the total number of individuals in the population. Hence, between times t and $t + 1$, the density of infected individuals due to infection of susceptible individuals is given by

$$S_t\left(1 - \left(1 - \frac{iI_t}{N}\right)^{N-1}\right),$$

which, in the limit $N \to \infty$ (i.e., in the limit of infinite-range interactions), is equal to

$$S_t\big(1 - (1 - \exp(-iI_t))\big).$$

See Boccara and Cheong [49].

a critical value. The initial conditions are such that $R_0 = 0$ and $I_0 \ll S_0$.

(iii) What is the behavior as a function of t of I_t if S_0 is either less than or greater than its critical value?

Exercise 4.4 (i) Extend the previous one-population SIR epidemic model to describe the spread of a sexually transmitted disease among heterosexual individuals.

(ii) Show that the limits S_∞^{m}, I_∞^{m}, R_∞^{m}, S_∞^{f}, I_∞^{f}, and R_∞^{f}, where the superscripts m and f refer, respectively, to the male and female populations, exist. What are the values of I_∞^{m} and I_∞^{f}?

(iii) According to the initial values S_0^{m}, I_0^{m}, S_0^{f}, and I_0^{f} and the ratios $r^{\mathrm{m}}/i^{\mathrm{m}}$ and $r^{\mathrm{m}}/i^{\mathrm{m}}$, discuss the existence of epidemics in one or both populations.

Exercise 4.5 Consider an SIS epidemic model; i.e., a model in which, after recovery, infected individuals again become susceptible to catch the disease (as, e.g., with common cold).

(i) Assuming, as in Exercise 4.3, that each susceptible individual can be equally infected by any infective individual, derive the recurrence equations satisfied by the densities of susceptible and infected individuals denoted, respectively, by S_t and I_t.[24]

(ii) Show that this model exhibits a transcritical bifurcation between an endemic state (i.e., a state in which the stationary density of infected individuals I_∞ is nonzero), and a disease-free state in which I_∞ is equal to zero.

Exercise 4.6 Determine the first period-doubling bifurcations of the one-parameter map $(n, r) \mapsto f(n, r) = -rn \log n$, which is a discrete version of the classical Gompertz map [147].

---

[24] See Boccara and Cheong [52].

Solutions

Solution 4.1 *Let f denote the one-parameter family of maps $(n, r) \mapsto rn^x(1-n)$. For intermediate values of n, the growth rate $rn^{x-1}(1-n)$ should be larger than 1; otherwise extinction is automatic. If $x > 1$, the value of the population density maximizing the growth rate is $(x-1)/x$, and the corresponding value of the growth rate is*

$$r\frac{(x-1)^{x-1}}{x^x}.$$

This quantity is larger than 1 if

$$r > \frac{x^x}{(x-1)^{x-1}}.$$

If, as a simple example, we assume that $x = 2$, then r has to be greater than 4, and, in this case, the recurrence equation $n_{t+1} = f(n_t, r)$ has three fixed points:

$$0 \quad and \quad \frac{\sqrt{r} + \sqrt{r-4}}{2\sqrt{r}},$$

which are asymptotically stable, and

$$\frac{\sqrt{r} - \sqrt{r-4}}{2\sqrt{r}},$$

which is always unstable (Figure 4.14).

The nonhyperbolic equilibrium point $(\frac{1}{2}, 4)$ is a saddle-node bifurcation point.

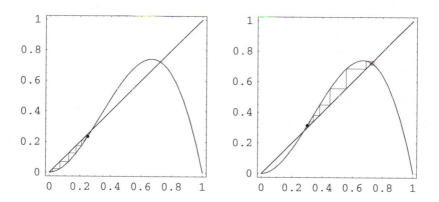

Fig. 4.14. *Cobwebs for the map $f : (n, r) \mapsto rn^x(1-n)$ for $x = 2$ and $r = 5$. The asymptotically stable fixed points are 0 and 0.723607, and the unstable fixed point is 0.276393. Starting from an initial population density $n_0 = 0.25$ (left cobweb), the population becomes extinct, while for an initial population density $n_0 = 0.3$, n converges to the high-density asymptotically stable fixed point.*

Solution 4.2 *If $0 < r < 4$, 0 is the only fixed point of the map $n \mapsto rn^2(1-n)$. If $r > 4$, there exist three fixed points 0 and $(\sqrt{r} \pm \sqrt{r-4})/2\sqrt{r}$; 0 and $(\sqrt{r} + \sqrt{r-4})/2\sqrt{r}$ are asymptotically stable, while $(\sqrt{r} - \sqrt{r-4})/2\sqrt{r}$ is unstable. The harvesting model makes sense only if $r > r_c = 4$.*

If $c > 0$ and $r - 4c - 4 > 0$, the map $n \mapsto rn^2(1-n) - cn$ has two nontrivial fixed points:

$$n_{\pm}^* = \frac{\sqrt{r} \pm \sqrt{r - 4c - 4}}{2\sqrt{r}};$$

n_+^* is asymptotically stable and n_-^* unstable. Therefore, a saddle-node bifurcation occurs at the nonhyperbolic point $(n^*, c^*) = (\frac{1}{2}, 1 - \frac{1}{4}r)$.

From an ecological point of view, if c is less than but close to its bifurcation value $(c \lesssim 1 - \frac{1}{4})$, there is a serious risk of extinction of the harvested species.

Solution 4.3 *(i) The three recurrence equations show that, as time t increases, S_t cannot increase and R_t cannot decrease. Since these quantities are bounded, the limits S_∞ and R_∞ exist. The second equation shows that I_∞ also exists and is equal to zero.*

(ii) By definition, there is an epidemic if $I_1 > I_0$.[25] According to the second equation, this occurs if

$$S_0\big(1 - \exp(-iI_0)\big) - rI_0 > 0.$$

If the initial conditions are $R_0 = 0$ and $I_0 \ll S_0$, then

$$S_0\big(1 - \exp(-iI_0)\big) - rI_0 = I_0(iS_0 - r) + O(I_0^2).$$

Hence, an epidemic occurs if, and only if,

$$S_0 > \frac{r}{i}.$$

This is the threshold phenomenon of Kermack and McKendrick (see Chapter 3, Remark 1).

(iii) Since, as t increases, S_t does not increase, if S_0 is less than r/i, for all values of t, S_t remains less than r/i. Hence, I_t goes monotonically to zero as t tends to ∞. If $S_0 > r/i$, the density of infective individuals increases as long as S_t is greater than r/i. Since S_t cannot increase, when it reaches the value r/i, I_t reaches its maximum value and then tends monotonically to zero as t tends to ∞.

Solution 4.4 *(i) Assuming that each susceptible individual can be equally infected by any infective individual (see Footnote 23), the equations describing the spread of a sexually transmitted disease among heterosexuals are*

$$S_{t+1}^m + I_{t+1}^m + R_{t+1}^m = \rho^m,$$
$$I_{t+1}^m = I_t^m + S_t^m\big(1 - \exp(-i^m I_t^f)\big) - r^m I_t^m,$$
$$R_{t+1}^m = R_t^m + r^m I_t^m,$$
$$S_{t+1}^f + I_{t+1}^f + R_{t+1}^f = \rho^f,$$
$$I_{t+1}^f = I_t^f + S_t^f\big(1 - \exp(-i^f I_t^m)\big) - r^f I_t^f,$$
$$R_{t+1}^f = R_t^f + r^f I_t^f,$$

---

[25] Refer to Chapter 3, Footnote 7.

where ρ^{m} and ρ^{f} are two constants representing the initial densities of males and females respectively. i^{m} (resp. i^{f}) is proportional to the probability for a male (resp. female) to be infected by a female (resp. male).

(ii) As for the one-population model, it follows that S_t^{m} and S_t^{f} cannot increase whereas R_t^{m} and R_t^{f} cannot decrease. Hence, all the limits S_∞^{m}, R_∞^{m}, S_∞^{f}, R_∞^{f} exist. The third and sixth equations show that the limits I_∞^{m} and I_∞^{f} also exist and are both equal to zero.

(iii) Due to the coupling between the two populations, a wider variety of situations may occur. For instance, if the initial conditions are

$$R_0^{\mathrm{m}} = R_0^{\mathrm{f}} = 0 \quad \text{and} \quad I_0^{\mathrm{m}} \ll S_0^{\mathrm{m}}, \;\; I_0^{\mathrm{f}} \ll S_0^{\mathrm{f}},$$

we have

$$I_1^{\mathrm{m}} - I_0^{\mathrm{m}} = i^{\mathrm{m}} S_0^{\mathrm{m}} I_0^{\mathrm{f}} - r^{\mathrm{m}} I_0^{\mathrm{m}} + O\big((I_0^{\mathrm{f}})^2\big),$$
$$I_1^{\mathrm{f}} - I_0^{\mathrm{f}} = i^{\mathrm{f}} S_0^{\mathrm{f}} I_0^{\mathrm{m}} - r^{\mathrm{f}} I_0^{\mathrm{f}} + O\big((I_0^{\mathrm{m}})^2\big).$$

Hence, according to the initial values S_0^{m}, S_0^{f}, I_0^{m}, and I_0^{f}, we may observe the following behaviors:

1. If $i^{\mathrm{m}} S_0^{\mathrm{m}} I_0^{\mathrm{f}} - r^{\mathrm{m}} I_0^{\mathrm{m}} < 0$ and $i^{\mathrm{f}} S_0^{\mathrm{f}} I_0^{\mathrm{m}} - r^{\mathrm{f}} I_0^{\mathrm{f}} < 0$, then $I_1^{\mathrm{m}} < I_0^{\mathrm{m}}$ and $I_1^{\mathrm{f}} < I_0^{\mathrm{f}}$. Since S_t^{m} and S_t^{f} are nonincreasing functions of time, I_t^{m} and I_t^{f} go monotonically to zero as t tends to ∞. No epidemic occurs.
2. If $i^{\mathrm{m}} S_0^{\mathrm{m}} I_0^{\mathrm{f}} - r^{\mathrm{m}} I_0^{\mathrm{m}} > 0$ and $i^{\mathrm{f}} S_0^{\mathrm{f}} I_0^{\mathrm{m}} - r^{\mathrm{f}} I_0^{\mathrm{f}} > 0$, then $I_1^{\mathrm{m}} > I_0^{\mathrm{m}}$ and $I_1^{\mathrm{f}} > I_0^{\mathrm{f}}$. The densities of infected individuals in both populations increase as long as the densities of susceptible and infective individuals satisfy the relations $i^{\mathrm{m}} S_t^{\mathrm{m}} I_t^{\mathrm{f}} - r^{\mathrm{m}} I_t^{\mathrm{m}} > 0$ and $i^{\mathrm{f}} S_t^{\mathrm{f}} I_t^{\mathrm{m}} - r^{\mathrm{f}} I_t^{\mathrm{f}} > 0$ and then tend monotonically to zero.
3. If $i^{\mathrm{m}} S_0^{\mathrm{m}} I_0^{\mathrm{f}} - r^{\mathrm{m}} I_0^{\mathrm{m}} < 0$ and $i^{\mathrm{f}} S_0^{\mathrm{f}} I_0^{\mathrm{m}} - r^{\mathrm{f}} I_0^{\mathrm{f}} > 0$, then $I_1^{\mathrm{m}} < I_0^{\mathrm{m}}$ and $I_1^{\mathrm{f}} > I_0^{\mathrm{f}}$. But, since I_{t+1}^{m} depends on I_t^{f}, the density of infected males does not necessarily goes monotonically to zero. After having decreased for a few time steps, due to the increase of the density of infected females, it may increase if $i^{\mathrm{m}} S_t^{\mathrm{m}} I_t^{\mathrm{f}} - r^{\mathrm{m}} I_t^{\mathrm{m}}$ becomes positive. The spread of the disease in the female population may trigger an epidemic in the male population, as shown in Figure 4.15. If, however, the increase of the density of infected females is not high enough, then the density of infected males will decrease monotonically, whereas the density of infected females will increase as long as the density of susceptible individuals S_t^{f} is greater than $r^{\mathrm{f}} I_t^{\mathrm{m}} / i^{\mathrm{f}} I_t^{\mathrm{m}}$ and then tend monotonically to zero. The disease spreads only in the female population, whereas no epidemic occurs among males.
4. If $i^{\mathrm{m}} S_0^{\mathrm{m}} I_0^{\mathrm{f}} - r^{\mathrm{m}} I_0^{\mathrm{m}} > 0$ and $i^{\mathrm{f}} S_0^{\mathrm{f}} I_0^{\mathrm{m}} - r^{\mathrm{f}} I_0^{\mathrm{f}} < 0$, our conclusions are similar to the case above.

Solution 4.5 (i) As in Exercise 4.3, assuming that each susceptible individual can be equally infected by any infective individual, we can write

$$S_{t+1} + I_{t+1} = \rho,$$
$$I_{t+1} = I_t + S_t\big(1 - \exp(-iI_t)\big) - rI_t,$$

where i is a positive real number proportional to the probability for a susceptible individual to catch the disease and r is the probability for an infected individual to recover.

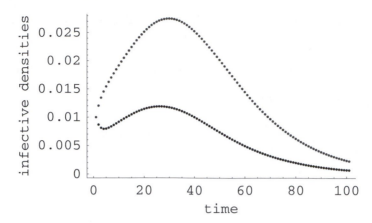

Fig. 4.15. *Time evolution of the densities of infected individuals for the two-population SIR model when the epidemic in one population triggers an epidemic in the other one.* $i^{\mathrm{m}} = 0.5$, $i^{\mathrm{f}} = 0.9$, $r^{\mathrm{m}} = 0.3$, $r^{\mathrm{f}} = 0.1$, $I_0^{\mathrm{m}} = 0.01$, $I_0^{\mathrm{f}} = 0.01$, $R_0^{\mathrm{m}} = 0$, $R_0^{\mathrm{f}} = 0$, $\rho^{\mathrm{m}} = 0.35$, $\rho^{\mathrm{f}} = 0.35$, *and number of iterations* $= 100$.

ρ *is the constant total density of individuals.*

(ii) Eliminating S_t from the second equation, we obtain

$$I_{t+1} = (1 - r)I_t + (\rho - I_t)\big(1 - \exp(-iI_t)\big).$$

The steady-state density of infected individuals I_∞ therefore satisfies the equation

$$I_\infty = (1 - r)I_\infty + (\rho - I_\infty)\big(1 - \exp(-iI_\infty)\big).$$

This equation has the obvious solution $I_\infty = 0$, which characterizes the disease-free state. In the vicinity of the bifurcation point, I_∞ is small, and the equation above can be written

$$I_\infty = (1 - r + \rho i)I_\infty - \big(i + \tfrac{1}{2}\rho i^2\big) I_\infty^2 + O\big((I_\infty)^3\big),$$

which shows that, in the (i, r)-plane, the transcritical bifurcation occurs along the line $\rho i - r = 0$. The disease-free state is stable if $\rho i < r$ and unstable if $\rho i > r$. The endemic state is stable in the latter case.

Solution 4.6 *For $n \in [0, 1]$ and $r > 0$, the map f has a quadratic maximum at $n = e^{-1}$. For $r = e$, the maximum reaches the value 1. There exist two fixed points: 0 and $e^{-1/r}$. The derivative of f is equal to $-r(1 + \log n)$, which shows that 0 is always unstable and $e^{-1/r}$ is asymptotically stable for $0 < r < 2$. Since*

$$f\left(\frac{1}{\sqrt{e}}, 2\right) = 1 \quad \text{and} \quad \frac{\partial f}{\partial n}\left(\frac{1}{\sqrt{e}}, 2\right) = -1,$$

the nonhyperbolic point $(1/\sqrt{e}, 2)$ may be a period-doubling bifurcation point. We have to check the remaining conditions. The second iterate of f is

$$(n, r) \mapsto f^2(r, n) = r^2 n \log n \log\big(-rn \log n\big),$$

and we have

$$\frac{\partial f^2}{\partial r}\left(\frac{1}{\sqrt{e}},2\right)=0,\qquad \frac{\partial^2 f^2}{\partial^2 n}\left(\frac{1}{\sqrt{e}},2\right)=0,$$

$$\frac{\partial^2 f^2}{\partial r \partial n}\left(\frac{1}{\sqrt{e}},2\right)=2,\qquad \frac{\partial^3 f^2}{\partial^3 n}\left(\frac{1}{\sqrt{e}},2\right)=-16e.$$

The one-parameter family of maps $(n,r)\mapsto -rn\log n$ *exhibits, therefore, a period-doubling bifurcation at the point* $(n,r)=(1/\sqrt{e},2)$.

It can be shown numerically that the 2-point cycle is asymptotically stable for $2<r<2.39536\ldots$. *For* $r=2.39536\ldots$, *there is a new period-doubling bifurcation, and we find that the 4-point cycle is asymptotically stable for* $2.39536\ldots<r<2.47138\ldots$. *There is actually an infinite sequence of period-doubling bifurcations. The first period-doubling bifurcations of this infinite sequence occurring for the parameter values:*

$$r_1=2,\quad r_2=2.39536\ldots,\quad r_3=2.47138\ldots,\quad r_4=2.48614\ldots$$

A few iterations for different values of r *are represented in Figure 4.16.*

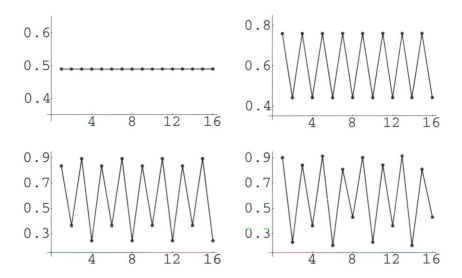

Fig. 4.16. *Iterations of the* $f:(n,r)\mapsto -rn\log n$ *for* $r=1.4$ *(fixed point),* $r=2.1$ *(2-point cycle),* $r=2.43$ *(4-point cycle), and* $r=2.483$ *(8-point cycle). 16 iterates are shown after 1000 have been discarded.*

5

Chaos

In 1963, Edward Lorenz [212] published a numerical analysis of a simplified model of thermal convection represented by three quadratic differential equations. He discovered that all nonperiodic solutions of his deterministic model were bounded but showed irregular fluctuations. Some thirty years later [213], here is how he described his discovery:

> ..., while conducting an extensive experiment in the theory of weather forecasting, I had come across a phenomenon that later came to be called "chaos"—seemingly random and unpredictable behavior that nevertheless proceeds according to precise and often easily expressed rules. Earlier investigators had occasionally encountered behavior of this sort, but usually under rather different circumstances. Often they failed to recognize what they had seen, and simply became aware that something was blocking them from solving their equations or otherwise completing their studies. My situation was unique in that, as I eventually came to realize, my experiment was doomed to failure unless I could construct a system of equations whose solutions behaved chaotically. Chaos suddenly became something to be welcomed, at least under some conditions, and in the ensuing years I found myself turning more and more toward chaos as a phenomenon worthy of study for its own sake.

The study of chaos can be traced back to Poincaré [290]. In *Science et Méthode*, first published in 1909, he already indicates the possibility for certain systems to be subject to *sensitive dependence on initial conditions*:

> *Si nous connaissions exactement les lois de la nature et la situation de l'univers à l'instant initial, nous pourrions prédire exactement la situation de ce même univers à un instant ultérieur. Mais, lors même que les lois naturelles n'auraient plus de secret pour nous, nous ne pourrons connaître la situation initiale qu'approximativement. Si cela nous permet de prévoir la situation ultérieure avec la même approximation, c'est tout ce qu'il nous faut, nous disons que le phénomène a*

été prévu, qu'il est régi par des lois ; mais il peut arriver que de petites différences dans les conditions initiales en engendrent de très grandes dans les phénomènes finaux ; une petite erreur sur les premières produirait une erreur énorme sur les derniers. La prédiction devient impossible et nous avons le phénomène fortuit.[1]

And a few lines below, he shows how these considerations allow an answer to the question:

Pourquoi les météorologistes ont tant de peine à prédire le temps avec quelque certitude ?[2]

More than 50 years later, computers were available, and in his seminal paper, Lorenz used almost identical terms to answer the same question:

When our results concerning the instability of nonperiodic flow are applied to the atmosphere, which is ostensibly nonperiodic, they indicate that prediction of the sufficiently distant future is impossible by any method unless the present conditions are known exactly. In view of the inevitable inaccuracy and incompleteness of weather observation, precise very-long forecasting would seem to be non-existent.

While the kind of complicated dynamics Lorenz discovered cannot exist in one- or two-dimensional systems of differential equations, this is not the case for recurrence equations. The modern theory of *deterministic chaos* started after the publication, in 1976, of a review article in which Robert May [230] called attention to the very complicated dynamics of some very simple one-population models:

Not only in research, but also in the everyday world of politics and economics, we would all be better off if more people realized that simple nonlinear systems do not necessarily possess simple dynamical properties.

---

[1] If we could know exactly the laws of nature and the situation of the universe at the initial instant, we should be able to predict exactly the situation of this same universe at a subsequent instant. But even when the natural laws should have no further secret for us, we could know the initial situation only *approximately*. If that permits us to foresee the subsequent situation *with the same degree of approximation*, this is all we require, we say that the phenomenon has been predicted, that it is ruled by laws. But this is not always the case; it may happen that slight differences in the initial conditions produce very great differences in the final phenomena; a slight error in the former would make an enormous error in the latter. Prediction becomes impossible and we have the fortuitous phenomenon. (English translation [290], p. 397.)

[2] Why have the meteorologists such difficulty in predicting the weather with any certainty? (English translation [290], p. 398.)

5.1 Defining chaos

In the previous chapter, we have seen that, in the case of the logistic map $(n, r) \mapsto f(n, r) = rn(1 - n)$, there exists an infinite sequence of parameter values $(r_k)_{k \in \mathbb{N}}$ for which f undergoes a period-doubling bifurcation, a 2^k-point cycle being asymptotically stable for all $r \in]r_k, r_{k+1}[$. The sequence (r_k) has a limit $r_\infty \approx 3.599692$. A natural question is: *What is the dynamics for $r > r_\infty$?* The answer is given by the bifurcation diagram represented in Figure 5.1. The bifurcation diagram has been computed for 300 parameter values equally spaced between 2.5 and 4. For each value of r, 300 iterates are calculated, but only the last 100 have been plotted.

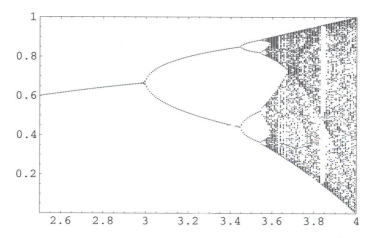

Fig. 5.1. *Bifurcation diagram of the logistic map $(n, r) \mapsto rn(1 - n)$. The parameter r, plotted on the horizontal axis, varies from 2.5 to 4, and the reduced population n, plotted on the vertical axis, varies between 0 and 1.*

For most values of $r \in]r_\infty, 4]$, the successive iterates of f seem to wander in an apparently random manner. For example, if $r = 4$, as illustrated in Figure 5.2, the trajectory through the initial point $n_0 = \sqrt{2} - 1$ appears to be dense[3] in $[0.1]$.

A numerical experiment cannot, of course, determine whether the trajectory converges to an asymptotically stable periodic orbit of very high period or is dense in an interval. If the trajectory is dense, say in $[0.1]$, we can divide this interval into a certain number of subintervals (bins) and count how many iterates fall into each subinterval. If the number of iterates is high enough, this

---

[3] Let I be an interval of $\mathbb{R}$. A subset J of I is dense in I if the closure of J coincides with I. In other words, any neighborhood of any point in I contains points in J. For example, the set of rational numbers $\mathbb{Q}$ is dense in $\mathbb{R}$.

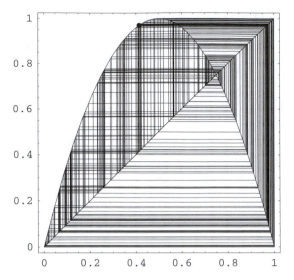

Fig. 5.2. *Cobweb showing a sequence of 250 iterates of the map $n \mapsto 4n(1-n)$. The initial value is $n_0 = \sqrt{2} - 1$ with a precision of 170 significant digits.*

could determine, if it exists, an approximate invariant measure,[4] as shown in Figure 5.3.

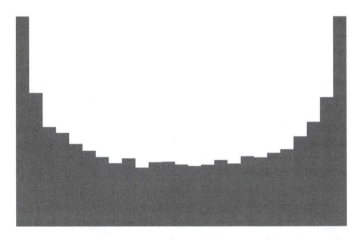

Fig. 5.3. *Histogram representing the approximate probability distribution of the iterates of the logistic map $f_4 : n \mapsto 4n(1-n)$. The histogram has been obtained from 75,000 iterates distributed in 25 bins.*

---

[4] Let f be a map defined on an interval I of $\mathbb{R}$; a measure μ on I is invariant for f if, for any measurable subset $E \subset I$, $\mu(E) = \mu(f^{-1}(E))$.

Remark 7. As already noticed by Poincaré, some maps may have a sensitive dependence on initial conditions. This feature, which is a necessary condition for a map to be chaotic (see Definition 20), implies that the precision—*i.e.*, number of significant digits—of each iterate is decreasing. In the case of the map $n \mapsto 4n(1-n)$, this precision decreases, on average, by 0.6 digit/iteration. In order to determine numerically a meaningful histogram, it is therefore essential to choose an initial value with a sufficient number of significant digits.

The considerations above are essentially heuristic, but, at least in the case of the map $f_4 : n \mapsto 4n(1-n)$, it is possible to completely understand its dynamics.

5.1.1 Dynamics of the logistic map f_4

In Example 25, we have seen that the logistic map f_4 and the binary tent map T_2, defined on $[0,1]$ by

$$f_4(x) = 4x(1-x) \quad \text{and} \quad T_2(x) = \begin{cases} 2x, & \text{if } 0 \le x < \frac{1}{2}, \\ 2 - 2x, & \text{if } \frac{1}{2} \le x \le 1, \end{cases}$$

are conjugate. This property greatly simplifies the study of the dynamics of f_4.

Let $0.x_1x_2x_3\ldots$ be the binary representation of $x \in [0,1]$, *i.e.*,

$$x = \sum_{i=1}^{\infty} \frac{x_i}{2^i},$$

where, for all $i \in \mathbb{N}$, $x_i \in \{0,1\}$. The binary representation of $T_2(x)$ is then given by

$$T_2(x) = \begin{cases} 0.x_2x_3x_4\ldots, & \text{if } 0 \le x \le \frac{1}{2}, \\ 0.(1-x_2)(1-x_3)(1-x_4)\ldots, & \text{if } \frac{1}{2} \le x \le 1. \end{cases}$$

These formulae are both correct for $x = \frac{1}{2}$. The binary representation of $\frac{1}{2}$ being either $0.1000\ldots$ or $0.01111\ldots$, the binary representation of $T_2(x)$ is, in both cases, $0.1111\ldots$, which is equal to 1.

The description of the iterates of x in terms of their binary representation leads to some remarkable results due to James Whittaker [344].

1- If the binary representation of x is finite (*i.e.*, if there exists a positive integer n such that $x_i = 0$ for all $i > n$), then after at most $n+1$ iterations the orbit of x will reach 0 and stay there. Hence, *there exists a dense set of points whose orbit reaches the origin and stays there.*

2- *If the binary representation of x is periodic with period p, then the orbit of x is periodic with a period equal to p or a divisor of p. For example, for $p = 2$,*

x is either equal to $0.010101\ldots = \frac{1}{3}$ or $0.101010\ldots = \frac{2}{3}$, and we verify that $T_2(\frac{1}{3}) = \frac{2}{3}$ and $T_2(\frac{2}{3}) = \frac{2}{3}$. In both cases, the orbit after, at most, one iteration reaches the fixed point $\frac{2}{3}$ (period 1). More generally, suppose that $x_{1+p} = x_1$. If after p iterations there have been an even number of conjugations (*i.e.*, changes of x_i in $1 - x_i$), we are back to the initial point $x = 0.x_1x_2x_3\ldots;$,but if, after p iterations, there have been an odd number of conjugations, we are not back to the initial point but to the point whose binary representation is $0.(1 - x_1)(1 - x_2)(1 - x_3)\cdots$. However, in this case,

$$T_2(0.(1 - x_1)(1 - x_2)(1 - x_3)\cdots) = T_2(0.x_1x_2x_3\ldots),$$

and after $p + 1$ iterations we are back to the first iterate $T_2(x)$. Here is an example. Consider $x = 0.001110011100111\ldots$, whose binary representation is periodic with period 5. We have

$$x = 0.001110011100111\ldots$$
$$T_2(x) = 0.01110011100111\ldots$$
$$T_2^2(x) = 0.1110011100111\ldots$$
$$T_2^3(x) = 0.001100011000\ldots \quad \text{conjugation}$$
$$T_2^4(x) = 0.01100011000\ldots$$
$$T_2^5(x) = 0.1100011000\ldots \quad \text{this is not } x$$
$$T_2^6(x) = 0.011100111\ldots \quad \text{but this is } T_2(x).$$

Using this method, we could show that the binary tent map has periodic orbits of all periods, and the set of all periodic points is dense in $[0.1]$. T_2 and f_4 being conjugate, the logistic map f_4 has the same property.

Regarding the existence of a periodic orbit, an amazing theorem, due to Šarkovskii, indicates which periods imply which other periods. First, define among all positive integers Šarkovskii's order relation by

$$3 \triangleright 5 \triangleright 7 \triangleright \cdots 2 \cdot 3 \triangleright 2 \cdot 5 \triangleright \cdots \triangleright 2^2 \cdot 3 \triangleright 2^2 \cdot 5 \triangleright \cdots$$
$$\triangleright 2^3 \cdot 3 \triangleright 2^3 \cdot 5 \triangleright \cdots\cdots \triangleright 2^3 \triangleright 2^2 \triangleright 2 \triangleright 1.$$

That is, first list all the odd numbers, followed by 2 times the odd numbers, 2^2 times the odd numbers, etc. This exhausts all the positive integers except the powers of 2 that are listed last in decreasing order. Since $\triangleright$ is an order relation, it is transitive (*i.e.*, $n_1 \triangleright n_2$ and $n_2 \triangleright n_3$ imply $n_1 \triangleright n_3$). Šarkovskii's theorem is[5]:

Theorem 10. *Let* $f : \mathbb{R} \to \mathbb{R}$ *be a continuous map. If* f *has a periodic orbit of period* n, *then, for all integers* k *such that* $n \triangleright k$, f *has also a periodic orbit of period* k.

---

[5] For a proof, see Štefan [326] or Collet and Eckmann [91].

We said above that the binary tent map T_2 and its conjugate, the logistic map f_4, have periodic orbits of all periods. Following Šarkovskii's theorem, to prove this result it suffices to prove that either T_2 or f_4 has a periodic orbit of period 3.

In order to generate an orbit of period 3 for f_4, we can start from $x = 0.100100100\ldots$, which generates an orbit of period 3 for the binary tent map. From the binary representation of x, we find that x satisfies the equation $8x = 4 + x$, thus $x = \frac{4}{7}$. In order to generate the corresponding period 3 orbit for f_4, we have to start from the point $h(x) = \sin^2(2\pi/7)$.

All these periodic orbits are unstable. Hence, periodic orbits computed with a finite precision will always present, after a number of iterations depending upon the precision, an erratic behavior. This feature is illustrated in Figure 5.4.

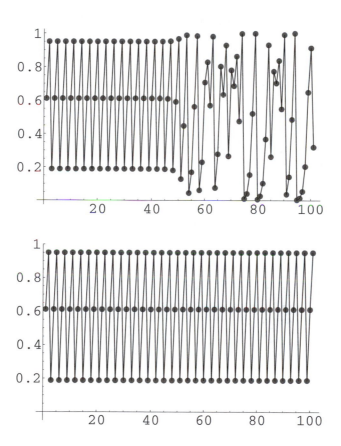

Fig. 5.4. *Unstable orbit of period 3 of the map $x \mapsto 4x(1-x)$ generated from an initial value with a precision equal to either 16 (top) or 60 (bottom).*

3- Now let us turn our attention to initial points whose binary representation is neither finite nor periodic such as, for instance, $x = 0.101001000100001\ldots$. It is readily verified that, given $\varepsilon > 0$ and $k = 0, 1, 2, \ldots$, there exists a positive integer t such that $|T_2^t(x) - 2^{-k}| < \varepsilon$.

Theorem 11. *Let $x_0 \in [0,1]$; then, arbitrarily close to x_0, there exists $x \in [0,1]$ such that the orbit through x $\{T_2^t(x) \mid t \in \mathbb{N}\}$, is dense in $[0,1]$.*

Denote $s_0 \in [0,1]$, a number whose finite binary representation agrees with sufficiently many initial digits of x_0 so that s_0 is as close as desired to x_0. Let (s_{ij}) be a sequence of numbers in $[0,1]$ with finite binary representations, where, for a given number s_{ij}, i is the finite number of digits of its binary representation and j its decimal value ($0 \le j \le 2^i - 1$). For example,

$$s_{20} = 0.00, \quad s_{21} = 0.01, \quad s_{22} = 0.10, \quad s_{23} = 0.11.$$

Now define recursively the binary representation of x as follows. First let $x = s_0$, then to the digits of s_0 add either the digits of each element of the sequence (s_{ij}) in the order $s_{10}, s_{11}, s_{20}, s_{21}, s_{22}, s_{23}, \ldots$ or the digits obtained by conjugation $\bar{s}_{10} = 0.1$, $\bar{s}_{11} = 0.0$, $\bar{s}_{20} = 0.11$, $\ldots$ according to the following rule. Once the digits of $n-1$ elements of either the sequence (s_{ij}) or the sequence $(\bar{s}_{ij})$ have been added, if the resulting finite representation contains an even number of conjugations, we add the digits of the nth element of (s_{ij}); otherwise we add the digits of the nth element of $(\bar{s}_{ij})$. The infinite binary representation so obtained gives the number x close to x_0. Now for any $y \in [0,1]$ and any positive ε, there is an s_{ij} such that $2^{-i} < \varepsilon/2$ and $|s_{ij} - y| < \varepsilon/2$. From this it follows that one of the iterates of x, say s, will have the same initial digits as s_{ij}, so

$$|s - y| \le |s - s_{ij}| + |s_{ij} - y| < \frac{\varepsilon}{2} + \frac{\varepsilon}{2} = \varepsilon,$$

which proves that there are iterates of x as close as we want to any given $y \in [0,1]$. The orbit of x is therefore dense in $[0,1]$.

The existence of dense orbits for the logistic map f_4 suggested by Figure 5.2 follows from Theorem 11 and the fact that the maps f_4 and T_2 are conjugate.

Theorem 11 also implies that, if J is an open interval of $[0,1]$, almost all points $x \in J$ will return to J infinitely many times under iteration of the binary tent map T_2. Under that form, Theorem 11 is a special case of the *Poincaré recurrence theorem*.

Definition 18. *A map $f : \mathcal{S} \to \mathcal{S}$ is said to be* topologically transitive *if for any pair of open sets (U, V) in $\mathcal{S}$ there exists a positive integer t such that $f^t(U) \cap V \neq \emptyset$.*

A topologically transitive map has points that eventually move under iteration from one arbitrarily small neighborhood to any other. If a map is topologically transitive, the phase space cannot be decomposed into two disjoint open sets that are invariant under the map.

Definition 19. *A map* $f : S \to S$ *is said to have* sensitive dependence on initial conditions *if there exists a positive real number* δ *such that, for any* $x \in S$ *and any neighborhood* $N(x)$ *of* x, *there exist* $y \in N(x)$ *and a positive integer* t *such that* $|f^t(x) - f^t(y)| > \delta$.

A map has sensitive dependence on initial conditions if there exist points arbitrarily close to x that eventually separate from x by at least δ under iteration of f. Note, however, that sensitive dependence on initial conditions requires that some points, but not necessarily all points, close to x eventually separate from x under iteration. If a map has sensitive dependence on initial conditions, one has to be very careful when drawing conclusions from numerical determinations of orbits.

As a consequence of Theorem 11, the binary tent map T_2 and its conjugate f_4 are topologically transitive and possess sensitive dependence on initial conditions.

5.1.2 Definition of chaos

We are now ready to define chaos. The following definition is one of the simplest. It is due to Devaney [102].

Definition 20. *Let* S *be a set. The mapping* $f : S \to S$ *is said to be* chaotic *on* S *if*

1. *f has sensitive dependence on initial conditions,*
2. *f is topologically transitive,*
3. *f has periodic points that are dense in* S.

And after having chosen this definition, Devaney adds the following comment:

> To summarize, a chaotic map possesses three ingredients: unpredictability, indecomposability, and an element of regularity. A chaotic system is unpredictable because of sensitive dependence on initial conditions. It cannot be broken down or decomposed into two subsystems (two invariant open subsets) which do not interact under f because of topological transitivity. And, in the midst of this random behavior, we nevertheless have an element of regularity, namely the periodic points which are dense.

According to Devaney's definition, the binary tent map T_2 and the logistic map f_4 are chaotic.

Actually, Devaney's definition of chaos is redundant. It can be shown [31] that sensitive dependence on initial conditions follows from topological transitivity and the existence of a dense set of periodic points. For general maps it is the only redundancy [12]. However, if one considers maps on an interval, finite or infinite, then continuity and topological transitivity imply the existence of a dense set of periodic points and sensitive dependence on initial conditions. The proof of this result, which uses the ordering of $\mathbb{R}$, cannot be extended to maps defined in $\mathbb{R}^n$ for $n > 1$ or to the unit circle S^1 [337].

Remark 8. The word "chaotic" to qualify the apparently random behavior of deterministic nonperiodic orbits of one-dimensional maps seems to have been used for the first time in 1975 by Li and York [204].

5.2 Routes to chaos

The infinite sequence of period-doubling bifurcations preceding the onset of chaos is called the *period-doubling route to chaos*. A route to chaos is a scenario in which a parameter-dependent system that has a simple deterministic time evolution becomes chaotic as a parameter is changed.

The period-doubling route is followed by a variety of deterministic systems, but it is not the only one. For example, the *intermittency* route, discovered by Pomeau and Manneville [292], is another scenario in which a deterministic system can reach chaos.

As described by the authors, for values of a control parameter r less than a critical transition value r_T, the attractor is a periodic orbit. For r slightly greater than r_T, the orbit remains apparently periodic during long time intervals (said to be *laminar phases*,) but this regular behavior seems to be abruptly and randomly disrupted by a *burst* of finite duration. These bursts occur at seemingly random times, much larger than—and not correlated with—the period of the underlying oscillations. As r increases substantially above r_T, bursts become more and more frequent, and regular oscillations are no longer apparent. Pomeau and Manneville have identified three types of intermittency transitions according to the nature of the bifurcations characterizing these transitions.

An example of an intermittent transition occurs for the logistic map $(n, r) \mapsto f(n, r) = rn(1 - n)$ for $r_T = 1 + 2\sqrt{2}$. If r is slightly greater than r_T, there exists a stable and an unstable 3-point cycle (Figure 5.5). These cycles coalesce at r_T through a saddle-node bifurcation for $f^3(n, r)$, and for r slightly less than r_T, we observe intermittent bursts, as shown in Figure 5.6. Numerical calculations were done with a precision of 200 significant digits on the initial value; 100 iterations have been discarded, and 170 are represented. For this type of intermittency, Pomeau and Manneville have shown that the average time between two bursts goes to infinity as r tends to r_T from below as $(r_T - r)^{-1/2}$.

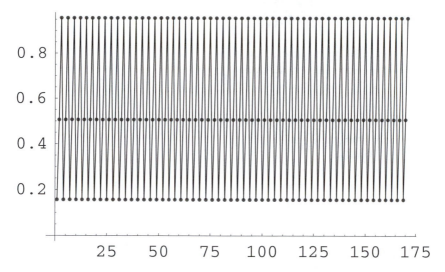

Fig. 5.5. *Stable 3-point cycle of the logistic map* $(n, r) \mapsto rn(1 - n)$ *for* $r = 1 + 2\sqrt{2} + 1/1000$. *Numerical calculations were done with a precision of 200 significant digits on the initial value. One hundred iterations have been discarded.*

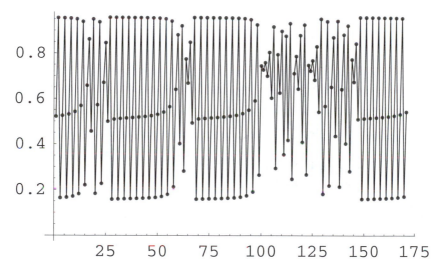

Fig. 5.6. *Intermittent bursts of the logistic map* $(n, r) \mapsto rn(1 - n)$ *for* $r = 1 + 2\sqrt{2} - 1/1000$. *Numerical calculations were done with a precision of 200 significant digits on the initial value. One hundred iterations have been discarded.*

5.3 Characterizing chaos

Computer studies have shown that chaos is *probably* present in many models. In the absence of a mathematical proof, when can we say that the time evolution of a system is chaotic? Some indications are given in the following sections.[6]

5.3.1 Stochastic properties

The histogram represented in Figure 5.3, while suggesting the existence of an invariant measure for the map f_4, shows that chaos implies deterministic randomness. In what follows, we describe the stochastic properties associated with the chaotic behavior of the binary tent map T_2 and the logistic map f_4. That is, for each map, we determine an invariant probability density ρ such that the probability $\rho(x)\,dx$ measures how frequently the interval $[x, x + dx]$ is visited by the dense orbit of a point $x_0 \in [0, 1]$.

Before going into any further detail, let us recall a few basic results of *ergodic theory*.[7]

Let $f : S \to S$ be a map. A subset A of S is *invariant* if $f(A) = A$. A measure μ is *invariant* for the map f if, for all measurable subsets A of S, $\mu(f^{-1}(A)) = \mu(A)$. The map f is *ergodic* with respect to the invariant measure μ if any measurable invariant subset A of S is such that either $\mu(A) = 0$ or $\mu(A) = \mu(S)$.

If f is ergodic with respect to the invariant probability measure μ[8] and $g : S \to \mathbb{R}$, an integrable function with respect to μ, then, for almost all $x_0 \in S$,

$$\lim_{t \to \infty} \frac{1}{t} \sum_{i=1}^{t} g \circ f^i(\mathbf{x_0}) = \int g(\mathbf{x})\,d\mu(\mathbf{x}).$$

That is, the time average is equal to the space average. If there exists a positive real function ρ such that $d\mu(x) = \rho(x)\,dx$, ρ is called an *invariant probability density*.

If the map $f : [0, 1] \to [0, 1]$ is such that any point x has k preimages $y_1, y_2, \ldots y_k$ by f (*i.e.*, for all $i = 1, 2, \ldots, k$, $f(y_i) = x$), the probability of finding an iterate of f in the interval $[x, x + dx]$ is then the sum of the probabilities of finding its k preimages in the intervals $[y_i, y_i + dy_i]$. Hence, from the relation

$$\rho(x)\,dx = \sum_{i=1}^{k} \rho(y_i)\,dy_i,$$

it follows that

---

[6] See Ruelle [303].

[7] See, for example, Shields [313].

[8] μ is a probability measure if $\mu(S) = 1$.

$$\rho(x) = \sum_{i=1}^{k} \frac{\rho(y_i)}{|f'(y_i)|}, \tag{5.1}$$

where we have taken into account that

$$\frac{dx}{dy_i} = f'(y_i).$$

Equation (5.1) is called the *Perron-Frobenius equation*. In the case of the binary tent map T_2, the Perron-Frobenius equation reads

$$\rho(x) = \tfrac{1}{2}\left(\rho\left(\tfrac{1}{2}x\right) + \rho\left(1 - \tfrac{1}{2}x\right)\right).$$

This equation has the obvious solution $\rho(x) = 1$. That is, the map T_2 preserves the Lebesgue measure. If $I_1 \cup I_2$ is the preimage by T_2 of an open interval I of $[0,1]$, we verify that

$$m(I_1 \cup I_2) = m(I_1) + m(I_2) = m(I),$$

where m denotes the Lebesgue measure. Using the relation

$$f_4 \circ h = h \circ T_2,$$

where $h : x \mapsto \sin^2(\tfrac{\pi}{2}x)$, the density ρ of the invariant probability measure for the map f is given by

$$\rho(x)\,dx = \frac{dh^{-1}}{dx}\,dx \tag{5.2}$$

$$= \frac{dx}{\pi\sqrt{x(1-x)}}. \tag{5.3}$$

The graph of ρ, represented in Figure 5.7, should be compared with the histogram in Figure 5.3.

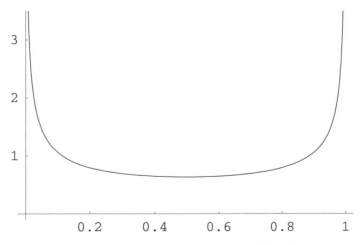

Fig. 5.7. *Invariant probability density $\rho : x \mapsto 1/\pi\sqrt{x(1-x)}$ for the logistic map f_4. Compare with Figure 5.3.*

5.3.2 Lyapunov exponent

Sensitive dependence on initial conditions is a necessary but not sufficient condition for a map f to be chaotic.[9] It is nonetheless interesting to be able to measure this essential feature of a chaotic map. Consider two neighboring initial points x_0 and $x_0 + \varepsilon$. If after t iterations of the map f the distance $|f^t(x_0 + \varepsilon) - f^t(x_0)|$ grows exponentially, we may define the exponent $\lambda(x_0)$ by

$$|f^t(x_0 + \varepsilon) - f^t(x_0)| \sim \varepsilon e^{t\lambda(x_0)},$$

or more precisely by

$$\lambda(x_0) = \lim_{t \to \infty} \lim_{\varepsilon \to 0} \frac{1}{t} \log \left| \frac{f^t(x_0 + \varepsilon) - f^t(x_0)}{\varepsilon} \right|$$

$$= \lim_{t \to \infty} \frac{1}{t} \log \left| \frac{df^t}{dx}(x_0) \right|. \tag{5.4}$$

$\lambda(x_0)$ is called the *Lyapunov exponent*.

$e^{\lambda(x_0)}$ is the average factor by which the distance between two neighboring points becomes stretched after one iteration.

Applying the chain rule to Relation (5.4), the Lyapunov exponent can also be expressed as

$$\lambda(x_0) = \lim_{t \to \infty} \frac{1}{t} \log \left| \prod_{i=0}^{t-1} f'(x_i) \right|$$

$$= \lim_{t \to \infty} \frac{1}{t} \sum_{i=0}^{t-1} \log |f'(x_i)|, \tag{5.5}$$

where $x_i = f^i(x_0)$, for $i = 0, 1, 2, \ldots, t-1$.

In the case of a chaotic map, except for a set of measure zero, the Lyapunov exponent does not depend upon x_0. If the map is ergodic with respect to an invariant measure μ, then the Lyapunov exponent expressed as a time average by (5.5) is also given by the space average:

$$\lambda = \int \log |f'(x)| d\mu(x). \tag{5.6}$$

Example 32. Lyapunov exponent of the logistic map f_4. The map f_4 is ergodic with respect to the invariant measure

$$d\mu(x) = \frac{dx}{\pi \sqrt{x(1-x)}},$$

---

[9] See Exercise 5.4 for an example of a map that has sensitive dependence on initial conditions and dense periodic points but is, however, not chaotic.

thus

$$\lambda = \frac{1}{\pi} \int_0^1 \frac{\log |4(1 - 2x)|}{\sqrt{x(1-x)}} \, dx$$

$$= \int_0^1 \log(4|\cos \pi u|) \, du$$

$$= \log 2.$$

Since a chaotic map has sensitive dependence on initial conditions, its Lyapunov exponent is necessarily positive. Notice that the result found for the logistic map f_4 could have been obtained without any calculation since f_4 is conjugate to T_2 and, for all $x \in [0, 1]$, $|T_2'(x)| = 2$.

5.3.3 "Period three implies chaos"

In their paper *Period Three Implies Chaos*, published in 1975, Li and York [204] gave the following theorem.

Theorem 12. *Let I be an interval and let $f : I \to I$ be a continuous map. Assume there is a point $x \in I$ such that its first three iterates satisfy*

$$\text{either} \quad f^3(x) \leq x < f(x) < f^2(x) \quad \text{or} \quad f^3(x) \geq x > f(x) > f^2(x). \quad (5.7)$$

Then, for all positive integers k, there exists in I a periodic point of period k. Furthermore, there exists an uncountable set $S \subset I$, containing no periodic or eventually periodic points,[10] such that, for all pairs of different points x_1 and x_2 in S,

$$\limsup_{t \to \infty} |f^t(x_1) - f^t(x_2)| > 0 \quad \text{and} \quad \liminf_{t \to \infty} |f^t(x_1) - f^t(x_2)| = 0. \quad (5.8)$$

For Li and York, "chaotic behavior" means existence of

1. infinitely many periodic points, and
2. an uncountable subset of I becoming highly folded under successive iterations.

According to Šarkovskii's theorem, the existence of infinitely many periodic points follows from the existence of a periodic point of either an odd period [205] or a period not equal to 2^n [274].

Condition (5.7) has been improved by Li, Misiurewicz, Pagani, and York [205], making the existence of chaotic behavior easier to verify. *If there exists an initial point x_0 such that*

$$f^n(x_0) < x_0 < f(x_0), \quad (5.9)$$

where n is an odd integer, then there exists a periodic orbit of odd period k with $1 < k \leq n$.

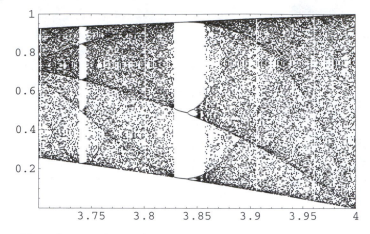

Fig. 5.8. *Magnification of the bifurcation diagram of Figure 5.1 around the period 3 window.*

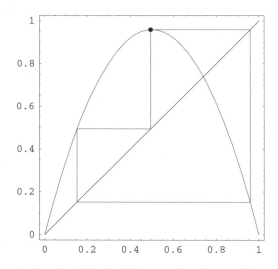

Fig. 5.9. *Three-point cycle of the logistic map $(n, r) \mapsto rn(1 - n)$ for $r = 3.835$. Fifty iterations are represented after 150 have been discarded.*

It is important to realize that, when the Li-York conditions of Theorem 12 are satisfied, it may happen that, for most initial values $x_0 \in I$, the orbit $\{f^t(x_0) \mid t \in \mathbb{N}\}$ is eventually periodic. For example, the magnification of the bifurcation diagram of the logistic map represented in Figure 5.8 shows the existence of 3-point cycles in a narrow range of values of the parameter r. Convergence to a typical cycle is shown in Figure 5.9. Such behavior is certainly not chaotic. However, according to the Li-York theorem there exists an

---

[10] A point is *eventually periodic* if it tends to a periodic point.

uncountable set of orbits that are chaotic. These orbits may, however, be rare and thus difficult to exhibit numerically. Therefore, time series obtained either numerically or experimentally are considered chaotic when apparently dense orbits in an invariant subset and sensitive dependence on initial conditions can be observed.

Example 33. Condition (5.9) is easily verified for the two discrete one-population models of Example 27.

For $n_{t+1} = n_t \exp(r(1 - n_t))$ with $r = 3$, we find

$$n_5 < n_0 < n_1, \quad \text{where} \quad n_0 = 0.3, \ n_1 = 2.44985, \ n_5 = 0.0877748.$$

For $n_{t+1} = rn_t(1 + n_t)^{-a}$ with $r = 40$ and $a = 8$, we find

$$n_7 < n_0 < n_1, \quad \text{where} \quad n_0 = 0.15, \ n_1 = 1.96141, \ n_7 = 0.0265951.$$

In Subsection 5.4.1 it is shown that apparent dense orbits are observed in both cases.

5.3.4 Strange attractors

A closed set A is an *attracting set* for a map $\mathbf{f}$ if there exists a neighborhood N of A such that

$$\lim_{t \to \infty} \mathbf{f}^t(N) = A.$$

If a domain D is such that $\mathbf{f}^t(D) \subset D$ for all positive t, D is called a *trapping region*. If we can find such a domain, then

$$A = \bigcap_{t \in \mathbb{N}} \mathbf{f}^t(D).$$

If an attracting set contains a dense orbit it is called an *attractor*. Attractors of chaotic systems often have complex structures. For example, a one-dimensional attractor can be an uncountable set of zero Lebesgue measure. In order to characterize such sets, we need to introduce a new type of measure. First, let us describe the typical classical example of such a set.

Example 34. The Cantor set [76]. Let (J_n) be a sequence of subsets of $[0, 1]$ defined by

$$J_0 = [0, 1],$$
$$J_1 = [0, \tfrac{1}{3}] \cup [\tfrac{2}{3}, 1],$$
$$J_2 = [0, \tfrac{1}{9}] \cup [\tfrac{2}{9}, \tfrac{1}{3}] \cup [\tfrac{6}{9}, \tfrac{7}{9}] \cup [\tfrac{8}{9}, 1],$$
$$\cdots\cdots$$

J_n is the union of 2^n pairwise disjoint closed intervals of length 3^{-n}. It is obtained from J_{n-1} by removing from each closed interval of length $3^{-(n-1)}$ an open interval of length 3^{-n} with the same center. The set

$$\mathcal{C} = \bigcap_{n=1}^{\infty} J_n$$

is called the *triadic Cantor set* or *ternary Cantor set*. Note that, for all $n \in \mathbb{N}$, the endpoints of all closed intervals, whose union at stage n is J_n, belong to $\mathcal{C}$.

$\mathcal{C}$ is closed since it is the intersection of closed sets. It contains no interval,[11] so its interior[12] is empty, that is, all its points are boundary points. Since any open set containing a point x of $\mathcal{C}$ contains at least another point of $\mathcal{C}$, all points of $\mathcal{C}$ are *limit points*.[13]

At each stage of the construction of $\mathcal{C}$, the open intervals that constitute the "middle thirds" of the closed intervals left at the previous stage are removed. Thus any element x of the Cantor ternary set can be written

$$x = \sum_{n=1}^{\infty} \frac{x_n}{3^n},$$

where, for all positive integers n, $x_n = 0$ or 2. In other words, the ternary expansion—that is, the expansion to the base 3—of an element $x \in \mathcal{C}$ does not contain the digit 1.

One might think that this statement is not correct since, for example, $7/9 = 2/3 + 1/3^2$ belongs to $\mathcal{C}$, and its ternary expansion, which is 0.21, contains the digit 1. There is indeed an infinite number of elements in $\mathcal{C}$ that have this property. The expansion of all of them contains only a finite number of terms, the last one being $1/3^{n_0}$. If this term is replaced by the series

$$\sum_{n=n_0+1}^{\infty} \frac{2}{3^n},$$

which is equal to $1/3^{n_0}$, the statement becomes correct. We shall always respect this convention and write $7/9$ as $0.2022\ldots$. Note that, with this convention, each point in $\mathcal{C}$ has a unique infinite expansion.

The Cantor set is not countable. If $\mathcal{C}$ were countable, there would exist a bijection $\varphi : \mathbb{N} \to \mathcal{C}$, and the set

$$\varphi(\mathbb{N}) = \{\varphi(n) \mid n \in \mathbb{N}\}$$

would coincide with $\mathcal{C}$. Let

---

[11] At stage n there is no interval of length greater than 2^n.

[12] Let $(X, \mathcal{T})$ be a topological space. A point x is an *interior point* of a subset $E \subset X$ if there exists an open set $U \in \mathcal{T}$ containing x such that $U \subset E$. The set of interior points of E is the *interior* of E. A point x is a *boundary point* of a set E if there exists an open set $U \in \mathcal{T}$ containing x that contains at least one point of E and one point of $X \backslash E$. The set of boundary points of E is called the *boundary* of E.

[13] Limit points are also called *accumulation points*.

$$(\forall n \in \mathbb{N}) \qquad \varphi(n) = \sum_{k=1}^{\infty} \frac{x_k^{(n)}}{3^k},$$

and consider the sequence (x_n) such that

$$x_n = \begin{cases} 0, & \text{if } x_n^{(n)} = 2, \\ 2, & \text{if } x_n^{(n)} = 0. \end{cases}$$

The number

$$x = \sum_{n=1}^{\infty} \frac{x_n}{3^n}$$

is clearly in $\mathcal{C}$ but not in $\varphi(\mathbb{N})$. Hence, there is no such bijection φ.[14]

The Cantor set is equipotent to the interval $[0, 1]$. To prove it, we have to exhibit a bijection $\psi : [0, 1] \to \mathcal{C}$. Let

$$x = \sum_{n=1}^{\infty} \frac{x_n}{2^n}$$

be an element of $[0, 1]$, where, for all $n \in \mathbb{N}$, $x_n = 0$ or 1, and define the *Cantor function* ψ by

$$\psi(x) = \sum_{n=1}^{\infty} \frac{2x_n}{3^n}.$$

It is clear that $\psi([0, 1]) = \mathcal{C}$. Furthermore, if

$$y = \sum_{n=1}^{\infty} \frac{y_n}{3^n},$$

is an element of $\mathcal{C}$, for all $n \in \mathbb{N}$, $y_n = 0$ or 2, and

$$x = \psi^{-1}(y) = \sum_{n=1}^{\infty} \frac{\frac{1}{2} y_n}{2^n}$$

exists and is unique. Then $\psi^{-1}(\mathcal{C}) = [0, 1]$, which finally proves that $\psi : [0, 1] \to \mathcal{C}$ is a bijection.

The Lebesgue measure of the Cantor set is zero. Since J_n is obtained from J_{n-1} by removing 2^{n-1} open intervals of length 3^{-n}, we have

$$m(\mathcal{C}) = 1 - \sum_{n=0}^{\infty} \frac{2^n}{3^{n+1}} = 0.$$

---

[14] This method of proof, called *Cantor's diagonal process*, is frequently used to demonstrate that a set is not countable.

To make a distinction between sets of zero Lebesgue measure, the Hausdorff dimension is a helpful notion [168]. Let A be a subset of $\mathbb{R}$; the *Hausdorff outer measure in dimension d* of A is

$$H_d^*(A) = \liminf_{\varepsilon \to 0} \sum_{j \in J} \left(m(I_j) \right)^d, \qquad (d > 0),$$

where the infimum is taken over all *finite* or *countable* coverings of A by open intervals I_j whose Lebesgue measure (here length) $m(I_j)$ is less than ε. The limit exists—but may be infinite—since the infimum can only increase as ε decreases.

1. If $H_{d_1}^*(A) < \infty$, then, for $d_2 > d_1$, $H_{d_2}^*(A) = 0$. Indeed, the condition $m(I_j) < \varepsilon$, for all $j \in J$, implies

$$\frac{\left(m(I_j) \right)^{d_2}}{\left(m(I_j) \right)^{d_1}} = \left(m(I_j) \right)^{d_2 - d_1} \le \varepsilon^{d_2 - d_1}.$$

Therefore,

$$\inf_{\varepsilon \to 0} \sum_{j \in J} \left(m(I_j) \right)^{d_2} \le \sum_{j \in J} \left(m(I_j) \right)^{d_2}$$

$$\le \varepsilon^{d_2 - d_1} \sum_{j \in J} \left(m(I_j) \right)^{d_1}.$$

Letting $\varepsilon \to 0$, the result follows.
2. If $0 < H_{d_1}^*(A) < \infty$, then, for $d_2 < d_1$, $H_{d_2}^*(A) = \infty$. This is a corollary of the preceding result.

Let A be a subset of $\mathbb{R}$; the number $\inf\{d \mid H_d^*(A) = 0\}$ is called the *Hausdorff dimension* of the subset A and denoted by $d_H(A)$.

The Hausdorff outer measure and Hausdorff dimension of a subset of $\mathbb{R}^k$ can be defined in a similar way using the Lebesgue measure (volume) m of open hypercubes.

Example 35. Hausdorff dimension of the ternary Cantor set. $\mathcal{C}$ is *self-similar*, that is,

$$3\mathcal{C} \cap [0, \tfrac{1}{3}] = \mathcal{C} \quad \text{and} \quad 3\mathcal{C} \cap [\tfrac{2}{3}, 1] - 2 = \mathcal{C},$$

where, for a given real number k, kA denotes the set $\{kx \mid x \in A\}$. But

$$H_d^*(\mathcal{C}) = H_d^*(\mathcal{C} \cap [0, \tfrac{1}{3}]) + H_d^*(\mathcal{C} \cap [\tfrac{2}{3}, 1]),$$

and

$$H_d^*(\mathcal{C} \cap [0, \tfrac{1}{3}]) = H_d^*(\mathcal{C} \cap [\tfrac{2}{3}, 1])$$

$$= \frac{1}{3^d} H_d^*(\mathcal{C}),$$

since, if $A = kB$, then $H_d^*(A) = k^d H_d^*(B)$. Finally,

$$H_d^*(\mathcal{C}) = \frac{2}{3^d} H_d^*(\mathcal{C}).$$

This relation shows that $H_d^*(\mathcal{C})$ takes a finite nonzero value if, and only if,

$$\frac{2}{3^d} = 1.$$

Hence, the Hausdorff dimension $d_H(\mathcal{C})$ of the ternary Cantor set $\mathcal{C}$ is given by

$$d_H(\mathcal{C}) = \frac{\log 2}{\log 3}.$$

Remark 9. There exist many different definitions of dimension. Let A be a subset of $\mathbb{R}^k$ and $N(\varepsilon, A)$ be the minimum number of open hypercubes of side ε needed to cover A. Then, the *capacity* of A is

$$d_C(A) = -\limsup_{\varepsilon \to 0} \frac{\log N(\varepsilon, A)}{\log \varepsilon}.$$

In the case of the ternary Cantor set, we have

$$d_C(\mathcal{C}) = \lim_{\varepsilon \to 0} \frac{\log 2^n}{\log 3^n} = \frac{\log 2}{\log 3}.$$

For the Cantor set $d_C(\mathcal{C}) = d_H(\mathcal{C})$. This is not always the case. In general, $d_C(A) \geq d_H(A)$ since in the definition of the Hausdorff dimension of a subset A of $\mathbb{R}^k$, we consider coverings of A by open hypercubes of variable sides.

Sets like the ternary Cantor set are called *fractals* by Mandelbrot [224, 225] who defined them as *sets for which the Hausdorff dimension strictly exceeds the topological dimension*. This definition, which has the advantage of being precise, is a bit restrictive. It excludes sets that can be accepted as fractals. Self-similarity, which is an essential characteristic of fractals, should be included in their definition.

In the following example, we study a two-dimensional map having a fractal attractor.

Example 36. The generalized baker's map. The three-parameter family of two-dimensional maps defined by

$$b_{r_1, r_2, a}(x, y) = \begin{cases} \left(r_1 x, \dfrac{1}{a} y \right), & \text{if } 0 \leq x \leq 1, \ 0 \leq y \leq a, \\[2ex] \left(r_2 x + \dfrac{1}{2}, \dfrac{1}{1-a}(y - a) \right), & \text{if } 0 \leq x \leq 1, \ a \leq y \leq 1, \end{cases}$$

where the parameters r_1, r_2, and a are positive numbers less than $\frac{1}{2}$, is called the generalized baker's map [120].[15] It can be verified that the map transforms the unit square $[0,1] \times [0,1]$ into the two rectangles

---

[15] The map models the process whereby a baker kneads dough by stretching and folding.

$$[0, r_1] \times [0, 1] \cup [\tfrac{1}{2}, \tfrac{1}{2} + r_2] \times [0, 1],$$

and, after a second iteration, we obtain four rectangles:

$$[0, r_1^2] \times [0, 1] \cup [\tfrac{1}{2} r_1, r_1(\tfrac{1}{2} + r_2)] \times [0, 1]$$
$$\cup [\tfrac{1}{2}, \tfrac{1}{2} + r_1 r_2] \times [0, 1] \cup [\tfrac{1}{2}(1 + r_2), \tfrac{1}{2} + r_2(\tfrac{1}{2} + r_2)] \times [0, 1].$$

At this stage it is important to note the following self-similarity property: if the interval $[0, r_1]$ on the x-axis is magnified by a factor $1/r_1$, the set

$$[0, r_1^2] \times [0, 1] \cup [\tfrac{1}{2} r_1, r_1(\tfrac{1}{2} + r_2)] \times [0, 1]$$

becomes a replica of the set obtained after the first iteration. Similarly, if the interval $[\tfrac{1}{2}, \tfrac{1}{2} + r_2]$, after a translation equal to $-\tfrac{1}{2}$, is magnified by a factor $1/r_2$, the set

$$[\tfrac{1}{2}, \tfrac{1}{2} + r_1 r_2] \times [0, 1] \cup [\tfrac{1}{2}(1 + r_2), \tfrac{1}{2} + r_2(\tfrac{1}{2} + r_2)] \times [0, 1]$$

also becomes a replica of the set obtained after the first iteration.

To determine the capacity and Hausdorff dimension of the attractor of the generalized baker's map, we first note that this attractor is the product of a Cantor set[16] along the x-axis and the interval $[0, 1]$ along the y-axis. Thus d_C and d_H are, respectively, equal to $1 + \bar{d}_C$ and $1 + \bar{d}_H$, where $\bar{d}_C$ and $\bar{d}_H$ are the capacity and the Hausdorff dimension of the attractor along the x-axis.

To find the capacity $\bar{d}_C(A)$ of the intersection of the attractor A and the interval $[0, 1]$ along the x-axis, we use the self-similarity property. We write

$$N(\varepsilon, A) = N_1(\varepsilon, A) + N_2(\varepsilon, A),$$

where $N_1(\varepsilon, A)$ is the number of intervals of length ε needed to cover the part of the attractor along the x-axis that lies in the interval $[0, r_1]$, and $N_2(\varepsilon, A)$ is the number of intervals of length ε needed to cover the part of the attractor along the x-axis that lies in the interval $[\tfrac{1}{2}, \tfrac{1}{2} + r_2]$. From the scaling property

$$N_1(\varepsilon, A) = N\left(\frac{\varepsilon}{r_1}, A\right) \quad \text{and} \quad N_2(\varepsilon, A) = N\left(\frac{\varepsilon}{r_2}, A\right).$$

Thus

$$N(\varepsilon, A) = N\left(\frac{\varepsilon}{r_1}, A\right) + N\left(\frac{\varepsilon}{r_2}, A\right).$$

If $\bar{d}_C(A)$ exists, $N(\varepsilon, A)$ should behave as $\varepsilon^{-\bar{d}_C}$ when ε is small enough. Hence,

$$1 = r_1^{\bar{d}_C} + r_2^{\bar{d}_C}.$$

If $r_1 = r_2 = \tfrac{1}{2}$, $\bar{d}_C = 1$, and for $r_1 = r_2 = r$, $\bar{d}_C$ decreases with r.

Again using self-similarity, we find that the Hausdorff dimension $d_H(A)$ of the attractor A is equal to its capacity $d_C(A)$.

---

[16] By "Cantor set" we mean a *closed set with no interior points, and such that all points are limit points.* A closed set that, as a Cantor set, coincides with the set of all its limit points is said to be *perfect*.

Chaotic attractors that are fractals are usually called *strange attractors*. More precisely, according to Ruelle [300, 301, 302]:

Definition 21. *A bounded set $A \subset \mathbb{R}^k$ is a strange attractor for a map* **f** *if there is a k-dimensional neighborhood N of A such that, for all $t \in \mathbb{N}$, $\mathbf{f}^t(N) \subset N$, and if, for all initial points in N, the map* **f** *has sensitive dependence on initial conditions.*

As for the generalized baker's map, strange attractors occur in dissipative systems[17] and in most cases result from a stretching and folding process.

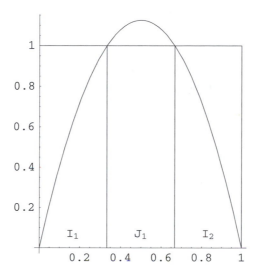

Fig. 5.10. *Graph of the logistic map $n \mapsto 4.5\,n(1-n)$.*

Example 37. A strange repeller: the logistic map for $r > 4$. Consider again the logistic map $(n, r) \mapsto f_r(n) = rn(1 - n)$ but, this time, for parameter values greater than 4. Since $f_r(0.5) > 1$, there exists an open interval J_1 centered at $n = \frac{1}{2}$ such that, for all $x \in J_1$, $\lim_{n \to \infty} f_r^n(x) = -\infty$. Hence, for the orbit of a point $x \in [0, 1]$ to be bounded, x has to belong to $[0, 1] \backslash J_1$, that is, to

$$I_1 \cup I_2 = [0, 1] \cap f_r^{-1}([0, 1]) = f_r^{-1}([0, 1]),$$

as shown in Figure 5.10. This condition, which is necessary but obviously not sufficient, can be rewritten

$$\bigcup_{i_1 = 1, 2} I_{i_1} = f_r^{-1}([0, 1]).$$

---

[17] That is, systems in which the determinant of the Jacobian matrix is less than 1. This implies that the Hausdorff dimension of a strange attractor is less than the dimension of the phase space.

More generally, for the orbit of a point $x \in [0,1]$ to be bounded, x has to belong to the union of 2^n disjoint closed intervals

$$\bigcup_{i_1,i_2,\ldots,i_n=1,2} I_{i_1,i_2,\ldots,i_n} = f_r^{-n}([0,1]).$$

If we let n go to infinity, we find that the orbit of a point $x \in [0,1]$ is bounded if, and only if, x belongs to the invariant set C_r defined by

$$C_r = \bigcap_{n=1}^{\infty} \bigcup_{i_1,i_2,\ldots,i_n=1,2} I_{i_1,i_2,\ldots,i_n}. \tag{5.10}$$

The set C_r is closed since it is the intersection of closed sets.

The Lebesgue measure (*i.e.*, the length) of the closed intervals $I_{i_1,n_2,\ldots,i_n}$ decreases as n increases. Denoting by $m(I)$ the Lebesgue measure of the interval I, we shall prove that

$$\lim_{n \to \infty} m(I_{i_1,n_2,\ldots,i_n}) = 0 \tag{5.11}$$

for $r > 2 + \sqrt{5}$.

First, let us show that, for all $x \in I_1 \cup I_2$, $|f_r'(x)| > 1$ if $r > 2 + \sqrt{5}$. Since $f_r'(x) = r - 2rx$ and $f_r''(x) = -2r < 0$, $|f_r'(x)|$ is minimum on $I_1 \cup I_2$ for x such that $f_r(x) = 1$, that is, for $x = (r \pm \sqrt{r^2 - 4r})/2r$. But

$$\left| f_r'\left(\frac{r \pm \sqrt{r^2 - 4r}}{2r}\right) \right| = \left| \sqrt{r^2 - 4r} \right|.$$

Thus, $\left| f_r'\left((r \pm \sqrt{r^2 - 4r})/2r\right) \right| > 1$ if

$$r^2 - 4r > 1; \quad \text{that is, if} \quad r > 2 + \sqrt{5}.$$

Let $\lambda_r = \inf_{x \in I_1 \cup I_2} |f_r'(x)|$. We shall prove by induction that

$$m(I_{i_1,n_2,\ldots,i_n}) \le \lambda_r^{-n}. \tag{5.12}$$

Consider the interval $I_i = [a_i, b_i]$, where $i = 1, 2$. The images by f_r of its endpoints are 0 and 1 for $i = 1$ and 1 and 0 for $i = 2$. By the mean value theorem, there exists a point $c_i \in I_i$ such that $f_r(b_i) - f_r(a_i) = f_r'(c_i)(b_i - a_i)$. Hence,

$$1 = |f_r(b_i) - f_r(a_i)| = |f_r'(c_i)| \, m(I_i) \ge \lambda_r m(I_i).$$

That is, $m(I_i) \le \lambda_r^{-1}$, which shows that (5.12) is true for $n = 1$.

Let $a_{i_1,i_2,\ldots,i_n}$ and $b_{i_1,i_2,\ldots,i_n}$ denote the endpoints of the interval $I_{i_1,i_2,\ldots,i_n}$. For $i_n = 1, 2$,

$$f_r(I_{i_1,i_2,\ldots,i_n}) = I_{i_1,i_2,\ldots,i_{n-1}}.$$

Hence,

$$m(I_{i_1,i_2,\ldots,i_{n-1}}) = |b_{i_1,i_2,\ldots,i_{n-1}} - a_{i_1,i_2,\ldots,i_{n-1}}|$$
$$= |f_r(b_{i_1,i_2,\ldots,i_n}) - f_r(a_{i_1,i_2,\ldots,i_n})|$$
$$= |f'_r(c_{i_1,i_2,\ldots,i_n})| \, |(b_{i_1,i_2,\ldots,i_n} - a_{i_1,i_2,\ldots,i_n})|$$
$$\geq \lambda_r \, m(I_{i_1,i_2,\ldots,i_n}),$$

and, if we assume that (5.12) is true for $n-1$, we finally obtain

$$m(I_{i_1,i_2,\ldots,i_n}) \leq \lambda_r \, m(I_{i_1,i_2,\ldots,i_{n-1}})$$
$$\leq \lambda_r^{-n}.$$

Thus, in the limit $n \to \infty$, Relation (5.11) is verified.

The set $\mathcal{C}_r$ is a Cantor set since it possesses the characteristic properties of such a set given in Footnote 16: it is closed with no interior points, and all its points are limit points. At the same time it is a repeller since any point in a neighborhood of $\mathcal{C}_r$ goes to $-\infty$. $\mathcal{C}_r$ is then a strange repeller.

5.4 Chaotic discrete-time models

5.4.1 One-population models

We have seen that, for continuous maps defined on an interval, topological transitivity implies the existence of a dense set of periodic points and sensitive dependence on initial conditions. To verify that a one-dimensional map is chaotic, it is therefore sufficient to prove that it is topologically transitive. It can be shown [299] that *if a map f is topologically transitive on an invariant set X, then the orbit of some point $x \in X$ is dense in X.* Except for very particular maps, it is, in general, almost impossible to determine a specific point that has a dense orbit. However, when analyzing models, a map is considered to be chaotic if it is possible to find numerically an orbit that "looks" dense.

In Figures 5.11 and 5.13 are represented the bifurcation diagrams of two one-population models. These bifurcation diagrams have been computed for 200 equally spaced parameter values. For each parameter value, 300 iterates are calculated, but only the last 100 have been plotted. Cobwebs corresponding to "apparently" dense orbits are represented in Figures 5.12 and 5.14.

5.4.2 The Hénon map

In order to find a simpler model than the Lorenz system (Equations (5.14) below) that nevertheless possesses a strange attractor, Hénon [169] considered

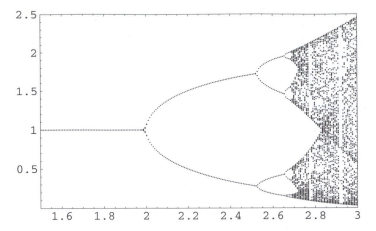

Fig. 5.11. *Bifurcation diagram of the one-population map* $(n, r) \mapsto n e^{r(1-n)}$. *The parameter* r, *plotted on the horizontal axis, varies from 1.5 to 3, and the population* n, *plotted on the vertical axis, varies between 0 and 2.5.*

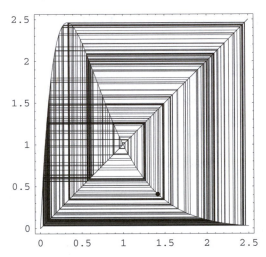

Fig. 5.12. *Cobweb showing a sequence of 250 iterates of the map* $(n, r) \mapsto n e^{r(1-n)}$ *for* $r = 3$. *The initial value is* $n_0 = \sqrt{2}$ *with a precision of 150 significant digits.*

the two-dimensional map[18]

$$\mathbf{H}_{a,b}(x, y) = (1 + y - ax^2, \ bx). \qquad (5.13)$$

The Jacobian matrix

$$D\mathbf{H}_{a,b}(x, y) = \begin{bmatrix} -2ax & 1 \\ b & 0 \end{bmatrix}$$

---

[18] It can be shown that the Hénon map is one of the reduced forms of the most general quadratic map, whose determinant of its Jacobian matrix is constant.

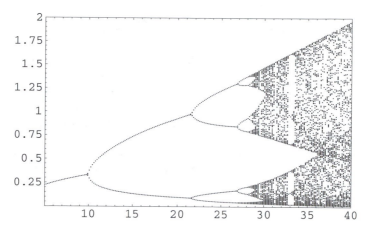

Fig. 5.13. *Bifurcation diagram of the one-population map* $(n, r) \mapsto rn(1+n)^{-8}$*. The parameter* r*, plotted on the horizontal axis, varies from 5 to 40, and the population* n*, plotted on the vertical axis, varies between 0 and 2.*

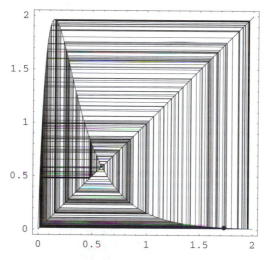

Fig. 5.14. *Cobweb showing a sequence of 250 iterates of the map* $(n, r) \mapsto rn(1+n)^{-8}$ *for* $r = 40$*. The initial value is* $n_0 = \sqrt{3}$ *with a precision of 150 significant digits.*

has a constant determinant equal to $-b$. The map is, consequently, a uniform contraction for $|b| < 1$ and area-preserving for $|b| = 1$. For $b \neq 0$, the map is invertible, and we have

$$\mathbf{H}_{a,b}^{-1}(x, y) = (b^{-1}y, \; x - 1 + ab^{-2}y^2).$$

If $a > a_0 = -\frac{1}{4}(1 - b)^2$, the map has two fixed points:

$$x^*_{1,2} = \frac{b - 1 \pm \sqrt{(1-b)^2 + 4a}}{2a}, \quad y^*_{1,2} = bx^*_{1,2}.$$

One is always unstable, while the other one is asymptotically stable if $a_1 = 3(1-b)^2/4 > a > a_0$. For $b = 0.3$, which is the value chosen by Hénon, $a_0 = -0.1225$ and $a_1 = 0.3675$. For a fixed value of b, as a increases, the map exhibits an infinite sequence of period-doubling bifurcations. Derrida, Gervois, and Pomeau [101] have determined the parameter values a_k of the first 11 period-doubling bifurcations and found that the rate of convergence is equal to the Feigenbaum number δ. Their estimate of a_∞ is 1.058048.... The universality of the period-doubling route to chaos is, therefore, valid for dissipative maps of dimensionality higher than 1.

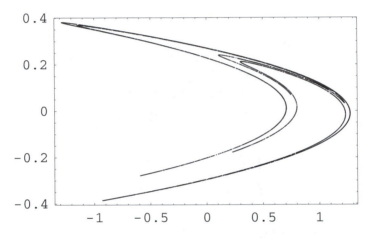

Fig. 5.15. *Ten thousand successive points obtained by iteration of the Hénon map starting from the unstable fixed point* $(0.631354\ldots, 0.189406\ldots)$.

After some preliminary numerical experiments, Hénon found a strange attractor for the parameter values $a = 1.4$ and $b = 0.3$. It is represented in Figure 5.15.

Two enlargements of the attractor are represented in Figures 5.16 and 5.17 showing its fine structure, which looks identical at all scales. Both frames contain the unstable fixed point $(0.631354\ldots, 0.189406\ldots)$, which apparently lies on the upper boundary of the attractor.

The Hausdorff dimension of the strange attractor of the Hénon map has been determined numerically by Russell, Hanson, and Ott [304][19], who found

---

[19] The authors divided the phase space in boxes of side ε and after many iterations counted how many boxes contained at least one iterate. The dimension is found assuming that, for a sufficiently small ε, the number of boxes behaves as ε^d. Strictly speaking, such a dimension approaches the value d_C of the capacity. However, for most strange attractors, $d_C = d_H$ (see Farmer, Ott, and York [120]).

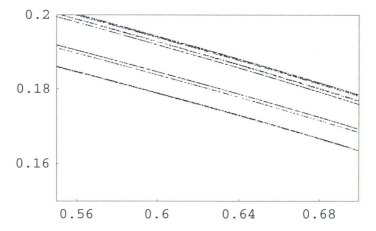

Fig. 5.16. *Enlargement of the Hénon attractor increasing the number of iterations to 10^5. The initial point in both cases is the unstable fixed point.*

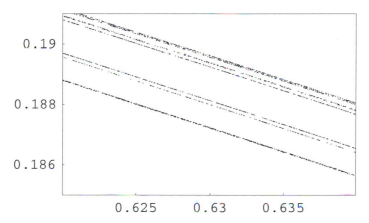

Fig. 5.17. *Enlargement of the Hénon attractor increasing the number of iterations to 10^6. The initial point is the unstable fixed point.*

$d_H = 1.261 \pm 0.003$ for $a = 1.4$ and $b = 0.3$, and $d_H = 1.202 \pm 0.003$ for $a = 1.2$ and $b = 0.3$.[20]

---

[20] All these numerical considerations do not constitute a proof of the existence of a strange attractor. Sometimes such a proof can be given. For the Hénon map, see Tresser, Coullet, and Arneodo [332].

5.5 Chaotic continuous-time models

5.5.1 The Lorenz model

In his historical paper, published in 1963 [212], Lorenz derived, from a model of fluid convection, a three-parameter family of three ordinary differential equations that appeared, when integrated numerically, to have extremely complicated solutions. These equations are

$$\dot{x} = \sigma(y - x)$$
$$\dot{y} = rx - y - xz \qquad (5.14)$$
$$\dot{z} = xy - bz,$$

where σ, r, and b are real positive parameters. The system is invariant under the transformation $(x, y, z) \to (-x, -y, z)$. Figures 5.18, 5.19 and 5.20 represent the projections of the Lorenz strange attractor, calculated for $\sigma = 10$, $r = 28$, and $b = 8/3$, on, respectively, the planes $x0y$, $x0z$, and $y0z$.

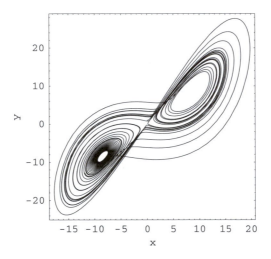

Fig. 5.18. *Projection on the $x0y$ plane of a numerical solution of the Lorenz equations for $t \in [0, 40]$ with $(x_0, y_0, z_0) = (0, 0, 1)$.*

The orbit is obviously not periodic. As t increases, the orbit winds first around the unstable nontrivial fixed point

$$(x^*, y^*, z^*) = (-8.48528\ldots, -8.48528\ldots, 27)$$

and then around the other unstable fixed point,

$$(x^*, y^*, z^*) = (8.48528\ldots, 8.48528\ldots, 27),$$

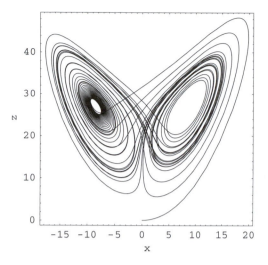

Fig. 5.19. *Projection on the x0z plane of a numerical solution of the Lorenz equations for $t \in [0, 40]$ with $(x_0, y_0, z_0) = (0, 0, 1)$.*

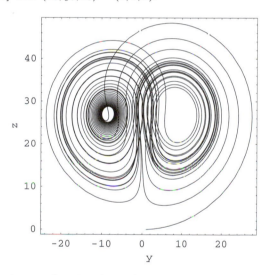

Fig. 5.20. *Projection on the y0z plane of a numerical solution of the Lorenz equations for $t \in [0, 40]$ with $(x_0, y_0, z_0) = (0, 0, 1)$.*

without ever settling down. Its shape does not depend upon a particular choice of initial conditions.

The divergence of the flow (trace of the Jacobian matrix) is equal to $-(\sigma + b + 1)$. Thus a three-dimensional volume element is contracted, as a function of time t, by a factor $e^{-(\sigma+b+1)t}$. It can be shown that there is a bounded ellipsoid $E \subset \mathbb{R}^3$ that all trajectories eventually enter.[21] Taken together, the existence of the bounded ellipsoid and the negative divergence of the flow imply that there exists a bounded set of zero Lebesgue measure inside the ellipsoid E towards which all trajectories tend.

Exercises

Exercise 5.1 *The converse of Šarkovskii's theorem is also true.[22] Show that the piecewise linear map defined on $[1,5]$ by*

$$f(x) = \begin{cases} 2x + 1, & \text{if } 1 \leq x \leq 2, \\ -x + 7, & \text{if } 2 \leq x \leq 3, \\ -2x + 10, & \text{if } 3 \leq x \leq 4, \\ -x + 6, & \text{if } 4 \leq x \leq 5, \end{cases}$$

has a periodic orbit of period 5 but no periodic orbit of period 3.

Exercise 5.2 *Consider the symmetric tent map defined by*

$$T_r(x) = \begin{cases} rx, & \text{if } 0 \leq x \leq \frac{1}{2}, \\ r(1-x), & \text{if } \frac{1}{2} \leq x \leq 1. \end{cases}$$

Find the values of the parameter r for which the map T_r has periodic orbits of period 3.

Exercise 5.3 *Find the condition under which the invariant probability density ρ is equal to 1 (i.e., the Lebesgue measure) for the asymmetric tent map defined by*

$$T_{a,b}(x) = \begin{cases} ax, & \text{if } 0 \leq x \leq \dfrac{b}{a+b}, \\ b(1-x), & \text{if } \dfrac{b}{a+b} \leq x \leq 1. \end{cases}$$

Exercise 5.4 *Let $f : \mathbb{R}_+ \to \mathbb{R}_+$ be the map defined by*

$$f(x) = \begin{cases} 3x, & \text{if } 0 \leq x \leq \frac{1}{3}, \\ -3x + 2, & \text{if } \frac{1}{3} \leq x \leq \frac{2}{3}, \\ 3x - 2, & \text{if } \frac{2}{3} \leq x \leq 1, \\ f(x-1) + 1, & \text{if } x \geq 1. \end{cases}$$

Show that f has sensitive dependence on initial conditions, has periodic points that are dense in $\mathbb{R}_+$, but is not chaotic.

Exercise 5.5 *Determine the bifurcation diagram of the one-parameter family of maps $(n, r) \mapsto -rn \log n$, which is a discrete version of the Gompertz model (see Exercise 4.6) for $1.8 \leq r \leq 2.7$.*

---

[21] For the most complete study of the Lorenz model, consult Sparrow [323].
[22] This example is taken from Li and York [204].

Exercise 5.6 Drosophila *population dynamics has been investigated to clarify if the observed fluctuations about the carrying capacity were the result of a chaotic behavior [251, 282]. Populations N_t and N_{t+1} at generation t and $t+1$, respectively, were fitted on the discrete theta model*

$$N_{t+1} = N_t \left(1 + r \left(1 - \left(\frac{N_t}{K} \right)^\theta \right) \right),$$

where K is the carrying capacity and r and θ are two positive parameters that measure the growth rate and its asymmetry. If we denote by $n_t = N_t/K$ the reduced population, find the first period-doubling bifurcation points and determine the bifurcation diagram of the two-parameter family of maps $(n, r, \theta) \mapsto f_{r,\theta}(n) = n + rn(1 - n^\theta)$.

Exercise 5.7 *Find the invariant probability density of the two-dimensional chaotic map*

$$(x_{t+1}, \, y_{t+1}) = (y_t, \, 4x_t(1 - x_t)).$$

Exercise 5.8 *Consider the two-parameter family of one-dimensional maps defined by*

$$f_{a,b}(x) = a + \frac{bx}{1 + x^2}.$$

Study its bifurcation diagram for $a \in [-5, 0]$ and $b \in [11, 12]$.

Solutions

Solution 5.1 *Figure 5.21 shows that the map f has a unique fixed point in the interval $[3, 4]$, solution of the equation $-2x + 10 = x$, i.e., $x = \frac{10}{3}$. This fixed point is unstable $(f'(\frac{10}{3}) = -2)$. Since f^3 has no other fixed point, f has no 3-point cycle, while it is readily verified that it has a 5-point cycle, namely $\{1, 3, 4, 2, 5\}$.*

Solution 5.2 *If $\{x_1^*, x_2^*, x_3^*\}$ is a 3-point cycle, we have to distinguish three possibilities.*

1. *Two points of the cycle (say x_1^* and x_2^*) are less than $\frac{1}{2}$, and the third one (x_3^*) is greater. In this case, the three points satisfy the equations*

$$x_2^* = rx_1^*, \quad x_3^* = rx_2^*, \quad x_1^* = r(1 - x_3^*).$$

Hence,

$$x_1^* = \frac{r}{1 + r^3}, \quad x_2^* = \frac{r^2}{1 + r^3}, \quad x_3^* = \frac{r^3}{1 + r^3}.$$

Since we assumed that

$$0 < x_1^* < \tfrac{1}{2}, \quad 0 < x_2^* < \tfrac{1}{2}, \quad \tfrac{1}{2} < x_3^* < 1,$$

the parameter r has to satisfy the condition

$$\tfrac{1}{2}(1 + \sqrt{5}) < r \leq 2.$$

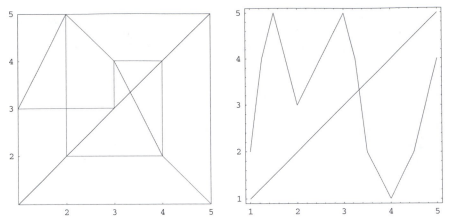

Fig. 5.21. *Exercise 5.1. Graphs of maps f (left) and f^3 (right). They show that f has a 5-point cycle $\{1, 3, 4, 2, 5\}$, but no 3-point cycle.*

2. One point of the cycle (say x_1^*) is less than $\frac{1}{2}$, and the other two (x_2^* and x_3^*) are greater. In this case, the three points satisfy the equations

$$x_2^* = rx_1^*, \quad x_3^* = r(1 - x_2^*), \quad x_1^* = r(1 - x_3^*).$$

Hence,

$$x_1^* = \frac{r}{1 + r + r^2}, \quad x_2^* = \frac{r^2}{1 + r + r^2}, \quad x_3^* = \frac{r(1 + r)}{1 + r + r^2}.$$

Since we assumed that

$$0 < x_1^* < \tfrac{1}{2}, \quad \tfrac{1}{2} < x_2^* < 1, \quad \tfrac{1}{2} < x_3^* < 1,$$

here again, the parameter r has to satisfy the condition

$$\tfrac{1}{2}(1 + \sqrt{5}) < r \le 2.$$

These two 3-cycles are shown in Figure 5.22.
3. If one point of the cycle (say x_2^*) is equal to $\frac{1}{2}$, then, among the other two, one (x_3^*) is greater than $\frac{1}{2}$, and the second one (x_1^*) is less than $\frac{1}{2}$. Then x_1^* and x_3^* have to satisfy the relation

$$rx_1^* = \tfrac{1}{2}, \quad \frac{r}{2} = x_3^*, \quad r(1 - x_3^*) = x_1^*.$$

In this case, r is the root of $r^3 - 2r^2 + 1 = 0$ in the semi-open interval $]1, 2]$; that is, $r = \frac{1}{2}(1 + \sqrt{5})$. This particular 3-cycle is represented in Figure 5.23.

Solution 5.3 *Let ρ be the invariant probability density for the asymmetric tent map $T_{a,b}$. If I is an open interval in $[0, 1]$, its preimage by $T_{a,b}^{-1}(I)$ is the union of two disjoint intervals I_1 and I_2 whose Lebesgue measure is*

$$m(I_1 \cup I_2) = m(I_1) + m(I_2) = \left(\frac{1}{a} + \frac{1}{b}\right) m(I).$$

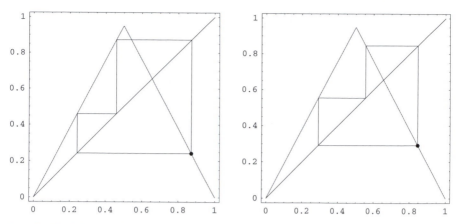

Fig. 5.22. *The two different types of 3-point cycles for the symmetric tent map T_r for $r = 1.9$.*

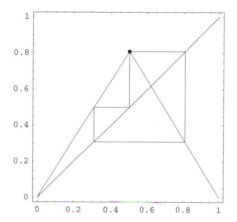

Fig. 5.23. *The 3-point cycle for the symmetric tent map T_r for $r = \frac{1}{2}(1 + \sqrt{5})$.*

Hence the map $T_{a,b}$ preserves the Lebesgue measure if

$$\frac{1}{a} + \frac{1}{b} = 1.$$

This relation is satisfied in particular by the symmetric binary tent map T_2 and by any asymmetric tent map $T_{a,b}$ with maximum

$$T_{a,b}\left(\frac{b}{a+b}\right) = \frac{ab}{a+b} = 1.$$

We could also have derived this result from the Perron-Frobenius equation, which, in the case of the asymmetric tent map, reads

$$\rho(x) = \frac{1}{a}\rho\left(\frac{x}{a}\right) + \frac{1}{b}\rho\left(1 - \frac{x}{b}\right).$$

This equation is satisfied by $\rho(x) = 1$ if

$$\frac{1}{a} + \frac{1}{b} = 1.$$

The graph of the asymmetric tent map for $a = 3$ and $b = 3/2$, which satisfy the condition above, is plotted in Figure 5.24.

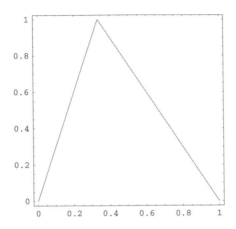

Fig. 5.24. *An example of an asymmetric tent map $T_{a,b}$ for $a = 3$ and $b = 3/2$ and invariant probability $\rho = 1$ (see the text).*

Solution 5.4 *The map f (see Figure 5.25) has sensitive dependence on initial conditions because $|f'(x)| = 3$ for all $x \in \mathbb{R}_+$. Between two consecutive integral values of x, f^n has $3^n - 2$ unstable fixed points. Since the distance between two consecutive fixed points is less than 3^{-n+1}, these periodic points are dense in $\mathbb{R}_+$. The map f, however, is not chaotic since it is not topologically transitive, because $f([k, k+1]) = [k, k+1]$, for any nonnegative integer k.*

Solution 5.5 *As for all bifurcation diagrams represented in this chapter, this one (see Figure 5.26) has been computed for 300 equally spaced parameter values, and for each parameter value, 300 iterates are calculated, but only the last 100 have been plotted.*

Solution 5.6 *The map $f_{r,\theta}$ has two fixed points: 0 and 1. 0 is always unstable $(f'_{r,\theta}(0) = 1 + r)$, and 1, which is the reduced carrying capacity, is asymptotically stable if $0 < r\theta < 2$ $(f'_{r,\theta}(1) = 1 - r\theta)$. For $r\theta = 2$, the family of maps undergoes a period-doubling bifurcation.*

While the stability of the fixed point 1 depends only on the product $r\theta$, this is not the case for the 2-point cycle. To simplify, we shall, therefore, fix the value of r and find the values of θ for which the family $f_{r,\theta}$ undergoes the first period-doubling bifurcations. For $r = 1$ the 2-point cycle is stable for $2 < \theta < 2.47887$ and the 4-point cycle is stable for $2.47887 < \theta < 2.5876$. The bifurcation diagram is represented in Figure 5.27.

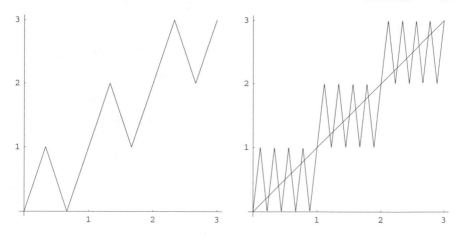

Fig. 5.25. *An example of a map in* $\mathbb{R}_+$ *(left figure) that has sensitive dependence on initial conditions and dense periodic points but that is not topologically transitive and, therefore, not chaotic. The right figure shows the graph of* f^2.

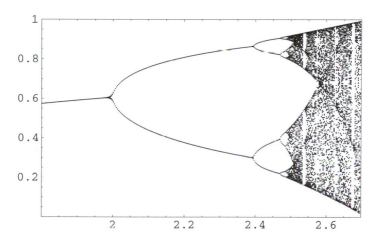

Fig. 5.26. *Bifurcation diagram of the discrete map* $(n, r) \mapsto -rn \log n$. *The parameter* r, *plotted on the horizontal axis, varies from* 1.8 *to* 2.7, *and the population* n, *plotted on the vertical axis, varies between* 0 *and* 1.

Solution 5.7 Let $\mathbf{f} : (x, y) \mapsto (y, 4x(1-x))$; then $\mathbf{f}^2 : (x, y) \mapsto (4x(1-x), 4y(1-y))$, which shows that

$$\mathbf{f}^2 : (x, y) \mapsto (f_4(x),\ f_4(y)),$$

where f_4 is the chaotic logistic map $x \mapsto 4x(1-x)$. Thus, taking into account the expression of the invariant probability density (5.3) found in Subsection 5.3.1, the invariant probability density of the two-dimensional map $\mathbf{f}$ is given by

$$\rho(x, y) = \frac{1}{\pi^2 \sqrt{xy(1-x)(1-y)}}.$$

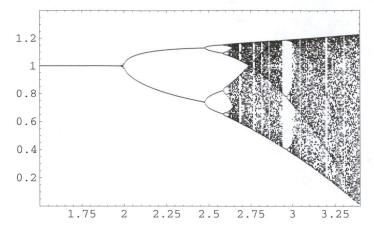

Fig. 5.27. *Bifurcation diagram of the discrete θ map $(n, r, \theta) \mapsto n + rn(1 - n^{\theta})$. The value of r is equal to 1, the parameter θ, plotted on the horizontal axis, varies from 1.5 to 3.4, and the population n, plotted on the vertical axis, varies between 0 and 1.4.*

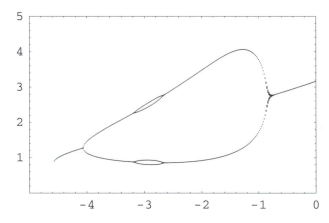

Fig. 5.28. *Bifurcation diagram of the two-parameter map $(x, a, b) \mapsto a + bx/(1 + x^2)$ for $b = 11$. The parameter a, plotted on the horizontal axis, varies from -5 to 0, and x, plotted on the vertical axis, varies between 0 and 5.*

Solution 5.8 *Figures 5.28 and 5.29 show the bifurcation diagram of the map $f_{a,b}$ for b equal to 11 and 12 respectively. For $b = 11$ and $a = -5$, the attractor is a stable fixed point. Then, as a increases from -5 to -3, we observe two period-doubling bifurcations, but as a increases further, the system undergoes two reverse period-doubling bifurcations and its attractor is again a fixed point. For $b = 12$ and $a = -5$, the attractor is, as for $b = 11$, a stable fixed point. But here, as a increases, the map becomes apparently chaotic after a sequence of period-doubling bifurcations, and as a further increases, through a reverse route, the attractor again becomes a fixed point. In the chaotic region, periodic windows are clearly visible.*

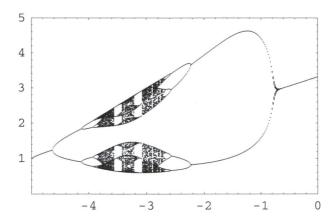

Fig. 5.29. *Bifurcation diagram of the two-parameter map* $(x, a, b) \mapsto a + bx/(1+x^2)$
for $b = 12$. *The parameter* a, *plotted on the horizontal axis, varies from* -5 *to* 0,
and x, *plotted on the vertical axis, varies between* 0 *and* 5.

Figure 5.30 shows the bifurcation diagram obtained for $a = -3$ and increasing b
from 11 to 12.

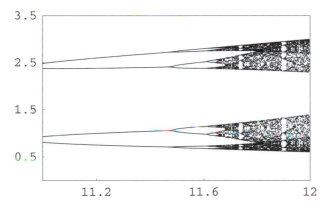

Fig. 5.30. *Bifurcation diagram of the two-parameter map* $(x, a, b) \mapsto a + bx/(1+x^2)$.
The parameter a *is equal to* -3, *the parameter* b, *plotted on the horizontal axis, varies*
from 11 *to* 12, *and* x, *plotted on the vertical axis, varies between* 0 *and* 3.5.

Part II

Agent-Based Models

In the preceding chapters, interacting individual actors have never been explicitly taken into account, we have only been interested in the evolution of average quantities. For instance, when modeling a predator-prey system, we only considered predator and prey densities and wrote down equations to determine how they varied as a function of time. In such an approach, each species is supposed to be homogeneously distributed in space. Even when space is taken into account, as in a model formulated in terms of partial differential equations, we have to assume that population densities are smooth functions of space, and an element of volume dx^3 is supposed to contain a large number of agents for "local densities" to be defined. Agent-based modeling is more ambitious; it deals directly with spatially distributed agents, which may be animals, people, or companies.

The agent-based models we shall discuss are formulated in terms of automata networks whose theory is not yet so well-developed. An *automata network* is a fully discrete dynamical system. It consists of a *graph* where each vertex takes states in a finite set.[1] The state of a vertex changes in time according to a rule that only depends upon the states of neighboring vertices in the graph. Automata networks are more precisely defined as follows.

Definition. *Let $G = (V, E)$ be a graph, where V is a set of vertices and E a set of edges. Each edge joins two vertices not necessarily distinct. An* automata network, *defined on V, is a triple $(G, Q, \{f_i \mid i \in V\})$, where G is a graph on V, Q a finite set of states and $f_i \colon Q^{|U_i|} \to Q$ a mapping, called the* local evolution rule *or* local transition rule *associated with vertex i. $U_i = \{j \in V \mid \{j, i\} \in E\}$ is the* neighborhood *of i (i.e., the set of vertices connected to i), and $|U_i|$ denotes the number of vertices belonging to U_i. The graph G is assumed to be* locally finite; *i.e., for all $i \in V$, $|U_i| < \infty$.*

Automata networks are ideal tools for *agent-based modeling*. Each agent has a position in space that is one of the vertices of the graph and an *individuality* represented by the state of the vertex where it is located. The interaction between the agents is described by the local evolution rule.

The states may just be integers. For instance, in a simple epidemic model, the state of a vertex could be equal to 0, 1 or 2 to represent, respectively, an empty site, a site occupied by a susceptible (*i.e.*, a healthy individual susceptible to catch the disease), or a site occupied by an infective (*i.e.*, an individual capable communicating the disease). Note that these numbers are just symbols. We could replace 0, 1, and 2 by, respectively, the letters e, s, and i. States may also be vectors, whose different components would represent, for example, an individual's sex, income, political affiliation, list of people met at previous times, etc. The word "vector" is not to be understood as an element of a linear space. It only denotes a list of symbols that are numbers, letters, or even lists of symbols.

---

[1] On graph theory, see Section 7.2.

In the following chapters, the notion of *critical behavior* will often play an important role. Critical behavior manifests itself in many-component systems and is characteristic of a cooperative behavior of the various components. This notion has been introduced in statistical physics for many-body systems such as ferromagnetic materials, alloys, or liquid helium, which exhibit *second-order phase transitions*; that is, a phase change as a function of a tuning parameter, such as temperature. A ferromagnetic material (*i.e.*, a system having a spontaneous nonzero magnetization), becomes, as its temperature is (in general) increased, paramagnetic (*i.e.*, its magnetization in the absence of an external magnetic field is equal to zero). In an ordered alloy such as β-brass—a 50% copper and 50% zinc alloy—the atoms of copper and zinc are located on two identical sublattices, one sublattice containing more copper and the other more zinc. As the temperature is increased, the alloy becomes disordered (*i.e.*, both sublattices contain equal fractions of copper and zinc). Liquid helium, which behaves as an ordinary liquid at temperatures above 2.19 K, becomes superfluid at lower temperatures. The temperature at which these phase transitions occur is called the *critical temperature*, and the system at the critical temperature—more generally at the *critical point*—is said to be in a *critical state*.

At the critical point, physical quantities such as entropy, volume, or magnetization that are first derivatives of the free energy are continuous, in contrast with second-order derivatives, such as the specific heat or the magnetic susceptibility, which are singular. These singular behaviors reflect the long-range nature of the correlations in the vicinity of the critical point.

Close to a second-order phase transition, correlation functions of fluctuating quantities (such as spins in the case of a para-ferromagnetic phase transition) at two different points decrease exponentially with a characteristic correlation length ξ. As the temperature T approaches the critical temperature T_c, ξ diverges as $(T - T_c)^{-\nu}$ if $T > T_c$ and $(T - T_c)^{-\nu'}$ if $T < T_c$.

This *cooperative effect* is characteristic of criticality. It implies that certain physical quantities either vanish or diverge as powers of $|T - T_c|$ as T approaches T_c. Today, when a many-agent system displays a power-law behavior for some observable, most researchers agree that this is a sign of some cooperative effect and a manifestation of the system complexity.

Despite the great variety of physical systems that exhibit second-order phase transitions, their critical behaviors, characterized by a set of *critical exponents*, fall into a small number of *universality classes* that only depend on the symmetry of the order parameter (such as the magnetization for a ferromagnet) and space dimension. Critical behavior is *universal* in the sense that it does not depend upon details whose characteristic size is much less than the correlation length, such as lattice structure, range of interactions (as long as this range is finite), spin length, etc. Moreover, for a given second-order phase transition, one needs to know only a rather small number of critical exponents to determine all other exponents. For instance, in the case of a para-ferromagnetic second-order phase transition, the specific heat at constant magnetic field C_B diverges as $(T - T_c)^{-\alpha}$ if $T > T_c$ and $(T - T_c)^{-\alpha'}$ if $T < T_c$, the magnetization M, which is identically equal to zero for $T > T_c$, goes to zero as $(T_c - T)^{\beta}$ if $T < T_c$, and the isothermal susceptibility χ_T diverges as $(T - T_c)^{-\gamma}$ if $T > T_c$ and $(T - T_c)^{-\gamma'}$ if $T < T_c$. If we assume that the free energy F, close to the critical point, is a generalized homogeneous function of $T - T_c$ and M that is a function satisfying, for all values of λ, the relation

$$F(\lambda(T - T_c), \lambda^{\beta} M) \equiv \lambda^{2-\alpha} F(T - T_c, M),$$

which implies that $F(T - T_c, M)$ can be written under the form

$$F(T - T_c, M) = |T - T_c|^{2-\alpha} f\left(\frac{M}{|T - T_c|^{\beta}}\right),$$

where f is a function of only one variable, it can be shown[2] that the critical exponents satisfy the following so-called *scaling relations*

$$\alpha = \alpha', \quad \gamma = \gamma', \quad \text{and} \quad \alpha + 2\beta + \gamma = 2.$$

An important distinguishing feature of the power-law behavior of physical quantities in the neighborhood of a critical point is that these quantities have no intrinsic scale. The function $x \mapsto e^{-x/\xi}$ has an intrinsic scale ξ, whereas the function $x \mapsto x^a$ has no intrinsic scale: power laws are *self-similar*.

Quantities exhibiting a power-law behavior have been observed in a variety of disciplines ranging from linguistics and geography to medicine and economics. As mentioned above, the emergence of such a behavior is regarded as the signature of a collective mechanism.

Second-order phase transitions are always associated with a broken symmetry. That is, the symmetry group of the ordered phase (the phase characterized by a nonzero value of the order parameter) is a subgroup of the disordered phase (the phase characterized by an order parameter identically equal to zero). To the order parameter, we can always associate a symmetry-breaking field. In the presence of such a field, the order parameter has a nonzero value, and, in this case, the system cannot exhibit a second-order phase transition.[3]

Depending upon the nature of the order parameter, a second-order phase transition exists only above a critical space dimensionality, called the *lower critical space dimension*. Critical exponents, which depend upon space dimensionality, have their mean-field values above another critical space dimensionality, called the *upper critical space dimension*. In the case of the Ising model, the lower and upper critical space dimensions are, respectively, equal to 1 and 4.

---

[2] See Boccara [45].

[3] The nature of the broken symmetry is not always obvious as, for instance, in the case of the normal-superfluid or normal-superconductor phase transitions. See Boccara [44].

6

Cellular Automata

Depending upon the nature of the graph G, there exist many types of automata networks. In this chapter, we focus on cellular automata in which the set of vertices, usually called *sites*, is either $\mathbb{Z}^n$, if the n-dimensional cellular automaton is infinite, or the torus $\mathbb{Z}_L^n$, if the n-dimensional cellular automaton is finite. $\mathbb{Z}_L$ denotes the set of integers modulo L. If the set of vertices is $\mathbb{Z}_L^n$, the cellular automaton is said to satisfy *periodic boundary conditions*. In what follows, we always assume that this is the case.

Cellular automata, constructed from many identical simple components but together capable exhibiting complex behavior, are ideal tools for developing models of complex systems [349].

6.1 Cellular automaton rules

Deterministic one-dimensional cellular automaton rules are defined as follows. Let $s(i, t) \in Q$ represent the state at site $i \in \mathbb{Z}$ and time $t \in \mathbb{N}$; a local evolution rule is a map $f : Q^{r_\ell + r_r + 1} \to Q$ such that

$$s(i, t+1) = f\big(s(i - r_\ell, t), s(i - \ell + 1, t), \ldots, s(i + r_r, t)\big), \qquad (6.1)$$

where the integers r_ℓ and r_r are, respectively, the *left radius* and *right radius* of the rule f; if $r_\ell = r_r = r$, r is called the *radius of the rule*. The local rule f, which is a function of $n = r_\ell + r_1 + 1$ arguments, is often said to be an *n-input rule*. The function $S_t : i \mapsto s(i, t)$ is the *state* of the cellular automaton at time t; S_t belongs to the set $Q^{\mathbb{Z}}$ of all *configurations*. Since the state S_{t+1} at $t + 1$ is entirely determined by the state S_t at time t and the local rule f, there exists a unique mapping $F_f : \mathcal{S} \to \mathcal{S}$ such that

$$S_{t+1} = F_f(S_t), \qquad (6.2)$$

which is called the *cellular automaton global rule* or the *cellular automaton evolution operator* induced by the local rule f.

In the case of two-dimensional cellular automata, there are several possible lattices and neighborhoods. If, for example, we consider a square lattice and a $(2r_1+1) \times (2r_2+1)$-neighborhood, the state $s\big((i_1, i_2), t+1\big) \in Q$ of site (i_1, i_2) at time $t+1$ is determined by the state of the $(2r_1 + 1) \times (2r_2 + 1)$-block of sites centered at (i_1, i_2) by

$$s\big((i_1, i_2), t+1\big)$$
$$= f \begin{pmatrix} s\big((i_1 - r_1, i_2 - r_2), t\big) & \cdots & s\big((i_1 + r_1, i_2 - r_2), t\big) \\ s\big((i_1 - r_1, i_2 - r_2 + 1), t\big) & \cdots & s\big((i_1 + r_1, i_2 - r_2 + 1), t\big) \\ \cdots & \cdots & \cdots \\ s\big((i_1 - r_1, i_2 + r_2), t\big) & \cdots & s\big((i_1 + r_1, i_2 + r_2), t\big) \end{pmatrix}, \quad (6.3)$$

where $f : Q^{(2r_1+1) \times (2r_2+1)} \to Q$ is the *two-dimensional cellular automaton local evolution rule*.

Most models presented in this chapter will be formulated in terms of *probabilistic cellular automata*. In this case, the image by the evolution rule of any $(r_\ell + r_r + 1)$-block, for a one-dimensional cellular automaton, or of any given $(2r_1 + 1) \times (2r_2 + 1)$-block, for a two-dimensional cellular automaton, is a discrete random variable with values in Q.

The simplest cellular automata are the so-called *elementary cellular automata* in which the finite set of states is $Q = \{0, 1\}$ and the rule's radii are $r_\ell = r_r = 1$. Sites in a nonzero state are sometimes said to be *active*. It is easy to verify that there exist $2^{2^3} = 256$ different elementary cellular automaton local rules $f : \{0, 1\}^3 \to \{0, 1\}$. The local rule of an elementary cellular automaton can be specified by its *look-up table*, giving the image of each of the eight three-site neighborhoods. That is, any sequence of eight binary digits specifies an elementary cellular automaton rule. Here is an example:

111	110	101	100	011	010	001	000
1	0	1	1	1	0	0	0

Following Wolfram [348], a *code number* may be associated with each cellular automaton rule. If $Q = \{0, 1\}$, this code number is the decimal value of the binary sequence of images. For instance, the code number of the rule above is 184 since

$$10111000_2 = 2^7 + 2^5 + 2^4 + 2^3 = 184_{10}.$$

More generally, the *code number* $N(f)$ of a one-dimensional $|Q|$-state n-input cellular automaton rule f is defined by

$$N(f) = \sum_{(x_1, x_2, \ldots, x_n) \in Q^n} f(x_1, x_2, \ldots, x_n) |Q|^{|Q|^{n-1} x_1 + |Q|^{n-2} x_2 + \cdots + |Q|^0 x_n}.$$

Figure 6.1 represents the first 50 iterations of the elementary cellular automaton rule 184. The cellular automaton size is equal to 50. The initial

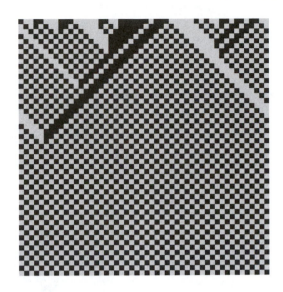

Fig. 6.1. *First 50 iterations of the elementary cellular automaton rule 184 starting from a random initial configuration with an equal number of 0s (light grey) and 1s (dark grey). Time is oriented downwards. Number of lattice sites: 50.*

configuration is random, and the density of active sites is exactly equal to $\frac{1}{2}$. As a result of the local rule, we observe the growth of a self-organized pattern. After a number of time steps of the order of the size of the cellular automaton, the initially disordered configuration becomes perfectly ordered, *i.e.*, the attractor consists of two configurations, namely

$$\ldots 01010101 \ldots \quad \text{and} \quad \ldots 10101010 \ldots .$$

In the case of cellular automata, the attractor is called the *limit set* and is defined by

$$\Lambda_F = \lim_{t \to \infty} F^t(\mathcal{S}) = \bigcap_{t \geq 0} F^t(\mathcal{S}),$$

where F is the global evolution rule and $\mathcal{S} = Q^{\mathbb{Z}}$ is the set of all configurations.

As mentioned in Chapter 1, such an emergent property is an essential characteristic of a complex system.

In the case of the elementary cellular automaton 184, the perfect order exists only if the density of active sites is exactly equal to $\frac{1}{2}$. If this is not the case, the limit set $\Lambda_{F_{184}}$, illustrated in Figure 6.2, is the set of all configurations consisting of finite sequences of alternating 0s and 1s separated by finite sequences of either 0s, if the density of active sites is less than $\frac{1}{2}$, or 1s, if the density of active sites is greater than $\frac{1}{2}$.

The evolution towards the limit set can be viewed as the elimination of *defects*, which are sequences of 0s or sequences of 1s. These defects can also be viewed as interacting *particlelike structures* evolving in a regular background.[1]

Fig. 6.2. *First 50 iterations of the elementary cellular automaton rule 184 starting from a random initial configuration. The density of active sites is exactly equal to 0.4 in the left figure and 0.6 in the right one. Time is oriented downwards. The number of lattice sites is equal to 50.*

6.2 Number-conserving cellular automata

Elementary cellular automaton 184 belongs to a special class of cellular automata. It is *number-conserving*; *i.e.*, it satisfies the condition

$$(\forall t \in \mathbb{N}) \qquad \sum_{i=1}^{L} s(i,t) = \text{constant},$$

where L is the cellular automaton size. It is not difficult to establish a necessary and sufficient condition for a one-dimensional $|Q|$-state n-input cellular automaton rule to be number-conserving [62]. An extension of these results to infinite lattices and higher dimensions can be found in Durand, Formenti, and Róka [112]. It can be shown that any one-dimensional cellular automaton can be simulated by a number-conserving cellular automaton; see Moreira [249].

---

[1] On interacting particlelike structures in spatiotemporal patterns representing the evolution of one-dimensional cellular automata, see Boccara, Nasser, and Roger [48].

Definition 22. *A one-dimensional* $|Q|$-*state n-input cellular automaton rule* f *is* number-conserving *if, for all cyclic configurations of length* $L \geq n$, *it satisfies*

$$f(x_1, x_2, \ldots, x_{n-1}, x_n) + f(x_2, x_3, \ldots, x_n, x_{n+1}) + \cdots$$
$$+ f(x_L, x_1 \ldots, x_{n-2}, x_{n-1}) = x_1 + x_2 + \cdots + x_L. \quad (6.4)$$

Theorem 13. *A one-dimensional* $|Q|$-*state n-input cellular automaton rule* f *is number-conserving if, and only if, for all* $(x_1, x_2, \ldots, x_n) \in Q^n$, *it satisfies*

$$f(x_1, x_2, \ldots, x_n) = x_1 + \sum_{k=1}^{n-1} \big(f(\underbrace{0, 0, \ldots, 0}_{k}, x_2, x_3, \ldots, x_{n-k+1})$$
$$- f(\underbrace{0, 0, \ldots, 0}_{k}, x_1, x_2, \ldots, x_{n-k}) \big). \quad (6.5)$$

To simplify the proof, we need the following lemma.

Lemma 1. *If* f *is a number-conserving rule, then*

$$f(0, 0, \ldots, 0) = 0. \quad (6.6)$$

Write Condition (6.4) for a cyclic configuration of length $L \geq n$ where all elements are equal to zero. □

To prove that Condition (6.5) is necessary, consider a cyclic configuration of length $L \geq 2n - 1$ that is the concatenation of a sequence $(x_1, x_2, \ldots, x_n)$ and a sequence of $L - n$ zeros, and express that the n-input rule f is number-conserving. We obtain

$$f(0, 0, \ldots, 0, x_1) + f(0, 0, \ldots, 0, x_1, x_2) + \cdots$$
$$+ f(x_1, x_2, \ldots, x_n) + f(x_2, x_3, \ldots, x_n, 0) + \cdots$$
$$+ f(x_n, 0, \ldots, 0) = x_1 + x_2 + \cdots + x_n, \quad (6.7)$$

where all the terms of the form $f(0, 0, \ldots, 0)$, which are equal to zero according to (6.6), have not been written. Replacing x_1 by 0 in (6.7) gives

$$f(0, 0, \ldots, 0, x_2) + \cdots + f(0, x_2, \ldots, x_n)$$
$$+ f(x_2, x_3, \ldots, x_n, 0) + \cdots + f(x_n, 0, \ldots, 0)$$
$$= x_2 + \cdots + x_n. \quad (6.8)$$

Subtracting (6.8) from (6.7) yields (6.5).

Condition (6.5) is obviously sufficient since, when summed on a cyclic configuration, all the left-hand side terms except the first cancel. □

Remark 10. The proof above shows that if we can verify that a cellular automaton rule f is number-conserving for all cyclic configurations of length $2n - 1$, then it is number-conserving for all cyclic configurations of length $L > 2n - 1$.

The following corollaries are simple necessary conditions for a cellular automaton rule to be number-conserving.

Corollary 1. *If f is a one-dimensional $|Q|$-state n-input number-conserving cellular automaton rule, then, for all $x \in Q$,*

$$f(x, x, \ldots, x) = x. \tag{6.9}$$

To prove (6.9), which is a generalization of (6.6), write Condition (6.5) for $x_1 = x_2 = \cdots = x_n = x$. $\square$

Corollary 2. *If f is a one-dimensional $|Q|$-state n-input number-conserving cellular automaton rule, then*

$$\sum_{(x_1, x_2, \ldots, x_n) \in Q^n} f(x_1, x_2, \ldots, x_n) = \frac{1}{2}(|Q| - 1)\,|Q|^n. \tag{6.10}$$

When we sum (6.5) over $(x_1, x_2, \ldots, x_n) \in Q^n$, all the left-hand-side terms except the first cancel, and the sum over the remaining terms is equal to $(0 + 1 + 2 + \cdots + (|Q| - 1))|Q|^{n-1} = \frac{1}{2}(|Q| - 1)\,|Q|^n$. $\square$

Number-conserving cellular automata can be used to model closed systems of interacting particles.

So far, we have only considered cellular automata evolving according to deterministic rules. In many applications it is often necessary to allow for probabilistic rules as defined on page 192.

Example 38. Highway car traffic. Vehicular traffic can be treated as a system of interacting particles driven far from equilibrium. The so-called *particle-hopping model* describes car traffic in terms of *probabilistic cellular automata*. The first model of this type was proposed by Nagel and Schreckenberg [257]. These authors consider a finite lattice of length L with periodic boundary conditions. Each cell is either empty (*i.e.*, in state e), or occupied by a car (*i.e.*, in state v), where $v = 0, 1, \ldots, v_{\max}$ denotes the car velocity (cars are moving to the right). If d_i is the distance between cars i and $i+1$, car velocities are updated in parallel according to the following subrules:

$$v_i(t + \tfrac{1}{2}) = \min(v_i(t) + 1, d_i(t) - 1, v_{\max}),$$

$$v_i(t + 1) = \begin{cases} \max(v_i(t + \tfrac{1}{2}) - 1, 0), & \text{with probability } p, \\ v_i(t + \tfrac{1}{2}), & \text{with probability } 1 - p, \end{cases}$$

where $v_i(t)$ is the velocity of car i at time t. Then, if $x_i(t)$ is the position of car i at time t, cars are moving according to the rule

$$x_i(t + 1) = x_i(t) + v_i(t + 1).$$

That is, at each time step, each car increases its speed by one unit (acceleration $a = 1$), respecting the safety distance and the speed limit. But, the model also includes some noise: with a probability p, each car decreases its speed by one unit.

Although rather simple, the model exhibits features observed in real highway traffic (*e.g.*, with increasing vehicle density, it shows a transition from laminar traffic flow to start-stop waves, as illustrated in Figure 6.3).

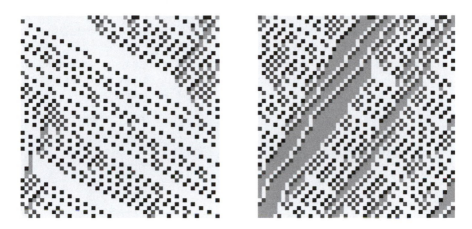

Fig. 6.3. *First 50 iterations of the Nagel-Schreckenberg probabilistic cellular automaton traffic flow model. The initial configuration is random with a density equal to 0.24 in the left figure and 0.48 in the right one. In both cases $v_{\max} = 2$ and $p = 0.2$. The number of lattice sites is equal to 50. Empty cells are very light grey while cells occupied by a car with velocity v equal to 0, 1, and 2 have darker shades of grey. Time increases downwards.*

When discussing road configurations evolving according to various illustrative rules, the knowledge of both car positions and velocities proves necessary. Therefore, although we are dealing with two-state cellular automaton rules, we shall not represent the state of a cell by its occupation number (*i.e.*, 0 or 1), but by a letter in the alphabet $\{e, 0, 1, \ldots, v_{\max}\}$ indicating that the cell is either empty (*i.e.*, in state e) or occupied by a car with a velocity equal to v (*i.e.*, in state $v \in \{0, 1, \ldots, v_{\max}\}$). Configurations of cells of this type will be called *velocity configurations*, or configurations for short.

When using velocity configurations, we shall not represent cellular automaton rules by their rule tables but make use of a representation that clearly exhibits the particle motion. This *motion representation*, or *velocity rule*, introduced by Boccara and Fukś [59], may be defined as follows. If the integer r is the rule's radius, list all the $(2r + 1)$-neighborhoods of a given particle represented by a 1 located at the central site of the neighborhood. Then, to each neighborhood, associate an integer v denoting the velocity of this particle, which is the number of sites this particle will move in one time step, with the convention that v is positive if the particle moves to the right and negative if it moves to the left. For example, rule 184 is represented by the radius 1 velocity rule:

$$\bullet 11 \to 0, \quad \bullet 10 \to 1. \tag{6.11}$$

The symbol $\bullet$ represents either a 0 or a 1 (*i.e.*, either an empty or occupied site). This representation, which clearly shows that a car can move to the next neighboring site on its right if, and only if, this site is empty, is shorter and more explicit.

Elementary cellular automaton rule 184 is a particular Nagel-Schreckenberg highway traffic flow model in which the probability p of decelerating is equal to zero and the maximum speed $v_{\max} = 1$. In this particular case, Fukś [133] has been able to determine the exact expression of the average car velocity $\langle v \rangle (\rho, t)$ as a function of car density ρ and time t. He found

$$\langle v \rangle (\rho, t) = \begin{cases} 1 - \dfrac{(4\rho(1-\rho))^t}{\sqrt{\pi t}}, & \text{if } \rho < \frac{1}{2}, \\ \dfrac{1-\rho}{\rho} \left(1 - \dfrac{(4\rho(1-\rho))^t}{\sqrt{\pi t}} \right), & \text{otherwise.} \end{cases}$$

In the limit $t \to \infty$, this yields

$$\langle v \rangle (\rho, \infty) = \begin{cases} 1, & \text{if } \rho < \frac{1}{2}, \\ \dfrac{1-\rho}{\rho}, & \text{otherwise.} \end{cases}$$

This deterministic model exhibits a second-order phase transition at $\rho = \frac{1}{2}$ between a *free-moving* phase and a *jammed* phase (see Figure 6.2). This phase transition may be characterized by the order parameter

$$m(\rho) = 1 - \langle v \rangle (\rho, \infty),$$

which is equal to zero in the free-moving phase.[2] The control parameter, whose role is usually played by the temperature in statistical physics, is here played by the car density.

It is straightforward to generalize the traffic flow model described by rule 184 to higher maximum velocities (see Figure 6.4). If, as before, d_i is the distance between car i and car $i + 1$, car velocities are updated in parallel according to the rule[3]

$$v_i(t+1) = \min(d_i(t) - 1, v_{\max}),$$

and car positions are updated according to the rule

---

[2] In phase transition theory, the order parameter is equal to zero in the phase of higher symmetry, its nonzero value characterizes the broken symmetry of the *ordered* phase.

[3] A similar model has also been studied by Fukui and Ishibashi [134].

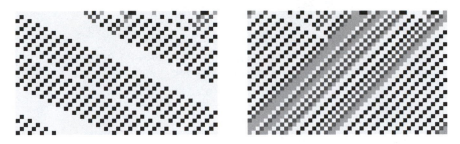

Fig. 6.4. *First 30 iterations of the generalized deterministic rule 184 traffic flow model. The initial configuration is random with a density equal to 0.24 in the left figure and 0.48 in the right one. In both cases $v_{\max} = 2$. The number of lattice sites is equal to 50. Empty cells are very light grey while cells occupied by a car with velocity v equal to 0, 1, and 2 have darker shades of grey. Time increases downwards.*

$$x_i(t+1) = x_i(t) + v_i(t+1).$$

In the steady state, the average car velocity $\langle v \rangle(\rho, \infty)$ is given, as a function of the car density ρ, by

$$\langle v \rangle(\rho, \infty) = \begin{cases} v_{\max}, & \text{if } \rho < \rho_c = \dfrac{1}{1 + v_{\max}}, \\ \dfrac{1 - \rho}{\rho}, & \text{otherwise}, \end{cases} \tag{6.12}$$

and the order parameter characterizing the second-order phase transition between the free-moving phase and the jammed phase is

$$m(\rho) = v_{\max} - \langle v \rangle(\rho, \infty), \tag{6.13}$$

or

$$m(\rho) = \begin{cases} 0, & \text{if } \rho < \rho_c, \\ \dfrac{\rho - \rho_c}{\rho \rho_c}, & \text{otherwise}. \end{cases}$$

In the jammed phase close to the critical car density ρ_c, the critical behavior of the order parameter is characterized by the exponent $\beta = 1$ defined by $m(\rho) \sim (\rho - \rho_c)^\beta$.

While the value of the critical exponent β can be found exactly, this is not the case for the other critical exponents. The exponents γ and δ have been determined by Boccara and Fukś [60]. They characterize, respectively, the critical behavior of the susceptibility and the response to the symmetry-breaking field at the critical point.

If $\rho < \rho_c$, any configuration in the limit set consists of *perfect tiles* of $v_{\max} + 1$ cells, as shown below.[4]

---

[4] Note that, in what follows, v is the velocity with which the car is going to move at the next time step.

$v_{\max}$	e	$\cdots$	e	e

in a sea of cells in state e (see Figure 6.4).

If $\rho > \rho_c$, a configuration belonging to the limit set only consists of a mixture of tiles containing $v + 1$ cells of the type

v	e	$\cdots$	e	e

where $v = 0, 1, \cdots, v_{\max}$.

If we introduce random braking, then, even at low density, some tiles become defective, which causes the average velocity to be less than $v_{\max}$. The random-braking probability p can, therefore, be viewed as a symmetry-breaking field conjugate to the order parameter m defined by (6.13). This point of view implies that the phase transition characterized by m will be smeared out in the presence of random braking as in ferromagnetic systems placed in a magnetic field.

The susceptibility, defined as

$$\chi(\rho) = \lim_{p \to 0} \frac{\partial m}{\partial p},$$

diverges, in the vicinity of ρ_c, as $(\rho_c - \rho)^{-\gamma}$ for $\rho < \rho_c$ and as $(\rho - \rho_c)^{-\gamma'}$ for $\rho > \rho_c$, and at $\rho = \rho_c$, the response function

$$\lim_{p \to 0} \frac{m(\rho_c, 0) - m(\rho_c, p)}{p}$$

goes to zero as $p^{1/\delta}$. For the model corresponding to $v_{\max} = 2$,[5] using numerical simulations and a systematic approximation technique, described in Example 42, Boccara and Fukś found $\gamma = \gamma' \approx 1$ and $\delta \approx 2$.

It is interesting to note that these values are found in equilibrium statistical physics in the case of second-order phase transitions characterized by nonnegative order parameters above the upper critical space dimension.

Close to the phase transition point, critical exponents obey scaling relations. If we assume that, in the vicinity of the critical point $(\rho = \rho_c, p = 0)$, the order parameter m is a generalized homogeneous function of $\rho - \rho_c$ and p of the form

$$m = |\rho - \rho_c|^\beta f\left(\frac{p}{|\rho - \rho_c|^{\beta\delta}}\right), \tag{6.14}$$

where the function f is such that $f(0) \neq 0$, then, differentiating f with respect to p and taking the limit $p \to 0$, we readily obtain

$$\gamma = \gamma' = (\delta - 1)\beta. \tag{6.15}$$

---

[5] For $v_{\max} = 2$, the cellular automaton local rule must have at least a left radius $r_\ell = 2$ and right radius $r_r = 1$. Using Wolfram codification, the corresponding 4-input rule code number is 43944.

These relations, verified by the numerical values, confirm the existence of a universal scaling function, which has also been checked directly [60].

Boccara [61] has shown that this highway traffic flow model satisfies, with other deterministic traffic flow models, a variational principle. For the sake of simplicity, in our discussion of this variational principle, it is sufficient to consider the case $v_{\max} = 2$; that is, the radius 2 velocity rule

$$\bullet\bullet\, 11\bullet \to 0, \quad \bullet\bullet\, 101 \to 1, \quad \bullet\bullet\, 100 \to 2. \tag{6.16}$$

According to velocity rule (6.16) and our choice of configuration representation, a cell occupied by a car with velocity v must be preceded by at least v empty cells. Each configuration is therefore a concatenation of the following four types of tiles:

$$\boxed{2}\ \boxed{e}\ \boxed{e}$$

$$\boxed{1}\ \boxed{e}\qquad \boxed{0}\qquad \boxed{e}$$

The first tile, which corresponds to cars moving at the speed limit $v_{\max} = 2$, is a *perfect tile*, the next two tiles, corresponding to cars with a velocity less than 2 (here 1 and 0), are *defective tiles*, and the last tile is a *free empty cell*; that is, an empty cell that is not part of either a perfect or a defective tile.

If $\rho < \rho_c = \frac{1}{3}$ (left part of Figure 6.4), only the first configurations contain defective tiles. After a few time steps, these tiles progressively disappear, and the last configurations contain only perfect tiles and free empty cells. Hence, all cars move at $v_{\max} = 2$, and the system is in the free-moving phase. If $\rho > \rho_c$ (right part of Figure 6.4), at the beginning, the same process of annihilation of defective tiles takes place, but, in this case, all defective tiles do not eventually disappear. A few cars move at $v_{\max}$, while other cars have either a reduced speed ($v = 1$) or are stopped ($v = 0$). The system is in the jammed phase.

To analyze the annihilation process of defective tiles in generalized rule 184 models, we need to define what we call a *local jam*.

Definition 23. *In deterministic generalized rule 184 models of traffic flow, a local jam is a sequence of defective tiles preceded by a perfect tile and followed by either a perfect tile or free empty cells.*

From this definition, it follows that:

Theorem 14. *In the case of deterministic generalized rule 184 models of traffic flow, the number of cars that belong to a local jam is a nonincreasing function of time.*

This result is a direct consequence of the fact that, by definition, a local jam is preceded by a car that is free to move, and, according to whether a new car joins the local jam from behind or not, the number of cars in the local jam remains unchanged or decreases by one unit. Note that *the jammed car just behind the free-moving car leading the local jam itself becomes free to move at the next time step.* ☐

In order to establish the variational principle for any value of $v_{\max}$, we will first prove the following lemma.

Lemma 2. *In the case of deterministic generalized rule 184 models of traffic flow, the number of free empty cells is a nonincreasing function of time.*

Let us analyze how the structure of the most general local jam changes in one time step. A local jam consisting of n defective tiles is represented below:

$$\cdots v_1 \underbrace{ee \cdots e}_{v_1} v_2 \underbrace{ee \cdots e}_{v_2} \cdots \cdots v_n \underbrace{ee \cdots e}_{v_n} v_{\max} \underbrace{ee \cdots e}_{v_{\max}} \cdots ,$$

where, for $i = 1, 2, \ldots, n$, $0 \le v_i < v_{\max}$. At the next time step, a car with velocity v_i located in cell k moves to cell $k + v_i$. Hence, if the local jam is followed by v_0 free empty cells, where $v_0 \ge 0$, we have to distinguish two cases:

1. If $v_0 + v_1 < v_{\max}$, then the number of jammed cars remains unchanged but the leftmost jammed car, whose velocity was v_1, is replaced by a jammed car whose velocity is $v_0 + v_1$:

   $$\cdots v_{\max} \overbrace{ee \cdots e}^{v_{\max} + v_0} v_1 ee \cdots e v_2 ee \cdots \ e \cdots \cdots v_n ee \cdots e v_{\max} ee \cdots e \cdots$$

   $$\cdots e (v_0 + v_1) ee \cdots e v_2 ee \cdots e \cdots \cdots v_n ee \cdots e v_{\max} ee \cdots e \cdots$$

2. If $v_0 + v_1 \ge v_{\max}$, then the local jam loses its leftmost jammed car:

   $$\cdots e v_1 ee \cdots e v_2 ee \cdots e \cdots \cdots v_n ee \cdots e v_{\max} ee \cdots e \cdots$$

   $$\cdots v_{\max} \underbrace{ee \cdots e}_{v_{\max} + v_0'} v_2 ee \cdots e \cdots \cdots v_n ee \cdots e v_{\max} ee \cdots e \cdots$$

   and, at the next time step, the local jam is followed by $v_0 + v_1 - v_{\max} = v_0' < v_0$ free empty cells.

If we partition the lattice in tile sequences whose endpoints are perfect tiles, then, between two consecutive perfect tiles, either there is a local jam, and the proof above shows that the number of free empty cells between the two perfect tiles cannot increase, or there is no local jam, and the number of free empty cells between the two perfect tiles remains unchanged. ☐

Remark 11. If a configuration contains no perfect tiles, then it does not contain free empty cells. Such a configuration belongs, therefore, to the limit set, and as we shall see below, the system is in its steady state. On a circular highway, if a configuration contains only one perfect tile, the reasoning above applies without modification.

Remark 12. The proof above shows that, at each time step, the rightmost jammed car of a local jam moves one site to the left. Local jams can only move backwards.

We can now prove the following variational principle:

Theorem 15. *In the case of deterministic generalized rule 184 models of traffic flow, for a given car density ρ, the average car flow is a nondecreasing function of time and reaches its maximum value in the steady state.*

If L is the lattice length, N the number of cars, and $N_{\text{fec}}(t)$ the number of free empty cells at time t, we have

$$N_{\text{fec}}(t) = L - N - \sum_{i=1}^{N} v_i(t)$$

since, at time t, car i is necessarily preceded by $v_i(t)$ empty cells. Dividing by L we obtain

$$\frac{N_{\text{fec}}(t)}{L} = 1 - \frac{N}{L} - \frac{N}{L} \frac{1}{N} \sum_{i=1}^{N} v_i(t).$$

Hence, for all t,

$$\rho_{\text{fec}}(t) = 1 - \rho - \rho \langle v \rangle (\rho, t), \tag{6.17}$$

where $\rho_{\text{fec}}(t)$ is the density of free empty cells at time t and $\langle v \rangle (\rho, t)$ the average car velocity at time t for a car density equal to ρ. This last result shows that, when time increases, since the density of free empty cells cannot increase, the average car flow $\rho \langle v \rangle (\rho, t)$ cannot decrease.   □

The annihilation process of defective tiles stops when there are either no more defective tiles or no more free empty cells. Since a perfect tile consists of $v_{\max} + 1$ cells, if the car density ρ is less than the critical density $\rho_c = 1/(v_{\max} + 1)$, there exist enough free empty cells to annihilate all the defective tiles, and all cars eventually become free to move. If $\rho > \rho_c$, there are not enough free empty cells to annihilate all defective tiles, and eventually some cars are not free to move at $v_{\max}$. At the end of the annihilation process, all subsequent configurations belong to the limit set, and the system is *in equilibrium* or *in the steady state*.

Note that the existence of a free-moving phase for a car density less than $\rho_c = 1/(v_{\max} + 1)$ can be seen as a consequence of Relation (6.17). When t goes to infinity, according to whether ρ_{fec} is positive or zero, this relation implies

$$\langle v \rangle (\rho, \infty) = \min \left(\frac{1 - \rho}{\rho}, v_{\max} \right).$$

If the system is finite, its state eventually becomes periodic in time, and the period is equal to the lattice size or one of its submultiples.

Since local jams move backwards and free empty cells move forwards, equilibrium is reached after a number of time steps proportional to the lattice size.

Remark 13. In the case of rule 184, the existence of a free-moving phase for a particle density ρ less than the critical density $\rho_c = \frac{1}{2}$ obviously implies that the average velocity $\langle v \rangle(\rho, t)$, as a function of time t, is maximum when $t \to \infty$. For $\rho > \frac{1}{2}$, this property is still true since the dynamics of the holes (empty sites) is governed by the conjugate of rule 184 (i.e., rule 226),[6] which describes exactly the same dynamics as Rule 184 but for holes moving to the left. Therefore, for all values of the particle density, the average velocity takes its maximum value in the steady state.

Recently, many papers on the application of statistical physics to vehicular traffic have been published. For a review see, for example, Chowdhury, Santen, and Schadschneider [87].

Example 39. Pedestrian traffic. While considerable attention has been paid to the study of car traffic since the early 1990s, comparatively very few cellular automaton models of pedestrian dynamics have been published. In contrast with cars, pedestrians can accelerate and brake in a very short time, and their average velocity is sharply peaked.

Pedestrian traffic models can provide valuable tools to plan and design a variety of pedestrian areas such as shopping malls, railway stations, or airport terminals.

One of the first pedestrian traffic models is due to Fukui and Ishibashi [135, 136], who studied pedestrians in a passageway moving in both directions. The passageway is represented as a square lattice of length L and width $W < L$ with periodic boundary conditions in which each cell is either occupied by one pedestrian or empty. Pedestrians are divided in two groups walking in opposite directions. Eastbound pedestrians may move only at odd time steps, while westbound pedestrians may move only at even time steps. At each odd time step, an eastbound pedestrian moves to his right-nearest cell when it is empty. If this cell is occupied by another eastbound pedestrian, he does not move, but if it is occupied by a westbound pedestrian, he tries to change lanes. At each even time step, a westbound pedestrian moves to his left-nearest cell when it is empty. If this cell is occupied by another westbound pedestrian, he does not move, but if it is occupied by an eastbound pedestrian, he tries to change lanes. When a pedestrian cannot move forward, he tries to change lanes according to the following rules:

1. In the first model, called the *sidestepping model,*
    (i) if both adjacent cells are empty, he moves to one of them chosen at random giving precedence, however, to pedestrians moving in the same direction without changing lanes,
    (ii) if only one cell is empty, he moves to this cell, here again giving

---

[6] If f is an n-input two-state deterministic cellular automaton rule, its conjugate, denoted Cf, is defined by

$$Cf(x_1, x_2, \cdots, x_n) = 1 - f(1 - x_1, 1 - x_2, \cdots, 1 - x_n).$$

precedence to a pedestrian moving without changing lanes, and
(iii) if none is empty, he does not move.
2. In a second model, called the *diagonal-stepping* model, instead of trying
to move to an adjacent cell, the pedestrian tries to move to one of the
cells in front of the adjacent cells—in the direction of motion—applying
the same rule as in the first model.

In their first paper, Fukui and Ishibashi study the problem of one westbound pedestrian walking across the passageway among a fixed density ρ of
eastbound pedestrians.

In the case of the sidestepping model, they find that, for $\rho \leq \rho_c = (W -
1)/2W$, the average flow of eastbound pedestrians grows linearly with ρ, for
$\rho_c \leq \rho \leq 1 - \rho_c$, the flow remains constant, and, for $1 - \rho_c < \rho \leq 1$, the flow
decreases linearly with ρ and becomes equal to zero for $\rho = 1$. In the limit
$W \to \infty$, as it should be, $\rho_c = \frac{1}{2}$.[7] Phase transitions at $\rho = \rho_c$ and $\rho = 1 - \rho_c$
are second order.

In the case of the diagonal-stepping model, for $\rho \leq 1 - \rho_c$, the average flow
of eastbound pedestrians grows linearly with ρ and then decreases linearly to
reach zero for $\rho > 1 - \rho_c$. At $\rho = 1 - \rho_c$ the phase transition is second order.

In their second paper, Fukui and Ishibashi consider an equal number of
westbound and eastbound pedestrians walking along a passageway who avoid
collisions trying first to move to one of the cells in front of the adjacent
cells, or, if these cells are occupied, trying to move to one of the adjacent
cells. Numerical simulations seem to indicate that, as a function of pedestrian
density, the system exhibits a first-order phase transition.

Adopting a slightly different point of view, it is possible to build up a much
simpler purely deterministic bidirectional pedestrian traffic model. Consider
a square lattice of length L and width $W < L$ with periodic boundary conditions in which N_w and N_e cells are, respectively, occupied by westbound
and eastbound pedestrians, with $N_w + N_e < LW$. As in the Fukui–Ishibashi
models, a pedestrian moves forward to the cell in front of him if it is empty.
If this cell is occupied by another pedestrian moving in the same direction,
the pedestrian does not move, but, if it is occupied by a pedestrian moving
in the opposite direction, the pedestrian moves to the cell in front of his right
adjacent cell (with respect to the moving direction), and if this cell is also
occupied, he moves to his right adjacent cell. If both cells are occupied, the
pedestrian does not move. In all cases, pedestrians who can move forward have
the right of way. Eastbound (resp. westbound) pedestrians move at odd (resp.
even) time steps. As a result of the lane-changing rule, to avoid collisions, the
local walking rule, which is of the form

$$s(i, j, t+1) = f \begin{pmatrix} s(i+1, j-1, t) & s(i+1, j, t) & s(i+1, j+1, t) \\ s(i, j-1, t) & s(i, j, t) & s(i, j+1, t) \\ s(i-1, j-1, t) & s(i-1, j, t) & s(i-1, j+1, t) \end{pmatrix},$$

---

[7] Refer to car traffic rule 184 and see Figure 6.1.

where $i \in \{1, L\}$ and $j \in \{1, W\}$, depends on a lesser number of variables and is not probabilistic. Figure 6.5 shows an example of spontaneous lane formation of pedestrians moving in the same direction.

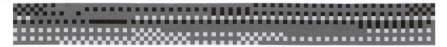

Fig. 6.5. *Multilane bidirectional pedestrian traffic. Lattice length: 100; lattice width: 10; number of pedestrians: 150 moving to the left (light grey cells) and 150 to the right (black cells); number of iterations: 200. Empty cells are grey.*

A few other lattice models of pedestrian traffic have been studied [252, 253, 254]. In the first paper, Muramatsu, Irie, and Nagatani consider a square lattice of length L and width $W < L$ whose sites are occupied by two types of biased random walkers. That is, eastbound walkers cannot move west and westbound walkers cannot move east, but both types may move either north or south. For each walker, the respective probabilities of moving to any of the three authorized neighboring sites depend upon the occupancy of these sites. For instance, if all three sites are empty, a walker will move forward with probability $D + (1 - D)/3$ and sideways with probability $(1 - D)/3$; if one lateral site is occupied, then the walker will move forward with probability $D + (1 - D)/2$ and laterally with probability $(1 - D)/2$. The probabilistic walking rules depend, therefore, upon a unique parameter D called the *drift strength*. Lateral walls are reflecting, and the right and left boundaries are open. A constant flow of walkers enters the system at each of these open boundaries. The walking rules are applied sequentially. In the case of a passageway, numerical simulations seem to indicate the existence of a first-order jamming phase transition depending upon D and W.

A promising model has been recently proposed by Burstedde, Klauck, Schadschneider, and Zittartz [75]. Pedestrians move on a square lattice with at most one pedestrian per cell. A 3×3 matrix M is associated with each pedestrian. The nine elements of this matrix give the respective probabilities for a pedestrian to move to one of the eight neighboring sites. The central element refers to the probability of not moving. In the simple problem of bidirectional traffic in a passageway, we only need two different types of matrices, one for pedestrians moving east and the other for pedestrians moving west. More complex problems could be handled using many more different types of matrices. One of the key features of this model is the existence of a *floor field*, which is a substitute for the long-range interactions necessary to take into account the geometry of a building, such as the existence of emergency exits. Actually, the authors introduce two floor fields: the *static* floor field matrix S, which does not depend upon time and the presence of pedestrians, is used to specify regions in space such as doors, in contrast with the *dynamic* floor field matrix D, which is modified by the presence of pedestrians and has its own dynamics and that is used to represent, for instance, the trace left by pedestrian footsteps. The expression of the probability for a pedestrian to move from site i to site j is then given by an expression of the form

$$p_{ij} = AM_{ij}S_{ij}D_{ij}(1 - n_{ij}),$$

where A is a normalization constant to ensure that, for all i, $\sum_{j=1}^{9} p_{ij} = 1$ and n_{ij} the occupation number of the neighboring cell j of cell i. The system is updated synchronously.

This model has been applied [188] to study a variety of situations such as the evacuation of a large room through one or two doors, the evacuation of a lecture hall, the formation of lanes in a long corridor, etc.

Remark 14. Cars and pedestrians are *self-propelled particles* that, as a result of local interactions, exhibit self-organized collective motion. There exist many other systems that display similar collective behaviors, such as, for instance, flocks of birds, schools of fish, or herds of quadrupeds. Cellular automata have rarely been used to model these systems. Researchers have mostly used so-called *off-lattice* models. Simple two-dimensional models, closely related to the XY model of ferromagnetism,[8] that exhibit *flocking behavior* have been proposed by different authors [331, 339]. Essentially, these models consist of a system of particles located in a plane, driven with a constant absolute velocity, whose direction is, at each time step, given by the average direction of motion of the particles in their neighborhood with the addition of some random perturbation. Such models undergo a second-order phase transition characterized by an order parameter breaking the symmetry of the two-dimensional rotational invariance of the average velocity of the particles. Extensive numerical simulations have been performed [98] on systems of 10^4 to 10^5 particles in a 100×100 plane with periodic boundary conditions.[9] Typically, the authors run simulations of 10^5 time steps. If the noise is uniformly distributed in the interval $[-\frac{1}{2}\eta, \frac{1}{2}\eta]$, the average velocity of the particles (the order parameter) behaves as $(\eta_c - \eta)^\beta$ with $\beta = 0.42 \pm 0.03$, where η_c is the critical noise amplitude. For a fixed lattice size, η_c behaves as a function of the particle density ρ as ρ^κ with $\kappa = 0.45 \pm 0.05$. When the absolute value of the particle velocity v_0 is equal to zero, this model coincides with a diluted XY model and does not exhibit long-range order. For a nonvanishing v_0, the model belongs to a new universality class of dynamic, nonequilibrium models.

6.3 Approximate methods

If we assume that the finite set of states Q is equal to $\{0, 1\}$, the simplest statistical quantity characterizing a cellular automaton configuration is the

---

[8] The XY model is a system of two-dimensional spins (or classical vectors) located at the sites of a d-dimensional lattice. Its Hamiltonian is $H = -J\sum_{ij} \mathbf{S}_i \cdot \mathbf{S}_j$, where the sum runs over all first-neighboring pairs of spins, and J is a positive or negative constant. According to the Mermin-Wagner theorem [240], such a system does not exhibit long-range order for $d \leq 2$.

[9] Particles are located anywhere in the plane, in particular, a lattice cell may contain more than one particle.

average density of active (nonzero) sites denoted by ρ. If the initial configuration is random (*i.e.*, if the states of the different cells are uncorrelated),
the iterative application of the cellular automaton rule introduces correlations
between these states.

Starting from an initial random configuration with a density of active sites
equal to $\rho(0)$, the simplest approximate method to determine the density as
a function of time t is the *mean-field approximation*. This method neglects
correlations, introduced by the application of the cellular automaton rule,
between the states of the cells. This approximation gives a simple expression
of the form

$$\rho(t+1) = f_{MFA}\big(\rho(t)\big),$$

where f_{MFA} is a map derived from the look-up table of the local rule f of the
cellular automaton assuming that the probability for a cell to be in state 1 at
time t is equal to $\rho(t)$.

Example 40. Elementary cellular automaton rule 184. Since the preimages of
1 by rule f_{184} are

$$111, \quad 101, \quad 100, \quad 011,$$

the probability $\rho(t+1)$ for a cell to be active at time $t+1$ is the sum of the
following terms:

$$\rho(t)^3, \quad 2\rho(t)^2(1-\rho(t)), \quad \rho(t)(1-\rho(t))^2.$$

The evolution of ρ described by the mean-field approximation for rule f_{184} is,
therefore, determined by the map

$$\rho \mapsto \rho^3 + 2\rho^2(1-\rho) + \rho(1-\rho)^2.$$

Taking into account the identity $\big(\rho + (1-\rho)\big)^2 = 1$, this yields

$$\rho(t+1) = \rho(t), \tag{6.18}$$

which is exact since rule 184 is number-conserving. It can be shown that, for
all number-conserving cellular automaton rules, the mean-field approximation
always gives Relation (6.18). This, obviously, does not mean that, for this set
of rules, the mean-field approximation is exact.

Example 41. Elementary cellular automaton rule 18. Since $18_{10} = 00010010_2$,
the look-up table for rule 18 is

111	110	101	100	011	010	001	000
0	0	0	1	0	0	1	0

The preimages of 1 by rule f_{18} being

$$100 \quad \text{and} \quad 001,$$

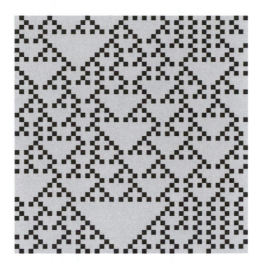

Fig. 6.6. *Limit set of the elementary cellular automaton rule 18. One hundred successive iterates of an initial random configuration have been discarded. The following 50 iterations are represented. Time is oriented downwards. Number of lattice sites: 50.*

the mean-field map is

$$\rho \mapsto 2(1 - \rho)^2.$$

Within the framework of the approximation, in the steady state, the density of active sites ρ_∞ is thus given by the solution to the equation

$$\rho_\infty = 2(1 - \rho_\infty)^2.$$

That is, $\rho_\infty = 1 - 1/\sqrt{2} = 0.292893\ldots$. This result is not exact. It can be shown [47] that the configurations in the limit set consist of sequences of 0s of odd lengths separated by isolated 1s (see Figure 6.6). More precisely, the structure of the configurations belonging to the limit set results from random concatenations of the 2-blocks 00 and 01 with equal probability. The exact density of active sites of such configurations is thus equal to $\frac{1}{4}$.

In order to build up better approximate methods than the mean field, we have to take explicitly into account the existence of correlations between the states of the cells. The *local structure theory* [162] is a systematic generalization of the mean-field approximation. In what follows, as a simplification, the theory is presented assuming that the set of cell states Q is $\{0, 1\}$.

An n-block B_n is a sequence $s_1 s_2 \ldots s_n$ of length n, where for $i = 1, 2, \ldots$, $s_i \in Q$. If $\mathcal{B}_n$ is the set of all n-blocks, an n-input rule f is a map from $\mathcal{B}_n$ into Q. The *truncation operators* L and R map any n-block B_n to an $(n-1)$-block by truncation from the left or right, respectively. That is, if $B_n = s_1 s_2 \ldots s_n$,

$$LB_n = s_2 s_3 \ldots s_n \quad \text{and} \quad RB_n = s_1 s_2 \ldots s_{n-1}.$$

L and R commute. Applied to a 1-block, L and R yield the null-block.

A map P_n from the set of all blocks of length less than or equal to n into $[0, 1]$ is a *block probability distribution of order* n if it satisfies the following self-consistency conditions:

1. For $i = 0, 1, 2, \ldots, n$, $P_n(B_i) \geq 0$.[10]
2. For $i = 0, 1, 2, \ldots, n$,

$$\sum_{B_i \in \mathcal{B}_i} P_n(B_i) = 1.$$

3. For all blocks B' of length $n - 1$,

$$P_n(B') = \sum_{B \text{ st } LB=B'} P_n(B).$$

4. For all blocks B' of length $n - 1$,

$$P_n(B') = \sum_{B \text{ st } RB=B'} P_n(B).$$

In $B \text{ st } LB = B'$ and $B \text{ st } RB = B'$, "st" stands for "such that."

Satisfaction of these last two conditions together implies that the probability of any block in the domain of P_n is the sum of the probabilities of blocks that contain the given block at a particular position.

For example, considering all 3-blocks

$$\mathcal{B}_3 = \{000, 001, 010, 011, 100, 101, 110, 111\},$$

the probability distribution P_3 must satisfy the conditions

$$P_3(00) = P_3(000) + P_3(001) = P_3(000) + P_3(100),$$
$$P_3(01) = P_3(010) + P_3(011) = P_3(001) + P_3(101),$$
$$P_3(10) = P_3(100) + P_3(101) = P_3(010) + P_3(110),$$
$$P_3(11) = P_3(110) + P_3(111) = P_3(011) + P_3(111).$$

These four constraints on P_3 imply that only four (2^2) parameters are needed to describe P_3 instead of eight (2^3). In general, for P_n, 2^{n-1} parameters are needed. Note that the conditions

$$P_3(0) = P_3(00) + P_3(01) = P_3(00) + P_3(10),$$
$$P_3(1) = P_3(10) + P_3(11) = P_3(01) + P_3(11),$$

do not further restrict the number of parameters.

The problem now is to construct block probability distributions satisfying the consistency relations. Given a block probability distribution P_n, we want to define a process to generate block probability distributions P_m for $m > n$.

---

[10] Since there is only one null-block, its probability is equal to 1.

It is reasonable to assume that, although the iterative application of a cellular automaton rule produces configurations in which nearby cells are correlated, these correlations die away with increasing cell separation. That is, for a block B that is long enough, the conditional probability of finding the block B augmented by a cell in state s on the right does not significantly depend upon the state of the left-most cell of B. Then

$$\frac{P(Bs)}{P(B)} = \frac{P(LBs)}{P(LB)}$$

or

$$P(Bs) = \frac{P(LBs)P(B)}{P(LB)}.$$

This last formula should be symmetric between the addition of cells to the right or left of B; changing Bs in B, B in RB, and LB in LRB, it can be written under the symmetric form

$$P(B) = \frac{P(LB)P(RB)}{P(LRB)}.$$

We shall use this relation to define an operator ϖ mapping an n-block probability distribution P_n to an $(n+1)$ probability distribution $\varpi(P_n)$, which gives the probability of $(n+1)$-blocks B by

$$\varpi\big(P_n(B)\big) = \frac{P_n(LB)P_n(RB)}{P_n(LRB)}. \tag{6.19}$$

Since $\varpi(P_n) = P_n$ for blocks of length n or less, we have to verify that the consistency relations are satisfied for blocks of length $n+1$.

Let B' and B be blocks of length n and $n+1$, respectively; then

$$\sum_{B \text{ st } LB=B'} \varpi(P_n)(B) = \sum_{B \text{ st } LB=B'} \frac{P_n(LB)P_n(RB)}{P_n(LRB)}$$

$$= \sum_{B \text{ st } LB=B'} \frac{P_n(B')P_n(RB)}{P_n(RB')}$$

$$= \frac{P_n(B')}{P_n(RB')} \sum_{B \text{ st } LB=B'} P_n(RB).$$

If $B'' = RB$, then $LB = B'$ implies $LRB = RB' = LB''$, and the relation

$$\sum_{B'' \text{ st } LB''=RB'} P_n(B'') = P_n(RB')$$

shows that

$$\sum_{B \text{ st } LB=B'} P_n(RB) = P_n(RB'),$$

which yields

$$\sum_{B \text{ st } LB=B'} \varpi(P_n)(B) = P_n(B').$$

Replacing the operator L by R, we also obtain

$$\sum_{B \text{ st } RB=B''} \varpi(P_n)(B) = P_n(B'').$$

Moreover,

$$\sum_{B \in \mathcal{B}_{n+1}} \varpi(P_n)(B) = \sum_{B' \in \mathcal{B}_n} \sum_{B \text{ st } LB=B'} \frac{P_n(RB)P_n(LB)}{P_n(LRB)}$$

$$= \sum_{B' \in \mathcal{B}_n} P_n(B')$$

$$= 1.$$

Example 42. Traffic flow models. In Example 38, we studied a cellular automaton model of traffic flow generalizing rule 184 for a speed limit $v_{\max} = 2$. Here we show how to apply the local structure theory to this model.

To construct a local structure approximation, it is more convenient to represent configurations of cars as binary sequences, where 0s represent empty spaces and 1s represent cars. Since for $v_{\max} = 2$ the speed of a car is determined by the states of, at most, two sites in front of it,[11] the minimal block size to obtain nontrivial results is 3 (the site occupied by a car plus two sites in front of it). In what follows, we limit our attention to order-three local structure approximation.

Using 3-block probabilities, we can write a set of equations describing the time evolution of these probabilities,

$$P_{t+1}(\sigma_2\sigma_3\sigma_4)$$

$$= \sum_{\substack{s_i \in \{0,1\} \\ i=0,1,\ldots,6}} w(\sigma_2\sigma_3\sigma_4 \mid s_0s_1s_2s_3s_4s_5s_6)P_t(s_0s_1s_2s_3s_4s_5s_6), \qquad (6.20)$$

where $P_t(\sigma_2\sigma_3\sigma_4)$ is the probability of block $\sigma_2\sigma_3\sigma_4$ at time t, and $w(\sigma_2\sigma_3\sigma_4 \mid s_0s_1s_2s_3s_4s_5s_6)$ is the conditional probability that the rule maps the seven-block $s_0s_1s_2s_3s_4s_5s_6$ into the three-block $\sigma_2\sigma_3\sigma_4$. States of lattice sites at time t are represented by s variables, while σ variables represent states at time $t + 1$, so that, for example, s_3 is the state of site $i = 3$ at time t, and σ_3 is the state of the same site at time $t + 1$. Conditional probabilities w are easily computed from the definition of the rule.

Equation (6.20) is exact. The approximation consists of expressing the seven-block probabilities in terms of three-block probabilities using Relation (6.19). That is,

---

[11] As a cellular automaton rule, it is the 4-input rule 43944. See Footnote 5.

$$P_t(s_0 s_1 s_2 s_3 s_4 s_5 s_6) =$$

$$\frac{P_t(s_0 s_1 s_2) P_t(s_1 s_2 s_3) P_t(s_2 s_3 s_4) P_t(s_3 s_4 s_5) P_t(s_4 s_5 s_6)}{P_t(s_1 s_2) P_t(s_2 s_3) P_t(s_3 s_4) P_t(s_4 s_5)}, \quad (6.21)$$

where $P_t(s_i s_{i+1}) = P_t(s_i s_{i+1} 0) + P_t(s_i s_{i+1} 1)$ for $i = 1, \ldots, 4$. Equations (6.20) and (6.21) define a dynamical system whose fixed point approximates three-block probabilities of the limit set of the cellular automaton rule. Due to the nonlinear nature of these equations, it is not possible to find the fixed point analytically. It can be done numerically.

From the knowledge of three-block probabilities $P_t(s_2 s_3 s_4)$, different quantities can be calculated.[12]

We have seen that the expression of the average velocity $\langle v \rangle (\rho, \infty)$ in the steady state as a function of the car density ρ can be determined exactly by (6.12). As a test of the local structure theory, we can check that the value of the average velocity as a function of ρ, which is given by

$$\langle v \rangle (\rho, \infty) = 2 P_\infty(100) + P_\infty(101),$$

agrees with the exact relation. The car density is given by

$$\rho = P_t(1) = P_t(100) + P_t(101) + P_t(110) + P_t(111).$$

More interestingly, we can determine the velocity probability distribution in the steady state; that is, the probabilities $P(v = 0)$, $P(v = 1)$, and $P(v = 2)$ for a car velocity to be equal either to 0, 1, or 2 as a function of car density in the limit $t \to \infty$. These probabilities are given by

$$P(v = 0) = P_\infty(110) + P_\infty(111),$$
$$P(v = 1) = P_\infty(101),$$
$$P(v = 2) = P_\infty(100).$$

For $0 < \rho \leq \rho_c = \frac{1}{3}$, $P(v = 0) = 0$, $P(v = 1) = 0$, and $P(v = 2) = 1$. For $\rho_c \leq \rho \leq 1$, the velocity probability distributions obtained from the expressions above, are represented in Figure 6.7.

6.4 Generalized cellular automata

In this section, we define a new class of evolution operators that contains as particular cases all cellular automaton global evolution rules.[13] In the following discussion, we suppose that the finite set of cell states Q coincides with

---

[12] The local structure theory and numerical simulations have been used by Boccara and Fukś [60] to determine critical exponents. See Example 38.

[13] This result had been derived first for a restricted set of rules by Boccara, Fukś, and Geurten [58].

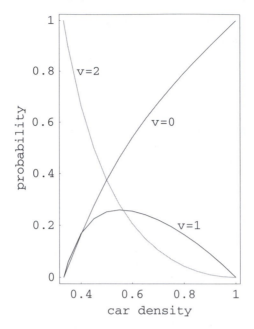

Fig. 6.7. *Velocity probability distributions for the deterministic car traffic flow model with $v_{\max} = 2$ determined using the local structure theory.*

$\{0, 1\}$. Moreover, we may always assume that the radii r_ℓ and r_r of the cellular automaton local rule f are equal by formally expressing f as a function of $2r + 1$ variables, where $r = \max\{r_\ell, r_r\}$. Under that form, f does not depend upon the values of the extra $|r_\ell - r_r|$ variables.[14]

Let $S_t : i \mapsto s(i, t)$ be the state of the system at time t, and put

$$\sigma(i, t) = \sum_{n=-\infty}^{\infty} s(i + n, t)p(n),$$

where p is a given probability measure on the set of all integers $\mathbb{Z}$;[15] that is, a nonnegative function such that

$$\sum_{n=-\infty}^{\infty} p(n) = 1.$$

---

[14] Note that the set of all radius r rules contains all n-input rules for $n \leq 2r + 1$. For example, a rule f with $r_\ell = 0$ and $r_r = 1$ may be represented as a function of three variables $(x_1, x_2, x_3) \mapsto f(x_1, x_2, x_3)$ satisfying, for $Q = \{0, 1\}$, the condition $f(x_1, x_2, x_3) = f(1 - x_1, x_2, x_3)$.

[15] The expression "probability measure" does not imply that we are considering stochastic cellular automata. A probability measure is just a finite measure such that the measure of the whole space is equal to 1.

Since $s(i,t) \in \{0,1\}$, for all $i \in \mathbb{Z}$ and $t \in \mathbb{N}$, it is clear that $\sigma(i,t) \in [0,1]$. The state S_{t+1} of the system at time $t+1$ is then determined by the function

$$i \mapsto s(i, t+1) = I_A(\sigma(i,t)),$$

where I_A is a given indicator function on $[0,1]$, that is, a function such that, for all $x \in [0,1]$,

$$I_A(x) = \begin{cases} 1, & \text{if } x \in A \subset [0,1], \\ 0, & \text{otherwise.} \end{cases}$$

Since the state S_{t+1} at time $t+1$ is entirely determined by the state S_t at time t, the probability measure p, and the subset A of $[0,1]$, there exists a unique mapping $F_{p,A} : \mathcal{S} \to \mathcal{S}$ such that

$$S_{t+1} = F_{p,A}(S_t).$$

$F_{p,A}$ is an evolution operator defined on the state space $\mathcal{S} = \{0,1\}^{\mathbb{Z}}$.

Theorem 16. *Let F_f be a cellular automaton global rule on $\mathcal{S} = \{0,1\}^{\mathbb{Z}}$ induced by a local rule f; then, there exists an evolution operator $F_{p,A}$ such that, for any configuration $x \in \mathcal{S}$, $F_{p,A}(x) = F_f(x)$.*

The set of all configurations with prescribed values at a finite number of sites is called a *cylinder set*. Cellular automaton rules being translation-invariant, we only need to consider cylinder sets centered at the origin. Since we assumed that both radii of the local rule are equal to r, we are interested in the 2^{2r+1} cylinder sets associated with the different $(2r+1)$-blocks $(x_{-r}, x_{-r+1}, \ldots, x_r)$. Each $(2r+1)$-block may be characterized by the binary number $\nu_r = x_0 x_{-1} x_1 x_{-2} x_2 \cdots x_{-r} x_r$; that is,

$$\nu_r = \sum_{n=1}^{r} x_{-n} 2^{2r-2n+1} + \sum_{n=0}^{r} x_n 2^{2r-2n}.$$

The cylinder set corresponding to a specific $(2r+1)$-block is denoted $C(r, \nu_r)$.

For any configuration x belonging to the cylinder set $C(r, \nu_r)$, the set of all numbers

$$\xi\big(C(r, \nu_r)\big) = \sum_{n=-\infty}^{\infty} x_n p(n)$$

belongs to the subinterval (called a *C-interval* in what follows)

$$\big[\xi_{\min}\big(C(r, \nu_r)\big), \xi_{\max}\big(C(r, \nu_r)\big)\big] \subset [0,1],$$

where

$$\xi_{\min}\big(C(r, \nu_r)\big) = \sum_{n=-r}^{r} x_n p(n)$$

and

$$\xi_{\max}\big(C(r,\nu_r)\big) = \xi_{\min}\big(C(r,\nu_r)\big) + \sum_{|n|=r}^{\infty} p(n).$$

There are 2^{2r+1} different $\xi_{\min}\big(C(r,\nu_r)\big)$ and as many different C-intervals.

If we want to define $2^{2^{2r+1}}$ different $F_{p,A}$ evolution operators, representing the $2^{2^{2r+1}}$ different cellular automaton local evolution rules f, we first have to find a probability measure p such that the 2^{2r+1} C-intervals are pairwise disjoint. Then, according to the local rule f to be represented, we choose a subset A of $[0,1]$ such that some of these C-intervals are strictly included in A, whereas the others have an empty intersection with A. The only problem is therefore to find the conditions to be satisfied by p.

Let

$$p(n) = \begin{cases} \dfrac{\beta-1}{\beta^{2n}}, & \text{if } n < 0, \\[2mm] \dfrac{\beta-1}{\beta^{2n+1}}, & \text{if } n \geq 0. \end{cases} \tag{6.22}$$

We verify that, for $\beta > 1$,

$$\sum_{n=-\infty}^{\infty} p(n) = 1.$$

In the trivial case of cellular automaton rules with $r = 0$, we have the following two C-intervals:

$$\xi_{\min}\big(C(0,0)\big) = 0, \quad \xi_{\max}\big(C(0,0)\big) = \frac{1}{\beta},$$

and

$$\xi_{\min}\big(C(0,1)\big) = \frac{\beta-1}{\beta}, \quad \xi_{\max}\big(C(0,1)\big) = 1.$$

These C-intervals are disjoint if

$$\xi_{\min}\big(C(0,1)\big) > \xi_{\max}\big(C(0,0)\big) \tag{6.23}$$

or

$$\frac{\beta-1}{\beta} > \frac{1}{\beta};$$

that is, for $\beta > 2$.

Increasing the radius by one unit, multiply the number of C-intervals by 4. That is, from each closed C-interval of the form

$$\left[\sum_{|n|\leq r} a_n p(n), \ \sum_{|n|\leq r} a_n p(n) + \sum_{|n|>r} p(n) \right],$$

where $(a_{-r}, a_{-r+1}, \cdots, a_{r-1}, a_r)$ is a specific $(2r+1)$-block, we generate four closed C-intervals, each one being associated with one of the following $(2r+3)$-blocks:

$$(0, a_{-r}, a_{-r+1}, \cdots, a_{r-1}, a_r, 0), \quad (0, a_{-r}, a_{-r+1}, \cdots, a_{r-1}, a_r, 1),$$
$$(1, a_{-r}, a_{-r+1}, \cdots, a_{r-1}, a_r, 0), \quad (1, a_{-r}, a_{-r+1}, \cdots, a_{r-1}, a_r, 1).$$

These new C-intervals are disjoint if, and only if, the following three conditions are satisfied:

$$p(r+1) > \sum_{|n|=r+2}^{\infty} p(n), \tag{6.24}$$

$$p(-r-1) > p(r+1) + \sum_{|n|=r+2}^{\infty} p(n), \tag{6.25}$$

$$p(-r-1) + p(r+1) > p(-r-1) + \sum_{|n|=r+2}^{\infty} p(n). \tag{6.26}$$

Conditions (6.24) and (6.26) are identical. Therefore, replacing $p(n)$ by its expression, and noting that

$$\sum_{|n|=r+2}^{\infty} p(n) = \frac{1}{\beta^{2r+3}},$$

we have finally to satisfy the conditions

$$\frac{\beta-1}{\beta^{2r+3}} > \frac{1}{\beta^{2r+3}} \quad \text{and} \quad \frac{\beta-1}{\beta^{2r+2}} > \frac{\beta-1}{\beta^{2r+3}} + \frac{1}{\beta^{2r+3}} = \frac{1}{\beta^{2r+2}}.$$

That is, β has to be greater than 2.

Therefore, the global evolution rule induced by any radius r cellular automaton local rule f can be represented as an evolution operator $F_{p,A}$, where the probability measure p is given by (6.22) for $\beta > 2$ and A a subset of $[0,1]$ that, according to f, includes some C-intervals, whereas the others have an empty intersection with it. A is not unique. □

Example 43. Radius 1 rules. The table below lists the eight different C-intervals for $\beta = 2.5$.

triplet	ν_r	C-interval	triplet	ν_r	C-interval
(0,0,0)	0	[0, 0.064]	(0,1,0)	4	[0.6, 0.664]
(0,0,1)	1	[0.096, 0.16]	(0,1,1)	5	[0.696, 0.76]
(1,0,0)	2	[0.24, 0.304]	(1,1,0)	6	[0.84, 0.904]
(1,0,1)	3	[0.336, 0.4]	(1,1,1)	7	[0.936, 1]

Using this table, we can represent any elementary cellular automaton global evolution rule. For example, the global rule induced by local rule 18 (see Example 41), defined by

$$f_{18}(x_1, x_2, x_3) = \begin{cases} 1, & \text{if } (x_1, x_2, x_3) = (0,0,1) \text{ or } (1,0,0), \\ 0, & \text{otherwise,} \end{cases} \tag{6.27}$$

may be represented by an evolution operator $F_{p,A}$, where p is given by (6.22) with $\beta = 2.5$, and A is any subset of $[0, 1]$ containing the union $[0.096, 0.16] \cup [0.24, 0.304]$ but whose intersection with all the other six C-intervals is empty. In particular, we could choose $A = [0.09, 0.31]$, illustrating the fact that A is not unique.

Remark 15. In the process described above, four new closed intervals, necessary to represent radius $(r + 1)$ rules, are obtained by removing three open intervals from each closed C-interval used to represent radius r rules. This process is reminiscent of the recursive construction of the triadic Cantor set (see Example 34). In the limit $r \to \infty$ (*i.e.*, when all open intervals have been removed), we obtain a Cantor-like set Σ; that is, a closed set with no interior points and such that all points are limit points (see page 166, Footnote 16).[16] This set consists of all the numbers represented by expansion of the form

$$\sigma(x) = \sum_{n=-\infty}^{\infty} x_n p(n), \tag{6.28}$$

where p is given by (6.22) for $\beta > 2$ and, for all $n \in \mathbb{Z}$, $x_n \in \{0, 1\}$. The right-hand side of (6.28) is the expansion of $\sigma(x)$ in (nonintegral) base β.[17]

The Lebesgue measure of Σ is obtained by subtracting from 1 the sum of the lengths of all the removed intervals. That is,

$$1 - \frac{\beta - 2}{\beta} - 2\sum_{r=0}^{\infty} \frac{\beta - 2}{\beta^{2r+2}} - 4\sum_{r=0}^{\infty} \frac{\beta - 2}{\beta^{2r+3}} = 1 - \frac{\beta - 2}{\beta} - \frac{2(\beta - 2)}{\beta^2 - 1} - \frac{4(\beta - 2)}{\beta(\beta^2 - 1)}$$

$$= \frac{6}{\beta(\beta^2 - 1)}.$$

As shown in Figure 6.8, the Lebesgue measure of Σ is equal to 1 for $\beta = 2$ (Σ coincides, in this case, with the set of all numbers of the interval $[0, 1]$ in base 2) and goes to zero as β tends to infinity. For a finite value of β, Σ has a nonzero Lebesgue measure (equal, for example, to 0.457143 for $\beta = 2.5$).[18]

Remark 16. An interesting feature of the definition of the space $\mathcal{F}_p$ of evolution operators for a given probability measure p is that it is possible to define the distance $d(F_{p,A_1}, F_{p,A_2})$ between two evolution operators F_{p,A_1} and F_{p,A_2}. The basic

---

[16] Let $x^{(1)}$ and $x^{(2)}$ be two elements of $\mathcal{S} = \{0, 1\}^{\mathbb{Z}}$; the distance between the two corresponding points $\sigma(x^{(1)})$ and $\sigma(x^{(2)})$ in Σ is defined by

$$d\big(\sigma(x^{(1)}), \sigma(x^{(2)})\big) = \sum_{n=-\infty}^{\infty} |x_n^{(1)} - x_n^{(2)}| p(n),$$

where $p(n)$ is given by (6.22).

[17] In the mathematical literature, these representations of a real number of the interval $[0, 1]$ are called *β-expansions*. For a review of their properties, see Blanchard [43].

[18] On Cantor-like sets, see Boccara [57].

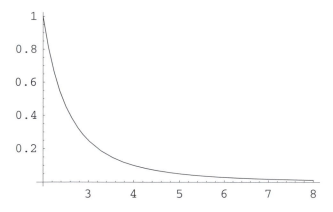

Fig. 6.8. *Lebesgue measure of the Cantor set Σ as a function of β.*

property of d, defined on $\mathcal{F}_p$, is that the distance between two different evolution operators (that is, two operators associated with two different subsets A_1 and A_2 of $[0,1]$) should be zero if the evolution operators F_{p,A_1} and F_{p,A_2} represent the same evolution rule (remember, the subset A defining $F_{p,A}$ is not unique).

The distance $d : \mathcal{F}_p \to [0,1]$ is defined by

$$d(F_{p,A_1}, F_{p,A_2}) = \int_0^1 |I_{A_1}(\sigma) - I_{A_2}(\sigma)|\, dF(\sigma), \qquad (6.29)$$

where $dF(\sigma)$ is a measure we have to define. The best way to understand the meaning of this measure is to use the language of probability theory. Let $(X_n)_{n\in\mathbb{Z}}$ be a doubly infinite sequence of identically distributed Bernoulli random variables such that, for all $n \in \mathbb{Z}$,

$$P(X_n = 0) = P(X_n = 1) = \tfrac{1}{2},$$

and consider the cumulative distribution function F of the random variable

$$\sum_{n=-\infty}^{\infty} X_n p(n),$$

that is, the function $F : [0,1] \to [0,1]$ such that[19]

$$F(\sigma) = P\left(\sum_{n=-\infty}^{\infty} X_n p(n) \le \sigma \right).$$

In other words, $F(\sigma)$ is the probability that the random variable $\sum_{n=-\infty}^{\infty} X_n p(n)$ does not exceed σ. F is a *singular measure* such that $F(0) = 0$ and $F(1) = 1$. It is represented in Figure 6.9. The meaning of the distance $d(F_{p,A_1}, F_{p,A_2})$ defined by (6.29) becomes clear. It is the measure of the *symmetrical difference* between the sets A_1 and A_2, that is, the set of all numbers $\sigma(x) \in [0,1]$ that belong to A_1 XOR A_2. Hence

---

[19] On probability theory, see, for example, Grimmett and Stirzaker [158].

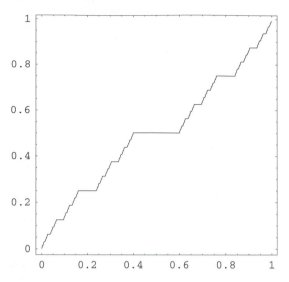

Fig. 6.9. *Cumulative distribution function* $F : [0,1] \to [0,1]$ *of the random variable* $\sum_{n=-\infty}^{\infty} X_n p(n)$ *for* $\beta = 2.5$.

$$d(F_{p,A_1}, F_{p,A_2}) = \int_{A_1 \triangle A_2} dF(\sigma),$$

where $A \triangle A_2$ is the symmetrical difference $(A_1 \cup A_2) \backslash (A_1 \cap A_2)$. For example, the distance between rule 18 defined by (6.27) and rule 22 defined by

$$f_{22}(x_1, x_2, x_3) = \begin{cases} 1, & \text{if } (x_1, x_2, x_3) = (0,0,1) \text{ or } (1,0,0) \text{ or } (0,1,0), \\ 0, & \text{otherwise,} \end{cases} \tag{6.30}$$

is, according to the table of C-intervals, the measure of any interval including the C-interval $[0.6, 0.664]$ but having an empty intersection with all the other seven C-intervals. By construction of the distribution function F, all C-intervals for a given radius are equally probable (have the same measure). Thus, the distance between rule 18 and rule 22 is equal to $\frac{1}{8}$. It does not depend upon the value of β for $\beta > 2$.

Remark 17. For a given p, the space $\mathbb{F}$ of all generalized cellular automaton evolution operators $\{F_{p,A} \mid A \subset [0,1]\}$ defined on $\mathbb{Z}$ is a metric space for the distance given by 6.29. It is easy to verify that the set of all cellular automaton global rules is countable and dense in $\mathbb{F}$, exactly as the set of positive rational numbers less than 1 is dense in the closed interval $[0,1]$. Thus, just as any irrational number in the interval $[0,1]$ can be approximated by a rational number in the same interval as close as we want, any generalized cellular automaton evolution operator $F_{p,A}$ can be approximated by a cellular automaton global rule (*i.e.*, with a finite radius) as close as we want. In other words, given $F_{p,A}$, we can always find a sequence of cellular automaton global rules (F_{f_n}) induced by the local rules f_n ($n \in \mathbb{N}$) that converges to $F_{p,A}$ for the metric d defined by (6.29).

6.5 Kinetic growth phenomena

Kinetic growth phenomena are the result of stochastic growth processes taking place in time. As a result, the systems we shall consider evolve according to a probabilistic growth rule. Strictly speaking, the percolation phenomenon described below is not a kinetic growth process, but it often plays an important role in many kinetic growth processes.[20] This is the reason why it is briefly described in this section.

Example 44. Percolation models. Percolation processes were defined in the late 1950s by Broadbent and Hammersley [72]. Discussing the spread of a fluid through a random medium, these authors distinguish *diffusion processes*, in which the random mechanism is ascribed to the fluid, from *percolation processes*, where, in contrast, the random mechanism is ascribed to the medium. Among the examples given to illustrate what is meant by percolation processes, Broadbent and Hammersley consider gas molecules (fluid) adsorbed on the surface of a porous solid (medium). The gas molecules move through all pores large enough to admit them. If p is the probability for a pore to be wide enough to allow for the passage of a gas molecule, they discovered that there exists a *critical probability* p_c below which there is no adsorption in the interior. Similarly, in a large orchard, if the probability p for an infectious disease to spread from one tree to its neighbors is less than a critical probability p_c, the disease will not spread through the orchard, only a small number of trees will be infected.

There exist two types of standard percolation models: *bond* and *site percolation models* (see Figures 6.10 and 6.11, respectively). In what follows, definitions are given for two-dimensional square lattices. Extension to higher dimensionalities and different lattice symmetries is straightforward.

On a square lattice, the so-called *Manhattan distance* is defined by

$$d(x, y) = |x_1 - y_1| + |x_2 - y_2|,$$

where $x = (x_1, x_2)$ and $y = (y_1, y_2)$ are two lattice sites. Viewed as a graph, a square lattice consists of a set of vertices $\mathbb{Z}^2$ (called sites) and a set of edges (called bonds) $\mathbb{E} = \{(x, y) \mid (x, y) \in \mathbb{Z}^2, |x - y| = 1\}$. That is, the set of edges is the set of all straight line segments joining nearest-neighboring vertices.

In the bond percolation model, we assume that each edge is either *open* with a probability p or *closed* with a probability $1 - p$, independently of all other edges. In contrast, in the site percolation model, we assume that each vertex is either *open* or *occupied* with a probability p or *closed* or *empty* with a probability $1 - p$, independently of all other vertices. An alternating sequence $(x_0, e_0, x_1, e_1, x_2, \ldots, x_{n-1}, e_{n-1}, x_n)$, of distinct vertices x_i and edges $e_{i-1} = (x_{i-1}, x_i)$ ($i = 0, 1, 2, \ldots, n$) is called a *path of length* n. For the bond (resp.

---

[20] Many examples showing the unifying character of the percolation concept can be found in de Gennes [141]).

site) percolation model, a path is said to be open (resp. closed) if all its edges (resp. vertices) are open (resp. closed). Two different sites are said to belong to the same *open cluster* if they are connected by an open path. It can be

Fig. 6.10. *Bond percolation model. In the left figure, the probability for a bond to be open (darker bond) is $p = 0.4$ and, in the right figure, this probability is $p = 0.7$.*

Fig. 6.11. *Site percolation model. In the left figure, the probability for a site to be occupied (darker site) is $p = 0.4$ and, in the right figure, this probability is $p = 0.7$.*

shown that if the dimension of the lattice is greater than or equal to 2, then the critical probabilities (or *percolation thresholds*) p_c^{bond} and p_c^{site} for the bond and site percolation models, respectively, satisfy the relation[21]

---

[21] See Grimmett [159].

$$0 < p_c^{\text{bond}} < p_c^{\text{site}} < 1. \tag{6.31}$$

Except for a very few lattice structures, critical probabilities can be determined using numerical simulations. For example, it can be shown that, for the square lattice, the bond critical probability p_c^{bond} is exactly equal to 0.5 [159] while the site critical probability p_c^{site} is only approximately equal to $0.592746\ldots$ [266].

While critical probabilities, for a given space dimensionality, depend on the symmetry of the lattice and whether bond or site percolation is considered, critical exponents are universal. Here are two exactly known results.[22]

1. The correlation length ξ diverges as $|p - p_c|^{-\nu}$ when p tends to p_c either from below or from above with $\nu = \frac{4}{3}$.
2. The *percolation probability*, which is the probability $P_\infty(p)$ that a given vertex belongs to an infinite cluster, is clearly equal to zero for $p < p_c$, but, for $p > p_c$, it behaves as $(p - p_c)^\beta$ with $\beta = \frac{5}{36}$.[23]

The percolation process we have described is not the only one. Another important process in applications is the *directed percolation* process. Consider the square lattice represented in Figure 6.12 in which open bonds are randomly distributed with a probability p. In contrast with the usual bond percolation problem, here bonds are directed downwards, as indicated by the arrows.

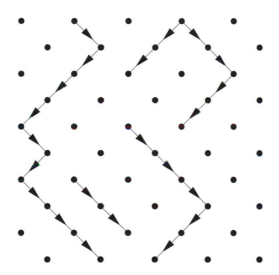

Fig. 6.12. *A configuration of directed bond percolation on a square lattice.*

---

[22] These exact results are obtained from the equivalence of both percolation models with Potts models; see Wu [350].

[23] For percolation, the lower and upper critical space dimensions are, respectively, equal to 1 and 6.

If we imagine a fluid flowing downwards from wet sites in the first row, one problem is to find the probability $P(p)$ that, following directed open bonds, the fluid will reach sites on an infinitely distant last row. There clearly exists a threshold value p_c above which $P(p)$ is nonzero. If the downwards direction is considered to be the time direction, the directed bond percolation process may be viewed as a two-input one-dimensional cellular automaton rule f such that

$$s(i, t+1) = f\big(s(i,t), s(i+1,t)\big)$$
$$= \begin{cases} 0, & \text{if } s(i,t) + s(i+1,t) = 0, \\ 1, & \text{with probability } p, \text{ if } s(i,t) + s(i+1,t) = 1, \\ 1, & \text{with probability } 1 - (1-p)^2, \text{ if } s(i,t) + s(i+1,t) = 2. \end{cases}$$

If ρ is the density of wet (active) sites,[24] the mean-field map is

$$\rho \mapsto 2p\rho(1-\rho) + (1 - (1-p)^2)\rho^2 = 2p\rho - p^2\rho^2.$$

This map shows the existence of a *directed bond percolation threshold*, or a *directed bond percolation probability* p_c^{DBP}, equal to $\frac{1}{2}$. That is, within the mean-field approximation, using the image of the flowing fluid, above p_c^{DBP}, the fluid has a nonzero probability of reaching an infinitely distant last row. These values are not exact. Numerical simulations show that p_c^{DBP} equals 0.6445 ± 0.0005 for the bond model.[25]

The most general two-input one-dimensional totalistic probabilistic cellular automaton rule,[26] called the *Domany-Kinzel cellular automaton rule* [103], may be written

$$s(i, t+1) = f\big(s(i,t), s(i+1,t)\big)$$
$$= \begin{cases} 0, & \text{if } s(i,t) + s(i+1,t) = 0, \\ 1, & \text{with probability } p_1, \text{ if } s(i,t) + s(i+1,t) = 1, \\ 1, & \text{with probability } p_2, \text{ if } s(i,t) + s(i+1,t) = 2. \end{cases}$$

Directed bond percolation corresponds to $p_1 = p$ and $p_2 = 2p - p^2$. The case $p_1 = p_2 = p$ is also interesting; it describes the *directed site percolation* process. In this case, numerical simulations show that the site percolation probability $p_c^{\text{DSP}} = 0.7058 \pm 0.0005$.

Example 45. General epidemic process. The general epidemic process (see Figure 6.13) describes, according to Grassberger [153, 77], who studied its critical properties, "the essential features of a vast number of population growth

---

[24] The density ρ, which is equal to $P(p)$, is the order parameter of the second-order phase transition.

[25] Refer to Kinzel [187] for critical exponents and scaling laws.

[26] A cellular automaton n-input rule is said to be totalistic if it only depends upon the sum of the n inputs.

phenomena." In its simplest version the growth process can be described as follows. Initially the cluster consists of the seed site located at the origin. At the next time step, a nearest-neighboring site is randomly chosen. This site is either added to the cluster with a probability p or rejected with a probability $1 - p$. At all subsequent time steps, the same process is repeated: a nearest-neighboring site of any site belonging to the cluster is selected at random, and it is either added to the cluster with a probability p or rejected with a probability $1 - p$. It is clear that there exists a critical probability p_c such that for $p > p_c$, the seed site has a nonzero probability of belonging to an infinite cluster.

Fig. 6.13. *Grassberger's general epidemic process. The cluster represents the spread of the epidemic to 4000 sites for $p = 0.6$.*

In order to determine the critical behavior of the general epidemic model, Grassberger performed extensive numerical simulations on a slightly different model that belongs to the same universality class and whose static properties are identical to a bond percolation model. In this model, every lattice site of a two-dimensional square lattice is occupied by only one individual, who cannot move away from it. The individuals are either susceptible, infected, or immune. At each time step, every infected individual infects each nearest-neighboring susceptible site with a probability p and becomes immune with probability 1. For this model, the critical probability p_c is exactly equal to $\frac{1}{2}$. If, at time $t = 0$, all the sites of one edge of the lattice are infected, among other results, Grassberger found that, at p_c, the average number of immune sites per row parallel to the initial infected row increases as a function of time as t^x, where the critical exponent x is equal to 0.807 ± 0.01.[27]

Example 46. Forest fires. A fire is characterized by its intensity,[28] its direction of travel, and its rate of spread.[29] MacKay and Jan [219] proposed a simple bond percolation model to describe the propagation of fire in a densely packed forest from a centrally located burning tree. The fire is supposed to be of low intensity so that its propagation can be viewed as a localized surface phenomenon (*i.e.*, a burning tree is only able to ignite its nearest neighbors).

Fig. 6.14. *Burning forest after 70 time steps. Burning, burnt and unburnt trees are, respectively, represented by white, black, and grey squares. The forest is a square lattice of size 101×101. In the left figure, the fire propagates isotropically with a probability $p = 0.6$. In the right figure, the fire propagates northward (resp. southward) with a probability $p_n = 0.65$ (resp. $p_s = 1 - p_n$) and eastward or westward with the same probability $p_e = p_w = 0.6$.*

More specifically, the authors assume that trees occupy the sites of a two-dimensional (square)[30] lattice. At each time step, the fire propagates from *burning* trees to their nearest neighbors with probability p, and burning trees become *burnt*. At $t = 0$, all sites are occupied by unburnt trees except for the

---

[27] For more detailed numerical results concerning this model, refer to Grassberger [153].

[28] Fire intensities are measured in kilocalories per second per meter of fire front. Low-intensity fires of about 800–900 kcal s^{-1} m^{-1} are two-dimensional phenomena, while high-intensity fires of about 15,000–25,000 kcal s^{-1} m^{-1} are three-dimensional phenomena.

[29] The rate of spread of a fire depends upon many factors: recent rainfall, oxygen supply, wind velocity, age and type of trees, etc.

[30] In their original paper, the authors assumed the lattice to be triangular.

central tree, which is burning. This growth model is identical to the model for the spread of blight in a large orchard mentioned by Broadbent and Hammersley (see Example 44).[31] It is also similar to the general epidemic process.

Precise numerical simulations [269] show that the critical probability $p_c = 0.5$. For $p = p_c$, the total number of burnt trees $M(t)$ grows as a function of time t as t^α, where $\alpha = 1.59$, while the radius of gyration $R(t)$ increases with time as $t^{2\beta}$, where $\beta = 0.79$.

It is possible to slightly modify the model above in order to take into account the influence of wind. We can, for example, suppose that the probability for a tree to be ignited by a neighboring burning tree is equal to p_n (resp. $p_s = 1 - p_n$) if it is located north (resp. south) of the burning tree and $p_e = p_w$ if it located either east or west of the burning tree (see Figure 6.14). The critical probability $p_{e,c} = p_{w,c}$ depends upon the value of the probability p_n. Von Niessen and Blumen [269] give the following values:

p_n	0.5	0.6	0.7	0.8	0.9
$p_{e,c}$	0.5	0.459	0.385	0.287	0.160

Example 47. Diffusion-limited aggregation. Cluster formation from a single particle is of interest in many domains such as, for instance, dendritic growth, flocculation, soot formation, tumor growth, and cloud formation. In 1981, Witten and Sander [347] proposed a model for random aggregates in which an initial seed particle is located at the center of a square lattice. A second particle is added at some random site at a large distance from the origin. This particle walks at random on the lattice until either it reaches a site adjacent to the seed and stops or first reaches the boundary of the lattice and is removed. A third particle is then introduced at a random distant point and walks at random until either it joins the cluster or reaches the boundary and is removed, and so forth. Figure 6.15 represents a (small) aggregate of 250 particles.

Numerical simulations showed that these structures have remarkable scaling and universal properties [237, 238]. In particular, these aggregates are fractals with a Hausdorff dimension equal to $5d/6$, where d is the Euclidean dimensionality.

In 1983 Meakin [239] and Kolb, Botet, and Jullien [191] investigated simultaneously a related model in which, in the initial state, N_0 lattice sites are occupied. Clusters—which may consist of only one particle—are picked at random and moved to a nearest-neighboring site selected at random. If a cluster contacts one or more clusters, these clusters merge to form a larger cluster. Clusters therefore grow larger and larger until one large cluster remains. Different variants of this model have been considered. Clusters may move with a probability independent of their size or with a probability inversely proportional to their size.

---

[31] See also Frish and Hammersley [131].

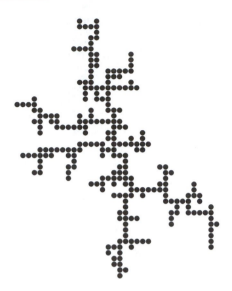

Fig. 6.15. *Diffusion-limited aggregate of 250 particles.*

6.6 Site-exchange cellular automata

A site-exchange cellular automaton is an automata network whose evolution rule consists of two subrules:

1. The first subrule is a probabilistic cellular automaton rule, such as, for example, a "diluted" one-dimensional rule f defined by

$$s(i, t+1) = Xf\big(s(i - r_\ell, t), s(i - r_\ell + 1, t), \ldots, s(i + r_r, t)\big), \qquad (6.32)$$

where f is a one-dimensional deterministic cellular automaton rule with left and right radii equal, respectively, to r_ℓ and r_r, and X is a Bernoulli random variable such that

$$P(X = 0) = 1 - p \quad \text{and} \quad P(X = 1) = p.$$

When $p = 1$, rule (6.32) coincides with the deterministic cellular automaton rule f.

2. The second subrule is a site-exchange rule defined as follows. An occupied site (*i.e.*, a site whose state value is 1) is selected at random and swapped with another site (either empty or occupied) also selected at random. The second site is either a nearest neighbor of the first site (short-range move) or any site of the lattice (long-range move). Between the application of the first subrule at times t and $t+1$, this operation is repeated $\lfloor m \times \rho(m, t) \times L \rfloor$ times, where m is a positive real number called the *degree of mixing*, $\rho(t)$ the density of occupied sites at time t, L the total number of lattice sites, and $\lfloor x \rfloor$ the largest integer less than or equal to x.

Site-exchange cellular automata were introduced by Boccara and Cheong [49, 52] and used in theoretical epidemiology in which infection by contact and recovery are modeled by two-dimensional probabilistic cellular automaton rules applied synchronously, whereas the motion of the individuals, which is an essential factor, is modeled by a sequential site-exchange rule.

While short-range moves are clearly diffusive moves on a finite-dimensional lattice, long-range moves may be viewed as diffusive motion of the individuals on an infinite-dimensional lattice. In order to help clarify the mixing process that results from the motion of the individuals, it might be worthwhile to characterize the mixing process as a function of m and the lattice dimensionality d.

Consider a random initial configuration $c(0, \rho)$ of walkers with density ρ on a d-dimensional torus $\mathbb{Z}_L^d$. Select sequentially $\lfloor m\rho L^d \rfloor$ walkers and move them to a neighboring site also selected at random if, and only if, the randomly selected neighboring site is vacant. L^d is the total number of sites of the torus (finite lattice), and m represents the *average number of tentative moves per random walker* since only a fraction of these moves are effective. The expression *at random* means that all possible choices are equally probable. Let $c(m\rho L^d, \rho)$ denote the resulting configuration.[32]

Theorem 17. *The d-dimensional sequential multiple random walkers model is equivalent to a one random walker model in which the walker has a probability $1/2d\rho L$ to move either to the left or to the right and a probability $1 - 1/d\rho L$ not to move, where ρ is the density of random walkers.*

To simplify the proof, consider a one-dimensional torus ($d = 1$), and let $P(m\rho L, \rho, j)$ be the probability that the site j is occupied after $m\rho L$ tentative moves. $P(m\rho L, \rho, j)$ is therefore the average value $\langle n(m\rho, \rho, j) \rangle_{\text{rw}}$ over all the possible random walks starting from the same initial configuration. In order to find the evolution equation of $P(m\rho L, \rho, j)$, consider the set of configurations in which $n(m\rho L, \rho, j-1)$, $n(m\rho L, \rho, j)$, and $n(m\rho L, \rho, j+1)$ have fixed values whereas, for all $i \neq j - 1, j, j + 1$, $n(m\rho L, \rho, i)$ takes any value with the restriction that the total density is equal to ρ. Let $p_{x_1 x_2 x_3}(m\rho L, \rho, j)$ denote the probability of a configuration such that

$$n(m\rho L, \rho, j - 1) = x_1, \quad n(m\rho L, \rho, j) = x_2, \quad n(m\rho L, \rho, j + 1) = x_3.$$

The probability $P(m\rho L + 1, \rho, j)$ for site j to be occupied by a random walker after $m\rho L + 1$ tentative moves is the sum of different terms.

1. If site j is empty after $m\rho L$ tentative moves, then at least one of its nearest-neighboring sites has to be occupied. The probability for a given site to be occupied being $1/\rho L$, and the probability to move either to the right or to the left being equal to $\frac{1}{2}$, the contribution to $P(m\rho L + 1, \rho, j)$ is in this case

---

[32] The following exact results were established by Boccara, Cheong, and Oram [54].

$$\frac{1}{2\rho L}\left(p_{001}(m\rho L,\rho,j)+p_{100}(m\rho L,\rho,j)\right)+\frac{1}{\rho L}p_{101}(m\rho L,\rho,j).$$

2. If site j is occupied by a random walker after $m\rho L$ tentative moves, the site has to remain occupied. The different contributions to $P(m\rho L+1,\rho,j)$ are, in this case, the sum of the following three terms:

$$\left(1-\frac{1}{\rho L}\right)p_{010}(m\rho L,\rho,j),$$

since the occupied site j should not be chosen to move,

$$\frac{1}{2\rho L}\left(p_{011}(m\rho L,\rho,j)+p_{110}(m\rho L,\rho,j)\right),$$

since the occupied site j should be selected to try to move to the neighboring occupied site, and

$$p_{111}(m\rho L,\rho,j).$$

Hence,

$$P(m\rho L+1,\rho,j)=\frac{1}{2\rho L}\left(p_{001}(m\rho L,\rho,j)+p_{100}(m\rho L,\rho,j)\right)$$
$$+\left(1-\frac{1}{\rho L}\right)p_{010}(m\rho L,\rho,j)$$
$$+\frac{1}{2\rho L}\left(p_{011}(m\rho L,\rho,j)+p_{110}(m\rho L,\rho,j)\right)$$
$$+\frac{1}{\rho L}p_{101}(m\rho L,\rho,j)+p_{111}(m\rho L,\rho,j).$$

Taking into account the relations

$$P(m\rho L,\rho,j-1)=p_{100}(m\rho L,\rho,j)+p_{101}(m\rho L,\rho,j)$$
$$+p_{110}(m\rho L,\rho,j)+p_{111}(m\rho L,\rho,j)$$
$$P(m\rho L,\rho,j)=p_{010}(m\rho L,\rho,j)+p_{011}(m\rho L,\rho,j)$$
$$+p_{110}(m\rho L,\rho,j)+p_{111}(m\rho L,\rho,j)$$
$$P(m\rho L,\rho,j+1)=p_{001}(m\rho L,\rho,j)+p_{011}(m\rho L,\rho,j)$$
$$+p_{101}(m\rho L,\rho,j)+p_{111}(m\rho L,\rho,j),$$

we finally obtain the following *discrete diffusion equation*:

$$P(m\rho L+1,\rho,j)=P(m\rho L,\rho,j)$$
$$+\frac{1}{2\rho L}\left(P(m\rho L,\rho,j+1)+P(m\rho L,\rho,j-1)-2P(m\rho L,\rho,j)\right). \quad (6.33)$$

For a d-dimensional torus, we would have obtained

$$P(m\rho L + 1, \rho, j) = P(m\rho L, \rho, j)$$

$$+ \frac{1}{2d\rho L} \left(\sum_{i \text{ nn } j} P(m\rho L, \rho, i) - 2dP(m\rho L, \rho, j) \right), \qquad (6.34)$$

where the summation runs over the $2d$ nearest neighbors i of j. This equation may also be written

$$P(m\rho L + 1, \rho, j) = \left(1 - \frac{1}{\rho L} \right) P(m\rho L, \rho, j)$$

$$+ \frac{1}{2d\rho L} \sum_{i \text{ nn } j} P(m\rho L, \rho, i), \qquad (6.35)$$

showing that the problem of ρL^d random walkers on a d-dimensional torus is equivalent to the problem of one random walker that may move to any single neighboring site with probability $1/2d\rho L^d$ or not move with probability $1 - 1/\rho L^d$. $\qquad \square$

To characterize the mixing process, consider the Hamming distance

$$d_H\left(c(0, \rho), c(n\rho L^d, \rho) \right) = \frac{1}{L^d} \sum_{j=1}^{L^d} \left(n(0, \rho, j) - n(m\rho L^d, \rho, j) \right)^2,$$

where $n(0, \rho, j)$ and $n(m\rho L^d, \rho, j)$ are the occupation numbers of site j in $c(0, \rho)$ and $c(m\rho L^d, \rho)$, respectively.

If m is very large, $c(0, \rho)$ and $c(m\rho L^d, \rho)$ are not correlated and the Hamming distance, which is the average over space of

$$\left(n(0, \rho, j) - n(m\rho L^d, \rho, j) \right)^2$$
$$= n(0, \rho, j) + n(m\rho L^d, \rho, j) - 2n(0, \rho, j)\, n(m\rho L^d, \rho, j),$$

is equal to $2\rho(1 - \rho)$.

Theorem 18. *In the limit $L \to \infty$, the average of the reduced Hamming distance δ_H, defined by*

$$\delta_H(m, d, \rho) = \lim_{L \to \infty} \left\langle \frac{d_H\left(c(0, \rho), c(n\rho L^d, \rho) \right)}{2\rho(1 - \rho)} \right\rangle$$

is a function of m and d only, i.e., it does not depend on ρ.

We shall first consider a complete graph (*i.e.*, a graph in which any pair of vertices are nearest neighbors). If the graph has N vertices, we have

$$P(m\rho N + 1, \rho, j) = P(m\rho N, \rho, j)$$

$$+ \frac{1}{\rho(N-1)N} \sum_{k=1}^{N} \left(P(m\rho N, \rho, k) - P(m\rho N, \rho, j) \right). \qquad (6.36)$$

Taking into account the relation

$$\sum_{k=1}^{N} P(m\rho N, \rho, k) = \rho N,$$

Equation (6.36) may also be written

$$P(m\rho N + 1, \rho, j) = \left(1 - \frac{1}{\rho(N-1)}\right) P(m\rho N, \rho, j) + \frac{1}{N-1}. \qquad (6.37)$$

Since this equation involves only one site—a typical result of mean-field equations—its solution is easy to obtain. We find

$$P(m\rho N, \rho, j) = \left(1 - \frac{1}{\rho(N-1)}\right)^{m\rho N} \left(n(0, \rho, j) - c\right) + \rho.$$

Initial configurations of numerical simulations are random; hence, only averages over all random walks and all initial configurations with a given density ρ of random walkers are meaningful quantities. Averaging over all random walks starting from the same initial configuration $c(0, \rho) = \{n(0, \rho, j) \mid j = 1, 2, \ldots, N\}$, the Hamming distance is

$$d_H\big(c(0, \rho), c(m\rho N, \rho)\big) =$$
$$\frac{1}{N}\sum_{j=1}^{N} \big(n(0, \rho, j) + P(m\rho N, \rho, j) - 2n(0, \rho, j)P(m\rho N, \rho, j)\big),$$

that is,

$$d_H\big(c(0, \rho), c(m\rho N, \rho)\big) = 2\rho - \frac{2}{N}\sum_{j=1}^{N} n(0, \rho, j)P(m\rho N, \rho, j).$$

Replacing $P(m\rho N, \rho, j)$ by its expression yields

$$n(0, \rho, j)P(m\rho N, \rho, j)$$
$$= \left(\left(1 - \frac{1}{\rho(N-1)}\right)^{m\rho N} \big(n(0, \rho, j) - c\big) + c\right) n(0, \rho, j)$$
$$= \left(1 - \frac{1}{\rho(N-1)}\right)^{m\rho N} \left(\big(n(0, \rho, j)\big)^2 - \rho n(0, \rho, j)\right) + \rho n(0, \rho, j),$$

and, taking the average over space, we obtain

$$\langle n(0, \rho, j)P(m\rho N, \rho, j)\rangle_{\text{space}}$$
$$= \left(1 - \frac{1}{\rho(N-1)}\right)^{m\rho N} \left(\langle\big(n(0, \rho, j)\big)^2\rangle_{\text{space}} - \rho^2\right) + \rho^2.$$

Since $\langle(n(0,\rho,j))^2\rangle_{\text{space}} = \langle n(0,\rho,j)\rangle_{\text{space}} = \rho$, letting N go to infinity, we finally obtain $2\rho(1-\rho)(1-e^{-m})$, *i.e.*,

$$\delta_H(m,\infty,\rho) = 1 - e^{-m},$$

where the symbol ∞ refers to the fact that, in the limit $N \to \infty$, a complete graph is equivalent to an infinite-dimensional lattice.

Consider now Equation (6.33), and replace $P(m\rho L, \rho, j)$ by its expression in terms of its Fourier transform $\widehat{P}(m\rho L, \rho, k)$ defined by

$$P(m\rho L, \rho, j) = \frac{1}{\sqrt{L}} \sum_{k=1}^{L} \widehat{P}(m\rho L, \rho, k) e^{2i\pi k/L}.$$

This yields

$$\widehat{P}(m\rho L + 1, \rho, k) = \left(1 - \frac{2}{\rho L}\sin^2\left(\frac{\pi k}{L}\right)\right)\widehat{P}(m\rho L, \rho, k)$$

and, therefore,

$$\widehat{P}(m\rho L, \rho, k) = \left(1 - \frac{2}{\rho L}\sin^2\left(\frac{\pi k}{L}\right)\right)^{m\rho L}\widehat{P}(0, \rho, k).$$

The average Hamming distance over all initial configurations is

$$\frac{1}{N}\sum_{j=1}^{L}\langle(P(0,\rho,j) + P(m\rho L,\rho,j) - 2P(0,\rho,j)P(m\rho L,\rho,j))\rangle_{\text{ic}}$$

$$= 2\rho - \frac{2}{L}\sum_{k=1}^{L}\langle\widehat{P}(0,\rho,k)\widehat{P}(0,\rho,-k)\rangle_{\text{ic}}\left(1 - \frac{2}{\rho L}\sin^2\left(\frac{\pi k}{L}\right)\right)^{m\rho L},$$

where $\langle f(j)\rangle_{\text{ic}}$ denotes the average over all initial configurations of $f(j)$. Since

$$\langle\widehat{P}(0,\rho,k)\widehat{P}(0,\rho,-k)\rangle_{\text{ic}}$$

$$= \frac{1}{N}\sum_{j_1=1}^{L}\sum_{j_2=1}^{L}\langle P(0,\rho,j_1)P(m\rho L,\rho,j_2)\rangle_{\text{ic}}e^{2i\pi k(j_1-j_2)/N},$$

we find

$$\frac{2}{L}\sum_{k=1}^{L}\langle\widehat{P}(0,\rho,k)\widehat{P}(0,\rho,-k)\rangle_{\text{ic}}\left(1 - \frac{2}{\rho L}\sin^2\left(\frac{\pi k}{L}\right)\right)^{m\rho L}$$

$$= \frac{1}{L^2}(L\rho + L(L-1)\rho^2) + (\rho - \rho^2)\frac{1}{L}\sum_{k=1}^{L-1}\left(1 - \frac{2}{\rho L}\sin^2\left(\frac{\pi k}{L}\right)\right)^{m\rho L}.$$

Finally, in the limit $L \to \infty$, the average Hamming distance over all random walks and all initial configurations is equal to

$$2\rho(1-\rho)\left(1 - \int_0^1 e^{-2m\sin^2\pi q}\, dq\right),$$

and the reduced Hamming distance is thus given by

$$\delta_H(m,1,\rho) = 1 - e^{-m}I_0(m),$$

where the function I_0, defined by

$$I_0(m) = \frac{1}{\pi}\int_0^\pi e^{m\cos u}\, du,$$

is the modified Bessel function of the first kind of order zero. $\delta_H(m,1,\rho)$ does not, therefore, depend upon ρ. For large values of m, $I_0(m)$ behaves as $e^m/\sqrt{2\pi m}$, hence, $\delta_H(m,1,\rho)$ approaches unity as m tends to infinity as $1/\sqrt{m}$.

Equation (6.34) could be solved in a similar way, and, in particular, we would find that $\delta_H(m,d,\rho)$ does not depend upon ρ and approaches unity as m tends to infinity as $1/\sqrt{m^d}$.

For small m, it is easy to verify that, for all d, the reduced Hamming distance behaves as m.   □

Example 48. Period-doubling route to chaos for a global variable of a one-dimensional totalistic cellular automaton. As a first example of a site-exchange cellular automaton, we shall study the behavior, as a function of the degree of mixing m in the case of long-range moves, of the infinite-time limit density of active sites $\rho(\infty,m)$ of the two-state one-dimensional radius r totalistic cellular automaton rule [33] defined by

$$s(i,t+1) = Xf\left(\sum_{j=-r}^{j=r} s(i+j,t)\right),$$

where

$$f(x) = \begin{cases} 1, & \text{if } S_{\min} \le x \le S_{\max}, \\ 0, & \text{otherwise.} \end{cases} \tag{6.38}$$

The spatiotemporal pattern of the cellular automaton evolving according to the radius 3 totalistic rule with $S_{\min} = 1$ and $S_{\max} = 6$ is represented in Figure 6.16. It can be shown [50] that the limit set of this cellular automaton consists of sequences of 0s and 1s whose lengths are multiples of 6. The distributions of 0s and 1s are identical, which implies that the asymptotic density

---

[33] Totalistic cellular automaton rules of this type were studied in detail by Bidaux, Boccara, and Chaté [42].

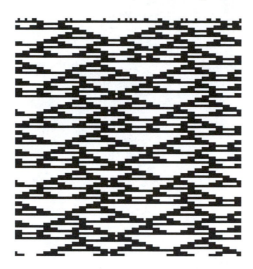

Fig. 6.16. *First 100 iterations of the radius 3 totalistic cellular automaton rule (6.38) (with $S_{\min} = 1$ and $S_{\max} = 6$) starting from a random initial configuration with 25 nonzero sites. 0s and 1s are, respectively, represented by white and black squares. Total number of lattice sites: 96.*

of nonzero sites is $\frac{1}{2}$. The average number of sequences of length $6n$ per site is $\frac{1}{3} \times 2^{n+3}$.

The prediction of the mean-field approximation is very bad. The mean-field map given by

$$\rho \mapsto 1 - \rho^7 - (1 - \rho)^7 = 7\rho(1 - \rho)(1 - 2\rho + 3\rho^2 - 2\rho^3 + \rho^4)$$

is chaotic.

Since the mean-field result should be correct when the number of tentative moves per active site m tends to infinity, the asymptotic density of active sites $\rho(\infty, m)$ should evolve chaotically when m is large enough. Interestingly, increasing m from 0, Boccara and Roger [50] observed for $\rho(\infty, m)$ a complete sequence of period-doubling bifurcations, as illustrated in Figure 6.17. In order to estimate the Feigenbaum number δ, some numerical simulations were done on lattices having $L = 1.6 \times 10^7$ sites to be able to locate the first four bifurcations with a reasonable accuracy. The parameter values of these bifurcation points are $m_1 = 0.340 \pm 0.005$, $m_2 = 0.655 \pm 0.005$, $m_3 = 0.780 \pm 0.001$, and $m_4 = 0.813 \pm 0.001$, yielding $\delta_1 = 2.5$ and $\delta_2 = 3.8$, where $\delta_n = (m_{n+1} - m_n)/(m_{n+2} - m_{n+1})$.

Taking into account the very small number of bifurcations, these results suggest that the period-doubling sequence has the universal behavior found for maps with a quadratic maximum. Plotting $\rho(t + 1, m)$ as a function of $\rho(t, m)$ for different values of m allowed verification of this hypothesis. Note that $\partial\rho(\infty, m)/\partial m$ is infinite at $m = 0$.

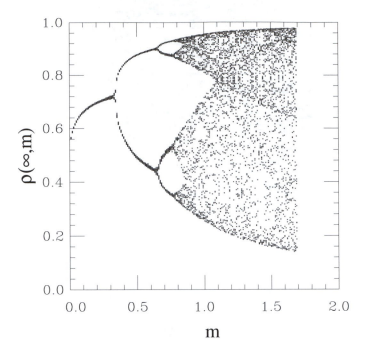

Fig. 6.17. *Bifurcation diagram for the asymptotic density of active sites $\rho(\infty, m)$ of the radius 3 totalistic cellular automaton rule (6.38) (with $S_{\min} = 1$ and $S_{\max} = 6$). The parameter m, which is the number of tentative moves per active site, is plotted on the horizontal axis and varies from 0 to 1.7*

Example 49. SIR epidemic model in a population of moving individuals. In an SIR epidemic model, individuals are divided into three disjoint groups:

1. *susceptible* individuals capable of contracting the disease and becoming infective,
2. *infective* individuals capable of transmitting the disease, and
3. *removed* individuals who have had the disease and are dead, have recovered and are permanently immune, or are isolated until recovery and permanent immunity occurs.

The possible evolution of an individual may, therefore, be represented by the transfer diagram

$$S \xrightarrow{p_i} I \xrightarrow{p_r} R,$$

where p_i denotes the probability for a susceptible to be infected and p_r the probability for an infective to be removed.

In a two-dimensional cellular automaton model, each site of a finite square lattice is either empty or occupied by a susceptible or an infective.

The spread of the disease is governed by the following rules:

1. Susceptibles become infective by contact (*i.e.*, a susceptible may become infective with a probability p_i if, and only if, it is in the neighborhood of an infective). This hypothesis neglects latent periods (*i.e.*, an infected susceptible becomes immediately infective).
2. Infectives are removed (or become permanently immune) with a probability p_r. This assumption states that removal is equally likely among infectives. In particular, it does not take into account the length of time the individual has been infective.
3. The time unit is the time step. During one time step, the two preceding rules are applied after the individuals have moved on the lattice according to a specific rule.
4. An individual selected at random may move to a site also chosen at random. If the chosen site is empty, the individual will move; otherwise, the individual will not move. The set in which the site is randomly chosen depends on the range of the move. To illustrate the importance of this range, two extreme cases may be considered. The chosen site may either be one of the four nearest neighbors (short-range move) or any site of the graph (long-range move). If N is the total number of individuals on $\mathbb{Z}_L^2$, mN individuals, where the positive real number m is the degree of mixing, are sequentially selected at random to perform a move. This sequential process allows some individuals to move more than others.

This model assumes that the population is closed. It ignores births, deaths by other causes, immigrations, and emigrations.

In order to have an idea of the qualitative behavior of the model, let us first study its mean-field approximation.

Denoting by S_t, I_t, and R_t the respective densities of susceptible, infective, and removed individuals at time t, we have

$$S_{t+1} = \rho - I_{t+1} - R_{t+1},$$
$$I_{t+1} = I_t + S_t\left(1 - (1 - p_i I_t)^4\right) - p_r I_t,$$
$$R_{t+1} = R_t + p_r I_t,$$

where ρ is a constant representing the total density of individuals.

From the equations above, it follows that, as functions of t, S_t is positive nonincreasing, whereas R_t is positive nondecreasing. Therefore, the infinite-time limits S_∞ and R_∞ exist. Since $I_t = \rho - S_t - R_t$, it follows also that I_∞ exists and satisfies the relation

$$R_\infty = R_\infty + p_r I_\infty,$$

which shows that $I_\infty = 0$. In such a model, there is no endemic steady state.

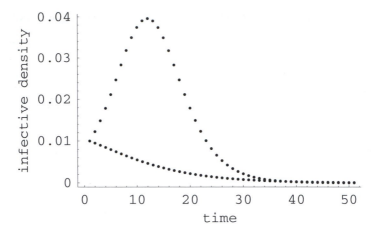

Fig. 6.18. *Time evolution of the density of infective individuals for the one-population SIR epidemic model using the mean-field approximation. The initial densities of susceptible and infective individuals are, respectively, equal to 0.69 and 0.01. For $p_r = 0.6$ and $p_i = 0.3$, an epidemic occurs, whereas there is no epidemic if, for the same value of p_r, $p_i = 0.2$.*

If the initial conditions are

$$R_0 = 0 \quad \text{and} \quad I_0 \ll S_0,$$

I_1 is small, and we have

$$I_1 - I_0 = (4p_i S_0 - p_r)I_0 + O(I_0^2).$$

Hence, according to the initial value of the density of susceptible individuals, we may distinguish two cases:

1. If $S_0 < p_r/4p_i$, then $I_1 < I_0$, and, since S_t is a nonincreasing function of time t, I_t goes monotonically to zero as t tends to ∞. That is, no epidemic occurs.
2. If $S_0 > p_r/4p_i$, then $I_1 > I_0$. The density I_t of infective individuals increases as long as S_t is greater than the threshold p_r/zp_i and then tends monotonically to zero. An epidemic occurred.

This shows that the spread of the disease occurs only if the initial density of susceptible individuals is greater than a threshold value. This threshold theorem was established for the first time by Kermack and McKendrick [182] using an epidemic model formulated in terms of a set of three differential equations (see Example 8, page 45). I_t being, in general, very small, the second recurrence equation of the model is well-approximated by

$$I_{t+1} = I_t + 4p_i S_t I_t - p_r I_t,$$

which shows that the mean-field approximation is equivalent to a time discrete formulation of the Kermack-McKendrick model. Figure 6.18 shows two typical time evolutions of the density of infective individuals.

In the numerical simulations done by Boccara and Cheong [49], the total density of individuals was equal to 0.6, slightly above the site percolation threshold in two dimensions for the square lattice (approximately equal to 0.592746 . . .) in order to be able to see cooperative phenomena. As the degree of mixing m increases, as expected, the density of infective individuals tends to the mean-field result. Long-range moves are a much more effective mixing process. For instance, in the case of an epidemic with permanent removal, simulation and mean-field results almost coincide for $m = 2$ for long-range moves compared to $m = 250$ for short-range moves. In the case of infective individuals recovering and becoming immune, the convergence to mean-field behavior is somewhat slower since the presence of the inert immune population on the lattice interferes with the mixing process.

As shown by Kermack and McKendrick [182], the spread of the disease does not stop for lack of susceptible individuals. For any given value of m, the stationary density of susceptible individuals $S_\infty(m)$ can be small but is always positive.

In the case of short-range moves and permanent removal, $S_\infty(m)$ tends to its mean-field value $S_\infty(\infty)$ as $m^{-\alpha}$ with $\alpha = 1.14 \pm 0.11$. If, instead of short-range moves, we consider long-range moves, the convergence to mean-field behavior is faster: the exponent α is equal to 1.73 ± 0.11. All these results were obtained for $p_i = 0.5$, $p_r = 0.3$, and with 100×100 lattices.

The results above show that the mean-field approximation is valid in the limit $m \to \infty$. A recent analysis of the spatiotemporal behavior of the spread of influenza for the epidemic of winter 1994–1995 [67] reveals a high degree of mixing of the population, making mean-field models appropriate. A possible explanation could be the change in transportation systems, allowing people to move over greater distances more easily and more frequently than in the past.

Example 50. SIS epidemic model in a population of moving individuals. In an SIS epidemic model, individuals are divided into two disjoint groups:

1. *susceptible* individuals capable of contracting the disease and becoming infective, and
2. *infective* individuals capable of transmitting the disease to susceptibles.

If p_i denotes the probability for a susceptible to be infected and p_r the probability for an infective to recover and return to the susceptible group, the possible evolution of an individual may be represented by the following transfer diagram:

$$S \xrightarrow{p_i} I \xrightarrow{p_r} S.$$

In a two-dimensional cellular automaton model, each site of a finite square lattice is either empty or occupied by a susceptible or an infective.

The spread of the disease is governed by the following rules.

1. Susceptible individuals become infected by contact (*i.e.*, a susceptible may become infective with a probability p_i if, and only if, it is in the neighborhood of an infective). This hypothesis neglects incubation and latent periods: an infected susceptible becomes immediately infective.
2. Infective individuals recover and become susceptible again with a probability p_r. This assumption states that recovery is equally likely among infective individuals but does not take into account the length of time the individual has been infective.
3. The time unit is the time step. During one time step, the two preceding rules are applied synchronously, and the individuals move on the lattice according to a specific rule.
4. As in the model above, an individual selected at random may perform either short-range or long-range moves.

Here again, this model assumes that the population is closed. It ignores births, deaths by other causes, immigrations, or emigrations.

Let us first study the qualitative behavior of the model using the mean-field approximation.

Denoting by S_t and I_t the respective densities of susceptible and infective individuals at time t, we have

$$S_{t+1} = \rho - I_{t+1},$$
$$I_{t+1} = I_t + S_t\left(1 - (1 - p_i I_t)^4\right) - p_r I_t,$$

where ρ is a constant representing the total density of individuals. Eliminating S_t in the second equation yields

$$I_{t+1} = (1 - p_r)I_t + (\rho - I_t)\left(1 - (1 - p_i I_t)^4\right).$$

In the infinite-time limit, the stationary density of infective individuals I_∞ satisfies the equation

$$I_\infty = (1 - p_r)I_\infty + (\rho - I_\infty)\left(1 - (1 - p_i I_\infty)^4\right).$$

$I_\infty = 0$ is always a solution to this equation. This value characterizes the *disease-free state*. It is a stable stationary state if, and only if, $4\rho p_i - p_r \leq 0$. If $4\rho p_i - p_r > 0$, the stable stationary state is given by the unique positive solution of the equation above. In this case, a nonzero fraction of the population is infected. The system is in the *endemic state*. For $4\rho p_i - p_r = 0$, the system, within the framework of the mean-field approximation, undergoes a transcritical bifurcation similar to a second-order phase transition characterized by a nonnegative order parameter, whose role is played, in this model, by the stationary density of infected individuals I_∞.

In the numerical simulations done by Boccara and Cheong [52], the total density of individuals was equal to 0.6, slightly above the site percolation threshold in two dimensions for the square lattice in order to be able to observe cooperative effects. Most simulations were performed on a 100×100 lattice and some on a 200×200 lattice to check possible size effects.

In the case of short-range moves, for given values of p_r and m there exists a critical value p_i^c of the probability for a susceptible to become infected. At this bifurcation point, the stationary density of infective individuals $I_\infty(m)$ behaves as $(p_i - p_i^c)^\beta$. When $m = 0$, the exponent β is close to 0.6, which is the value obtained for the two-dimensional directed percolation. When $m = \infty$, that is for the mean-field approximation, $\beta = 1$.

Because it neglects correlations, which play an essential role in the neighborhood of a second-order phase transition, the mean-field approximation cannot correctly predict the critical behavior of short-range interaction systems [45]. For standard probabilistic cellular automata, this is also the case [42, 53].

For a given value of p_r, the variations of β and p_i^c as functions of m are found to exhibit two regimes reminiscent of crossover phenomena. In the small m regime (i.e., for $m \lesssim 10$), p_i^c and particularly β have their $m = 0$ values. In the large m regime (i.e., for $m \gtrsim 300$), p_i^c and β have their mean-field values. In agreement with what is known in phase transition theory, the exponent β does not seem to depend upon p_r; i.e., its value does not change along the second-order transition line.

The fact that, for $m = 0$, the value of β, for this model, is equal to the two-dimensional directed percolation value strongly suggests that the critical properties of this model are universal (i.e., model-independent).

For given values of p_i and p_r, the asymptotic behaviors of the stationary density of infective individuals $I_\infty(m)$ for small and large values of m may be characterized by the exponents

$$\alpha_0 = \lim_{m \to 0} \frac{\log \left(I_\infty(m) - I_\infty(0) \right)}{\log m},$$

$$\alpha_\infty = \lim_{m \to \infty} \frac{\log \left(I_\infty(\infty) - I_\infty(m) \right)}{\log m}.$$

It is found that $\alpha_0 = 0.177 \pm 0.15$ and $\alpha_\infty = -0.945 \pm 0.065$.

The fact that α_0 is rather small shows the importance of motion in the spread of a disease. The stationary number of infective individuals increases dramatically when the individuals start to move. In other words, the response $\partial I_\infty(m)/\partial m$ of the stationary density $I_\infty(m)$ to the degree of mixing m tends to ∞ when m tends to 0.

In the case of long-range moves, for a fixed value of p_r, the variations of p_i^c and β are very different from those for short-range moves. Whereas for short-range moves β and p_i^c do not vary in the small-m regime, for long-range moves, on the contrary, the derivatives of β and p_i^c with respect to m tend to ∞ as m tends to 0. For small m, the asymptotic behaviors of β and p_i^c may therefore be characterized by the exponents

$$\alpha_\beta = \lim_{m \to \infty} \frac{\log \big(\beta(m) - \beta(0)\big)}{\log m},$$

$$\alpha_{p_i^c} = \lim_{m \to 0} \frac{\log \big(p_i^c(m) - p_i^c(0)\big)}{\log m}.$$

Both exponents are found to be close to 0.5.

Example 51. Rift Valley Fever in Senegal. The Rift Valley Fever is a viral disease infecting sheep, cattle, and humans in contact with infected animals. The virus is transmitted among susceptible hosts by mosquitoes such as the *Aedes vexans* and *Culex pipiens*. The eggs may remain in *diapause* (*i.e.*, a period of reduced metabolic activity), until the damp soil on which the eggs are laid is flooded to form a pool suitable for the larvae. A model for the propagation of the disease in Senegal explaining the appearance of endemicity in subsahelian regions has been given by Dubois, Favier, and Sabatier [110]. In order to take into account seasonal migration of herds and shepherds, the authors consider a two-dimensional square lattice in which a viremic herd at a given site can infect, with a probability α, herds located in its von Neumann neighborhood and with a probability β a randomly selected distant site. This model is not, strictly speaking a site-exchange cellular automaton. A herd infecting a distant site does so by moving back and forth between two consecutive time steps, *i.e.*, a given herd always stays at the same location. A time step corresponds to a full year during which the seasonal migration took place. It is clear that interesting results are expected only for nonzero values of both α and β. While for $\alpha \neq 0$ and $\beta = 0$ we have a classical cellular automaton epidemic model, for $\alpha \neq 0$ and $\beta \neq 0$ we have what is called a small-world model (see next chapter), which may exhibit an explosive propagation of the disease.

When the motion of the individuals is an important factor of an agent-based model, instead of a sequential site-exchange rule, one could use a synchronous moving rule defined, for example, as follows. Consider N random walkers on a d-dimensional hypercubic lattice, each lattice site being occupied by at most one random walker. At each time step, each walker decides to move to one of his $2d$ neighboring sites with a probability equal to $1/2d$. If this neighboring site is empty, the walker performs the move; otherwise, he does not. If two or more random walkers choose to move to the same empty site, then one can imagine many rules to resolve this conflict. For example, *timorous walkers* could agree not to move, while *amiable walkers* could agree on randomly selecting one of them to move to the empty site.[34] The multiple amiable random walkers should have, when the number of iterations is small, a degree of mixing slightly higher than the timorous model (see Exercise 6.6). The essential difference between sequential site-exchange rules and multiple

---

[34] A cellular automaton model implementing these rules has been published by Nishidate, Baba, and Gaylord [270].

random walker cellular automaton rules is that the degree of mixing, measured by the reduced Hamming distance between two configurations, can be continuously increased from zero in the first case, whereas it jumps to a finite value after only one iteration in the second case.

6.7 Artificial societies

Lately there has been a growing interest in agent-based modeling in the social sciences.[35] Following Epstein and Axtell, agent-based models of social processes are referred to as *artificial societies*. The aim of these models, which may include [117] "trade, migration, group formation, combat, interaction with an environment, transmission of culture, propagation of disease, and population dynamics," is "to discover fundamental local or micro mechanisms that are sufficient to *generate* the macroscopic social structures and collective behaviors of interest."

Example 52. Spatial segregation. In probably the first attempt at agent-based modeling in the social sciences, Schelling [306, 307] devised a very simple model of spatial segregation in which individuals prefer that a minimum fraction of their neighbors be of their own "color."

Consider a two-dimensional cellular automaton on $\mathbb{Z}_L^2$ (*i.e.*, an $L \times L$ square lattice with periodic boundary conditions). Each site is either occupied by an individual with a probability ρ or empty with a probability $1 - \rho$. Individuals are of two different types. The state of an empty site is 0, while the state of an occupied site is an integer $a \in \{1, 2\}$ representing the individual type (race, gender, social class). The local evolution rule consists of the following steps.

1. Each individual counts how many individuals in his Moore neighborhood (eight sites) share with him the same type.
2. If this number is larger than or equal to four, the individual stays at the same location; if this is not the case, the individual tries to move to a randomly selected nearest-neighboring site.
3. Individuals who decide to move do so according to the multiple random walkers model described on page 242 in which when two or more random walkers are facing the same empty site they agree not to move (timorous walkers).

Starting from a random configuration, Figure 6.19 shows the formation of segregated neighborhoods. Since an unsatisfied individual moves to a randomly selected location, he may dissatisfy his new neighbors who, consequently, will try to move to a new location. Long-range moves not being permitted, some unsatisfied individuals who would like to move to new locations cannot do so if they are trapped in a neighborhood fully occupied

---

[35] See, in particular, Epstein and Axtell [117] and Axelrod [17].

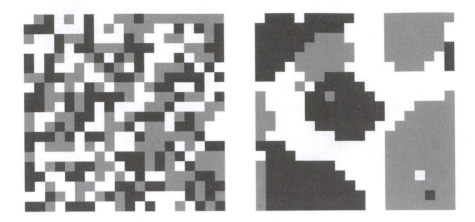

Fig. 6.19. *Spatial segregation. Lattice size:* 20 × 20; *number of individuals: 238 (138 of type 1 and 130 of type 2 represented by light grey and dark grey squares, respectively). The left figure represents the initial random configuration, and the right one shows the formation of segregated neighborhoods after 2000 iterations. Note the existence of two individuals trapped in Moore neighborhoods fully occupied by individuals of a different type.*

by individuals of a different type. Because individuals move away from other individuals of different type, they also move away from empty sites that have no type. As a consequence, empty sites also segregate.

Remark 18. If we increase the number of types, say from two to four (see Figure 6.20), as expected, it takes more iterations to observe well-segregated neighborhoods.

Since individuals were only permitted to perform short-range moves, this model is not a very realistic model of spatial segregation. However, it is rather realistic if we view it as a model of clustering of people, say by gender or generation, at a party. In a more realistic model of spatial segregation, we should allow unsatisfied individuals to perform long-range moves to reach a distant preferable location. A model of spatial segregation allowing individuals to perform long-range moves has been proposed by Gaylord and D'Andria in an interesting book [140] intended for the reader who wishes to create his own computer programs to model artificial societies. As in the model above, individuals of two different types occupy a fraction ρ of the sites of a finite square lattice with periodic boundary conditions. The state of an empty site is 0, and the state of an occupied site is either 1 or 2 according to the individual type. The evolution rule consists of the following steps.

1. Build up the list of the individuals of each type who want to move using the rule that an individual wants to move if his Moore neighborhood contains fewer individuals of his own type than the other type.

Fig. 6.20. *Spatial segregation. Lattice size:* 20×20; *number of individuals: 268 (70, 69, 61, and 68 of, respectively, types 1, 2, 3, and 4, represented by darker shades of grey from light grey to black). The left figure represents the initial random configuration, and the right one shows the formation of segregated neighborhoods after 10,000 iterations.*

2. List the positions of all individuals of type 1 who want to move.
3. List the positions of all individuals of type 2 who want to move.
4. List the positions of all empty sites that are suitable for individuals of type 1.
5. List the positions of all empty sites that are suitable for individuals of type 2.
6. If the length of the list of a given type of individual who wants to move is not equal to the list of empty sites suitable to that type of individual, match the lengths of the two lists by randomly eliminating elements of the longest list.
7. Carry out the moves by swapping the positions of individuals of a given type who want to move with the positions of the empty sites suitable to that type of individual.

As expected, it takes much fewer iterations to observe the formation of segregated neighborhoods when individuals are permitted long-range moves. Figure 6.21 shows the initial random configuration and the resulting configuration after 200 iterations. Note that in this model an individual does take into account empty sites in deciding to stay or to move. This is why, in contrast with the previous model, there is no segregation of empty sites.

Example 53. Dissemination of culture. Axelrod [16] has proposed a model for the dissemination of culture in which an individual's culture is described by a list of *features* such as political affiliation or dress style. For each feature there is a set of *traits*, which are the alternative values the feature may have, such

Fig. 6.21. *Spatial segregation allowing long-range moves. Lattice size:* 20×20*; number of individuals: 254 (125 of type 1 and 129 of type 2 represented by light grey and dark grey squares, respectively). The left figure represents the initial random configuration, and the right one shows the formation of segregated neighborhoods after 200 iterations.*

as the different colors of a piece of clothing. If we assume that there are five features and each feature can take on any one of ten traits, then the culture of an individual is represented by a sequence of five digits such as $(4, 5, 9, 0, 2)$. The *degree of cultural similarity* between two individuals is defined as the percentage of their features that have the identical trait. For example, the degree of cultural similarity between $(4, 5, 9, 0, 2)$ and $(4, 7, 5, 0, 2)$ is equal to 60%, for these two sequences share traits for three of the five cultural features.

The basic idea is that individuals who are similar to each other are likely to interact and then become even more similar. This process of social influence is implemented by assuming that the probability of interaction between two neighboring individuals is equal to their degree of similarity.

The local evolution rule consists of the following steps.

1. Select at random a site to be active and one of its neighbors.
2. Select at random a feature on which the active site and its neighbor differ, and change the active site's trait on this feature to the neighbor's trait on this feature with a probability equal to the degree of similarity between the two sites. If the two sites have a degree of similarity equal to 100%, nothing happens.

Applying this evolution rule repeatedly, Axelrod found that cultural features tend to be shared over larger and larger regions. Since two neighboring sites with completely different cultures cannot interact, the process eventually stops with several surviving cultural regions.

An interesting result concerns the influence of the number of features and number of traits per feature on the number of surviving cultural regions:

The number of surviving cultural regions is much smaller if there are many cultural features with few traits per feature than if there are few cultural features with many traits per feature.

Increasing the number of cultural features increases the interaction probability since there is a greater chance that two neighboring sites will have the same trait on at least one feature. On the contrary, increasing the number of traits per feature has the opposing effect since, in this case, there is a smaller chance that two neighboring sites will have the same trait.

In the original Axelrod model, there is no movement and the sites may be viewed as villages. Gaylord and D'Andria [140] proposed a modified version of the Axelrod model incorporating random motion of the individuals[36] and bilateral, instead of unilateral, cultural exchange. Their model is a two-dimensional cellular automaton on $\mathbb{Z}_L^2$. Each site is either occupied by an individual with a probability ρ or empty with a probability $1 - \rho$. The state of an empty site is 0, while the state of an occupied site is an ordered pair (i, j). The first element i is a list of s specific traits of cultural features, each trait having an integer value between 1 and m. The second element j, randomly selected at each time step in the set $\{n, e, s, w\}$, indicates the direction faced by the individual (*i.e.*, north, east, south or west).

The local evolution rule consists of two subrules.

1. The first subrule is a bilateral cultural exchange rule: two individuals facing each other interact if they have some, but not all, traits in common. The interaction is carried out by randomly selecting one feature with traits having different values and changing these traits to a common value for both individuals with a probability equal to the degree of cultural similarity between the two individuals. The new value taken by the trait is randomly selected between the two differing values. For example, if a north-facing individual's traits list is $(1, 4, 2, 7, 9)$ and her south-facing neighbor's traits list is $(1, 8, 2, 7, 5)$, these individuals will change either their second or fifth feature trait with a probability equal to 0.6. If the fifth trait is randomly chosen, then the new lists will be $(1, 4, 2, 7, x)$ and $(1, 8, 2, 7, x)$, where x is an integer selected at random in the set $\{5, 6, 7, 8, 9\}$.

2. The second rule is the multiple random walkers moving rule defined on page 242 in which when two or more random walkers are facing the same empty site they agree not to move (timorous walkers).

---

[36] Mobility had already been included in Axelrod's model and was shown to result in fewer stable cultural regions. See Axtell, Axelrod, Epstein, and Cohen [18]. See an adapted version in Axelrod [17].

Fig. 6.22. *Dissemination of culture among individuals, each of whom has two possible traits for each of their two features. Empty sites are white, and the traits lists* $(1,1)$, $(1,2)$, $(2,1)$, *and* $(2,2)$ *are darker shades of grey from light grey to black. Lattice size:* 20×20; *number of individuals: 335. The left figure represents the initial random configuration, and the right one is the resulting configuration after 10,000 iterations.*

Figure 6.22 shows the dissemination of culture among 335 individuals on a 20×20 lattice. Each individual has two possible traits for each of their two features. The table below gives the number of individuals for each traits list for the initial random configuration and for the configuration obtained after 10,000 time steps.

traits list	$(1,1)$	$(1,2)$	$(2,1)$	$(2,2)$
initial configuration	102	89	70	74
final configuration	211	0	114	10

The traits list $(1,2)$ has completely disappeared, and only ten individuals having the traits list $(2,2)$ remain. The dominant "cultures" are $(1,1)$ and $(2,1)$.

In the social sciences, the interaction between members of a population is often modeled as a *game*. Game theory was originally designed by John von Neumann and Oskar Morgenstern [258] to solve problems in economics.

In a two-person game, the players try to outsmart one another by anticipating each other's strategies. The "stone-paper-scissors" game played by most children makes clear what is meant by *strategy*. The game's rules are:

1. scissors defeats paper,
2. paper defeats stone,
3. stone defeats scissors.

Assuming that the winner wins one unit from the loser, the outcomes are listed in the following table:

player B

	stone	*paper*	*scissors*
stone	$(0, 0)$	$(-1, 1)$	$(1, -1)$
player A *paper*	$(1, -1)$	$(0, 0)$	$(-1, 1)$
scissors	$(-1, 1)$	$(1, -1)$	$(0, 0)$

Each player has three strategies: "stone," "paper," and "scissors." The number of units received by players A and B adopting, respectively, strategies i and j are given by the ordered pair (x_{ij}, y_{ij}). For example, if A adopts strategy "scissors" and B strategy "paper," A wins one unit and B loses one unit, represented by $(1, -1)$. The entries in the table above are referred to as *payoffs*, and the table of entries is the *payoff matrix*. Usually, only the payoffs received by player A from player B are listed, and the payoff matrix of the "stone-paper-scissors" game is

$$\begin{bmatrix} 0 & -1 & 1 \\ 1 & 0 & -1 \\ -1 & 1 & 0 \end{bmatrix}. \tag{6.39}$$

In the "stone-paper-scissors" game, the gain of player A equals the loss of player B, and vice versa. Such a game is said to be a *zero-sum game*. Since the game is entirely determined by the payoff matrix, it is called a *matrix game*.

Suppose players A and B have, respectively, m and n strategies, and let a_{ij} be the payoff A receives from B. The payoff matrix is

$$A = \begin{bmatrix} a_{11} & a_{12} & \cdots & a_{1n} \\ a_{21} & a_{22} & \cdots & a_{2n} \\ \cdots & \cdots & \cdots & \cdots \\ a_{m1} & a_{m2} & \cdots & a_{mn} \end{bmatrix}.$$

If player A adopts strategy i, his payoff is at least

$$\min_{1 \le j \le n} a_{ij}.$$

Since player A wishes to maximize his payoff, he should choose strategy i so as to receive a payoff not less than

$$\max_{1 \le i \le m} \min_{1 \le j \le n} a_{ij}.$$

If player B adopts strategy j, his loss is at most

$$\max_{1 \le i \le m} a_{ij}.$$

Since player B wishes to minimize his loss, he should choose strategy j so as to lose not more than

$$\min_{1 \le j \le n} \max_{1 \le i \le m} a_{ij}.$$

Theorem 19. *Let the payoff matrix of a matrix game be* $[a_{ij}]$; *then*

$$\max_{1 \leq i \leq m} \min_{1 \leq j \leq n} a_{ij} \leq \min_{1 \leq j \leq n} \max_{1 \leq i \leq m} a_{ij}. \tag{6.40}$$

Since, for all $i = 1, 2, \ldots, m$ and for all $j = 1, 2, \ldots, n$, we have

$$\min_{1 \leq j \leq n} a_{ij} \leq a_{ij} \quad \text{and} \quad a_{ij} \leq \max_{1 \leq i \leq m} a_{ij},$$

for all (i, j), we have

$$\min_{1 \leq j \leq n} a_{ij} \leq \max_{1 \leq i \leq m} a_{ij}.$$

The left-hand side of the inequality above does not depend upon j; hence, taking the minimum with respect to j of both sides yields

$$\min_{1 \leq j \leq n} a_{ij} \leq \min_{1 \leq j \leq n} \max_{1 \leq i \leq m} a_{ij}.$$

This result being true for all i, taking the maximum with respect to i of both sides, we finally obtain inequality (6.40). $\square$

If the elements of the payoff matrix $[a_{ij}]$ satisfy the relation

$$\max_{1 \leq i \leq m} \min_{1 \leq j \leq n} a_{ij} = \min_{1 \leq j \leq n} \max_{1 \leq i \leq m} a_{ij}, \tag{6.41}$$

this common value is called the *value of the game*, and there exist i_s and j_s such that

$$\min_{1 \leq j \leq n} a_{i_s, j} = \max_{1 \leq i \leq m} \min_{1 \leq j \leq n} a_{ij}$$

and

$$\max_{1 \leq i \leq m} a_{i, j_s} = \min_{1 \leq j \leq n} \max_{1 \leq i \leq m} a_{ij}.$$

Thus,

$$\min_{1 \leq j \leq n} a_{i_s, j} = \max_{1 \leq i \leq m} a_{i, j_s}.$$

But

$$\min_{1 \leq j \leq n} a_{i_s, j} \leq a_{i_s j_s} \leq \max_{1 \leq i \leq m} a_{i, j_s}.$$

Hence,

$$\min_{1 \leq j \leq n} a_{i_s, j} = a_{i_s j_s} = \max_{1 \leq i \leq m} a_{i, j_s}.$$

Therefore,

$$a_{i_s j} \geq a_{i_s j_s} \geq a_{i j_s}, \tag{6.42}$$

which shows that $a_{i_s j_s}$ is equal to the value of the game. Relation (6.42) shows that whichever strategy is adopted by player B, if player A adopts strategy i_s, the payoff cannot be less than the value of the game, and whichever strategy is adopted by player A, if player B adopts strategy j_s, the payoff cannot be greater than the value of the game. This is the reason why strategies i_s and j_s are said to be *optimal strategies* for, respectively, players A and B. The pair (i_s, j_s) is called a *saddle point of the game*, and $(i, j) = (i_s, j_s)$ is a *solution of the game*.

The existence of a saddle point ensures that Relation (6.41) is satisfied. In this case, the element $a_{i_s j_s}$ of the payoff is, at the same time, the minimum of its row and the maximum of its column. Using this remark, we verify that the "stone-paper-scissors" game has no saddle point.

Saddle points are not necessarily unique. If (i_{s_1}, j_{s_1}) and (i_{s_2}, j_{s_2}) are two different saddle points, then the relations

$$a_{ij_{s_1}} \le a_{i_{s_1} j_{s_1}} \le a_{i_{s_1} j} \quad \text{and} \quad a_{ij_{s_2}} \le a_{i_{s_2} j_{s_2}} \le a_{i_{s_2} j},$$

which are valid for all i and j, imply

$$a_{i_{s_1} j_{s_1}} \le a_{i_{s_1} j_{s_2}} \le a_{i_{s_2} j_{s_2}} \le a_{i_{s_2} j_{s_1}} \le a_{i_{s_1} j_{s_1}};$$

that is,

$$a_{i_{s_1} j_{s_1}} = a_{i_{s_1} j_{s_2}} = a_{i_{s_2} j_{s_2}} = a_{i_{s_2} j_{s_1}} = a_{i_{s_1} j_{s_1}}.$$

Hence, if (i_{s_1}, j_{s_1}) and (i_{s_2}, j_{s_2}) are two different saddle points, (i_{s_1}, j_{s_2}) and (i_{s_2}, j_{s_1}) are also saddle points.

When saddle points exist, suitable strategies are clear, but if, as in the "stone-paper-scissors" game, there are no saddle points—that is, when the elements of the payoff matrix $[a_{ij}]$ are such that

$$\max_{1 \le i \le m} \min_{1 \le j \le n} a_{ij} \le \min_{1 \le j \le n} \max_{1 \le i \le m} a_{ij}$$

—the game has no solution in the sense given above. In such a game, players cannot select optimal strategies; they are led to adopt *mixed strategies*. Let $[a_{ij}]_{\substack{1 \le i \le m \\ 1 \le j \le n}}$ be the payoff matrix of a matrix game. Mixed strategies for players A and B, respectively, are sequences $(x_i)_{1 \le i \le m}$ and $(y_j)_{1 \le j \le m}$ of nonnegative numbers such that

$$\sum_{i=1}^m x_i = 1 \quad \text{and} \quad \sum_{j=1}^n y_j = 1. \tag{6.43}$$

If player A adopts strategy i with probability x_i and player B adopts strategy j with probability y_j, then the expected payoff of player A is

$$\sum_{i=1}^m \sum_{j=1}^n a_{ij} x_i y_j.$$

If S_m (resp. S_n) denotes the set of all sequences $X = (x_i)_{1 \le i \le m}$ (resp. $Y = (y_j)_{1 \le j \le n}$) of all nonnegative numbers satisfying Conditions (6.43), then the expected payoff of player A using the mixed strategy $X = (x_i) \in S_m$ is at least

$$\min_{Y \in S_n} \sum_{i=1}^{m} \sum_{j=1}^{n} a_{ij} x_i y_j,$$

and, to maximize this quantity, he should choose a mixed strategy $X \in S_m$ such that his expected payoff would be at least equal to

$$\max_{X \in S_m} \min_{Y \in S_n} \sum_{i=1}^{m} \sum_{j=1}^{n} a_{ij} x_i y_j.$$

Similarly, player B should choose a mixed strategy $Y \in S_n$ to prevent player A from receiving an expected payoff greater than

$$\min_{Y \in S_n} \max_{X \in S_m} \sum_{i=1}^{m} \sum_{j=1}^{n} a_{ij} x_i y_j.$$

Proceeding as we did to prove Theorem 19, it can be shown that

$$\max_{X \in S_m} \min_{Y \in S_n} \sum_{i=1}^{m} \sum_{j=1}^{n} a_{ij} x_i y_j \leq \min_{Y \in S_n} \max_{X \in S_m} \sum_{i=1}^{m} \sum_{j=1}^{n} a_{ij} x_i y_j.$$

Actually, John von Neumann proved in 1928 that for all finite[37] two-person zero-sum games[38]

$$\max_{X \in S_m} \min_{Y \in S_n} \sum_{i=1}^{m} \sum_{j=1}^{n} a_{ij} x_i y_j = \min_{Y \in S_n} \max_{X \in S_m} \sum_{i=1}^{m} \sum_{j=1}^{n} a_{ij} x_i y_j. \qquad (6.44)$$

This important result is known as the *minimax theorem*. The pair of sequences $(X_s, Y_s) \in S_m \times S_n$ is said to be a solution of the game, and X_s and Y_s are optimal mixed strategies for, respectively, players A and B. The common value of both sides of Relation (6.44) is called the *value of the game*.

In a two-person zero-sum game, the players have conflicting interests: what is gained by one player is lost by the other. In a nonzero-sum game, this is not necessarily the case. Consider, for example, the celebrated *prisoner's dilemma*. Two prisoners, A and B, suspected of committing a serious crime are isolated in different cells and each is asked whether the other is guilty. Both prisoners want, naturally, to get the shortest prison sentence, and both know the consequences of their decisions. Each prisoner has two strategies: to accuse or not to accuse the other.

1. If both defect—that is, each of them accuses the other—both go to jail for five years.

---

[37] That is, all matrix games having a finite number of strategies.
[38] See von Neumann and Morgenstern [258] and, for an alternative simple proof, Jianhua [180].

2. If both cooperate—that is, neither of them accuses the other—both go to jail for one year, say, on account of carrying concealed weapons.
3. If one defects while the other does not, the former goes free and the latter goes to jail for 15 years.

For both prisoners, the only rational choice is to defect. Although each prisoner ignores what the other will do, if he defects, he will either get five years instead of 15 if the other also defects or go free if the other does not defect. The paradox is that when both prisoners act selfishly and do not cooperate, they both get a five-year sentence, whereas both could have been sentenced only one year if they had cooperated.

More formally, in the two-person prisoner's dilemma game, if both players cooperate, they receive a payoff R (reward); if both defect, they receive a lesser payoff P (punishment); and if one defects while the other cooperates, the defector receives the largest payoff T (temptation), while the cooperator receives the smallest S (sucker), as shown in the following table

	cooperation	*defection*
cooperation	(R, R)	(S, T)
defection	(T, S)	(P, P)

where $T > R > P > S$ and $R > \frac{1}{2}(T + S)$, so that, in an iterated game, in which players can modify their strategies at each time step, each player gains on average more per game by agreeing to repeatedly cooperate (R) than by agreeing to alternately cooperate and defect out-of-phase $\left(\frac{1}{2}(T + S)\right)$.

As emphasized by Axelrod and Hamilton in their 1981 paper [15], if two players can meet only once, the best strategy is to defect, but if

> *an individual can recognize a previous interactant and remember some aspects of the previous outcomes, then the strategic situation becomes an iterated Prisoner's Dilemma with a much richer set of possibilities. A strategy would take the form of a decision rule which determined the probability of cooperation or defection as a function of the history of the interaction so far.*

Here are a few examples of evolving strategies adopted by players on a square lattice with periodic boundary conditions.

Example 54. Iterated prisoner's dilemma game: individuals have memory of past encounters. On an $L \times L$ lattice with periodic boundary conditions $\mathbb{Z}_L^2$, each site is either occupied by an individual with probability ρ or empty with probability $1 - \rho$. The state of an empty site is 0, while the state of an occupied site is a five-tuple where:

1. the first element is an integer representing the individual's name;
2. the second element is either d or c, indicating whether the individual is a defector or a cooperator;

3. the third element is a list of defectors who have interacted with the individual in previous encounters;
4. the fourth element is an integer representing the sum of the payoffs from previous time steps;
5. the last element, randomly selected at each time step in the set $\{n, e, s, w\}$, indicates the direction faced by the individual (*i.e.*, north, east, south or west).

The local evolution rule consists of the following two subrules.

1. Two facing individuals decide to interact if neither of them is on the other individual's list of previously encountered defectors.
2. The second rule is the multiple random walkers moving rule defined on page 242 in which when two or more random walkers are facing the same empty site they agree not to move (timorous walkers).

In agreement with what is found in the literature, we used the payoff values $T = 5$, $R = 3$, $P = 1$, and $S = 0$. Figure 6.23 shows that, in the long run, cooperators do better than defectors.

In this model, information about defectors is kept private. If individuals are allowed to share information, the average gain of defectors eventually declines.[39]

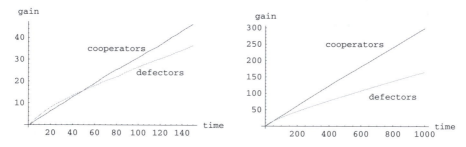

Fig. 6.23. *Average gain of defectors and cooperators in an iterated prisoner's dilemma game in which individuals refuse to interact with previously encountered defectors. Lattice size: 30×30, number of cooperators: 360, number of defectors: 355, total number of iterations: 1000. The left figure represents the average gain for each population for the first 150 iterations, showing that at the beginning the average gain of defectors is higher.*

In Example 54, each individual always adopted the same strategy; and if the cooperators' average gain was higher in the long run than the defectors' average gain, this was due to the fact that the former refused to interact with defectors they had previously encountered.

---

[39] See Gaylord and D'Andria [140], p. 57.

While defection is the unbeatable strategy if individuals know they will never meet again, when after the current encounter the two opponents will meet again with a nonzero probability w, strategic choices become more complex. To find out what type of strategy was the most effective, Axelrod in 1980 conducted two computer tournaments [13, 14]. The strategies were submitted by a variety of people, including game theorists in economics, sociology, and political science or professors of evolutionary biology, physics, and computer science. In both cases, the strategies were paired in a round-robin tournament, and in both cases the winner was Anatol Rapoport,[40] who submitted the simplest strategy, called *tit for tat*. This strategy consists in cooperating on the first encounter (*i.e.*, being nice) and then doing whatever the opponent did in the previous encounter (*i.e.*, being provoking and forgiving).

Axelrod and Hamilton [15] argued that, if two individuals have a probability w of meeting again greater than a threshold value, tit for tat is an *evolutionarily stable strategy*, which means that *a population of individuals using that strategy cannot be invaded by a rare mutant adopting a different strategy*.[41] They determined the w threshold value as follows.

1. If two players adopt the tit-for-tat strategy, every one receives a payoff R at each time step, that is, a total payoff[42]

$$R + Rw + Rw^2 + \cdots = \frac{R}{1-w}.$$

2. If a player adopts the tit-for-tat strategy against another player who always defects, the latter receives a payoff T the first time and P thereafter, so it cannot invade a population of tit-for-tat strategists if

$$\frac{R}{1-w} \geq T + \frac{Pw}{1-w}$$

or

$$w > \frac{T-R}{T-P}. \tag{6.45}$$

3. If a player adopts the tit-for-tat strategy against another player who alternates defection and cooperation, the latter receives the sequence of payoffs $T, S, T, S, \ldots$, and so cannot invade a population of tit-for-tat strategists if

$$\frac{R}{1-w} \geq \frac{T+Sw}{1-w^2}$$

---

[40] An interesting early discussion of the prisoner's dilemma game "addressed specifically to experimental psychologists and generally to all who believe that knowledge about human behavior can be gained by the traditional method of interlacing theoretical deductions with controlled observations" can be found in Rapoport and Chammah with the collaboration of Carol Orwant [293].

[41] On the concept of stability in evolutionary game theory, see Maynard Smith [235] and Cressman [96].

[42] The game is supposed to go on forever.

or

$$w > \frac{T - R}{R - S}.$$ (6.46)

From these two results, it follows that no mutant adopting one of these two strategies can invade a population of tit-for-tat strategists if w is such that

$$w > \max\left(\frac{T - R}{T - P}, \frac{T - R}{R - S}\right).$$ (6.47)

According to Axelrod and Hamilton [15], if w satisfies (6.47), a population of tit-for-tat strategists cannot be invaded not only by a mutant who either always defects or alternates defection and cooperation but also, as a consequence, by any mutant adopting another strategy. This result, which would imply that the tit-for-tat strategy is evolutionarily stable, has been questioned by Boyd and Lorberbaum [70] (see also May [232]). Essentially, the argument of these authors is that Axelrod and Hamilton proved that, if tit for tat becomes common in a population where w is sufficiently high, it will remain common because such a strategy is *collectively stable*. A strategy S_e is collectively stable if, for any possible strategy S_i,

$$V(S_e \mid S_e) \geq V(S_i \mid S_e),$$

where $V(S_1 \mid S_2)$ is the expected fitness of individuals who use strategy S_1 when interacting with individuals using strategy S_2. But collective stability does not imply evolutionary stability.[43]

Tit for tat is nevertheless a very successful strategy, and there have been a few attempts to test tit for tat as a model to explain cooperative behavior among animals in nature. It seems that the first test was reported by Lombardo [211], who examined the interactions between parent and nonbreeding tree swallows (*Tachycineta bicolor*) and showed that parents and nonbreeders seem to resolve their conflicting interests by apparently adopting a tit-for-tat strategy. He used stuffed model nonbreeders and simulated acts of defection by nonbreeders by making as though nonbreeders had killed nestlings. A more sophisticated test has been reported by Milinski [244], who studied the behavior of three-spined sticklebacks (*Gasterosteus aculeatus*) when approaching a potential predator to ascertain whether it poses a threat or not. Using a system of mirrors, single three-spined sticklebacks "inspecting" a predator were provided with either a simulated cooperating companion or a simulated defecting one. In both cases, the test fish adopted the tit-for-tat strategy.

Example 55. Iterated prisoner's dilemma game: tit-for-tat strategists against defectors. Consider a population of mobile players in which most individuals adopt the tit-for-tat strategy and the rest always defect. As illustrated in Figure 6.24, the average gain of individuals adopting the tit-for-tat strategy is eventually higher than the average gain of a small fraction of individuals who always defect.

---

[43] See Exercise 6.10.

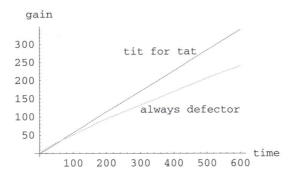

Fig. 6.24. *Average gain of individuals adopting the tit-for-tat strategy or being always the defector in an iterated prisoner's dilemma game. Lattice size: 25 × 25; number of individuals who always defect: 62; number of individuals adopting the tit-for-tat strategy: 459; total number of iterations: 600.*

Example 56. Iterated prisoner's dilemma game: imitating the most successful neighbor. Nowak and May [271] have obtained fascinating results concerning the evolution of cooperation among individuals located at the sites of a two-dimensional square lattice. The individuals interact with their neighbors through simple deterministic rules and have no memory. The evolution produces constantly changing surprising spatial patterns reminiscent of Persian carpets, in which both cooperators and defectors persist forever.

Consider an $L \times L$ square lattice in which initially all sites are occupied by cooperators except a few sites located in the middle of the lattice and disposed symmetrically that are occupied by defectors. At each time step, each individual plays with all the individuals located in his Moore neighborhood, including himself. He then compares his payoff to the payoffs of his neighbors and adopts at the next time step the strategy of the most successful neighbor. The outcome depends on the initial configuration and the value of the single parameter b characterizing the reduced payoff values:

$$S = P = 0, \quad R = 1, \quad T = b > 1.$$

For $b = 2$, successive patterns are shown in Figure 6.25. Note that regions occupied by individuals who defect or cooperate twice in a row are separated by narrow boundaries of individuals who just changed their strategy.

Since the early 1980s, the iterated prisoner's dilemma game has become the most important model for studying the evolution of cooperation among a population of selfish individuals, and a very large number of articles have been published on the subject.

Fig. 6.25. *Successive patterns generated from an initial state in which all individuals located at the sites of a 60×60 square lattice are cooperators except the four sites $(28, 28)$, $(28, 32)$, $(32, 28)$, and $(32, 32)$, who are defectors. A site is colored based on the strategies adopted at time $t - 1$ and t by the individual located at that site according to the following code: if the individual defects at times $t - 1$ and t, the site is black; if the individual defects at $t - 1$ and cooperates at t, the site is dark grey; if the individual cooperates at $t - 1$ and defects at t the site is light grey, and if the individual cooperates at $t - 1$ and t, the site is very light grey.*

Exercises

Exercise 6.1 *We have seen that configurations in the limit set of elementary cellular automaton rule 18 consist of sequences of 0s of odd lengths separated by isolated 1s as illustrated in Figure 6.6. The evolution towards the limit set can be viewed as interacting particle-like defects.[44] Defects in the case of rule 18 are only of one type, namely two*

---
[44] Boccara, Nasser, and Roger [48]

consecutive 1s, which generate a sequence of an even number of 0s. These defects (kinks) were first studied by Grassberger [154], who showed that they perform a random walk. When two defects meet, they annihilate, and this implies, in the case of an infinite lattice, that the density of these defects decreases as $t^{-1/2}$, where t is the number of iterations.

Illustrate this annihilation process starting from an initial configuration containing only two defects.

Exercise 6.2 Let $\mathcal{S} = \{0,1\}^{\mathbb{Z}}$ be the set of all configurations and M a surjective mapping on $\mathcal{S}$.[45]

(i) If F is a global one-dimensional cellular automaton rule of radius r, show that there exists a unique global cellular automaton rule Φ of radius not greater than r such that[46]

$$\Phi \circ M = M \circ F \qquad \text{(R)}$$

if, and only if, for any configuration $c \in \mathcal{S}$, the image by $M \circ F$ of the set $M^{-1}(c)$ contains only one element. When Φ exists, it is called the transform of F by M. Prove that, if Φ exists, it is unique.

(ii) It can be shown that an n-input mapping M is surjective if, and only if, each n-block has exactly the same number of preimages. Using this result, show that the 3-input mapping M_{30}, whose Wolfram code number is 30 (Wolfram code numbers are defined on page 192), is surjective.

(iii) Determine all radius 1 rules F that have a transform Φ under the 3-input surjective mapping 30.

Exercise 6.3 Consider a string of 0s and 1s of finite length L; the density classification problem is to find a two-state cellular automaton rule f_ρ such that the evolution of the string according to f_ρ (assuming periodic boundary conditions) converges to a configuration that consists of 1s only (resp. 0s only) if the density of 1s (resp. 0s) is greater than (resp. less than) a given value ρ. It has been found that there exists no such rule [199]. The problem can, however, be solved using two cellular automaton rules. Following Fukś [133], if $\rho = \frac{1}{2}$, the solution is first to let the string evolve according to radius 1 rule 184 for a number of time steps of the order of magnitude of half of the string length and then again to let the last iterate evolve according to radius 1 rule 232 for a number of time steps of the order of magnitude of half the string length. Justify this result.

Exercise 6.4 A two-lane one-way car traffic flow consists of two interacting lanes with cars moving in the same direction with the possibility for cars blocked in their lane to shift to the adjacent lane. If we assume that the road is a circular highway with neither entries nor exits, such a system may be modeled as a one-dimensional cellular automaton with periodic boundary conditions in which the state of cell i is a two-dimensional vector

$$\mathbf{s}_i = (s_i^1, s_i^2), \quad \text{where} \quad (s_i^1, s_i^2) \in \{0,1\} \times \{0,1\},$$

---

[45] A mapping M on $\mathcal{S}$ is surjective if, for all $c \in \mathcal{S}$, the set of all preimages $M^{-1}(c)$ of c by M is not empty. That is, each configuration $c \in \mathcal{S}$ has at least one preimage.

[46] M has to be surjective because the domain of definition of Φ has to be the whole space $\mathcal{S}$.

s_i^1 and s_i^2 representing, respectively, the occupation numbers of lane 1 and lane 2 at location i, as shown below.

...	s_{i-1}^1	s_i^1	s_{i+1}^1	...
...	s_{i-1}^2	s_i^2	s_{i+1}^2	...

If the speed limit is $v_{\max} = 1$, we may adopt the following moving rule:

1. If the site ahead of a site occupied by a car is empty, the car moves to that site.
2. If the site ahead of a site occupied by a car is occupied, the car shifts to the adjacent site if this site and the site behind are both empty.

This rule is a radius 1 rule of the form

$$(s_{i,1}, s_{i,2})_{t+1} = \mathbf{f}\big((s_{i-1,1}, s_{i-1,2})_t, (s_{i,1}, s_{i,2})_t, (s_{i+1,1}, s_{i+1,2})_t, \big).$$

In traffic engineering, the car flow—that is, the product of the car density by the average car velocity—as a function of the car density is called the fundamental diagram. Write down explicitly traffic rule $\mathbf{f}$, and determine the fundamental diagram of this model.

Exercise 6.5 Let

$$\mathbf{Q} = \underbrace{\{0, 1, 2, \ldots, q-1\} \times \cdots \times \{0, 1, 2, \ldots, q-1\}}_{\ell}.$$

An n-input one-dimensional cellular automaton ℓ-vectorial rule

$$(\mathbf{x}_1, \mathbf{x}_2, \cdots, \mathbf{x}_n) \mapsto \mathbf{f}(\mathbf{x}_1, \mathbf{x}_2, \cdots, \mathbf{x}_n)$$

where, for $i = 1, 2, \ldots, n$, $\mathbf{x}_i \in \mathbf{Q}$, is number-conserving if, for all cyclic configurations of length $L \geq n$,

$$\sum_{j=1}^{\ell} \Big(f_j(\mathbf{x}_1, \mathbf{x}_2, \ldots, \mathbf{x}_n) + f_j(\mathbf{x}_2, \mathbf{x}_3, \ldots, \mathbf{x}_{n+1}) + \cdots + f_j(\mathbf{x}_L, \mathbf{x}_1 \ldots, \mathbf{x}_{n-1}) \Big)$$

$$= \sum_{j=1}^{\ell} (x_{1,j} + +x_{2,j} + \cdots + x_{L,j}),$$

where $(f_1, f_2, \ldots, f_\ell)$ are the components of $\mathbf{f}$ and $(x_{i,1}, x_{i,2}, \ldots, x_{i,\ell})$ the components of $\mathbf{x}_i$. Extend the proof of Theorem 13 and find the necessary and sufficient condition for vectorial rule $\mathbf{f}$ to be number-conserving.

Exercise 6.6 In order to evaluate the degree of mixing of the multiple random walkers cellular automaton model described at the end of Section 6.6 (Nishidate-Baba-Gaylord model), study, as a function of the number of time steps, the reduced Hamming distance between the initial configuration and the configuration obtained after t time steps.

Exercise 6.7 The Eden model [114] is one of the simplest growth models. Consider a finite square lattice with free boundary conditions, and suppose that, at time t, some sites are occupied while the other sites are empty. The growth rule consists of selecting at random an empty site, the nearest neighbor of an occupied site, and occupying that site at $t + 1$. Assume that at $t = 0$ only the central site of the lattice is occupied.

According to Eden, this process may reasonably represent the growth of bacterial cells or tissue cultures of cells that are constrained from moving. For example, the common sea lettuce (Ulva lactuca) grows as a sheet only two cells thick at its periphery. Write a small program to generate Eden clusters.

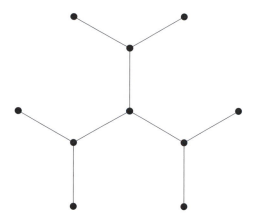

Fig. 6.26. *Bethe lattice for* $z = 3$.

Exercise 6.8 *In most lattice models, the existence of closed paths (also called* loops*) causes difficulties in finding exact solutions. In statistical physics, special lattices with no loops have been very helpful in understanding critical phenomena. Such lattices, called* Bethe lattices, *are undirected graphs in which all vertices have the same degree* z. *An example for* $z = 3$ *is represented in Figure 6.26.*

(i) Find the site percolation critical probability of a Bethe lattice, and show that the critical exponent β, *characterizing the critical behavior of the site percolation probability* $P_\infty^{\mathrm{site}}(p)$ *for* $p > p_c^{\mathrm{site}}$, *is equal to 1 for all values of* z. *Examine the case* $z = 3$.

(ii) Answer the same questions for the bond percolation model.

Exercise 6.9 *Motivated by the problem of one fluid displacing another from a porous medium under the action of capillary forces, Wilkinson and Willemsen [345] considered a new kind of percolation problem, which may describe any kind of invasion process proceeding along a path of least resistance. The invasion percolation model considered here consists of a two-dimensional square lattice in which a random number, uniformly distributed in the interval* $[0, 1]$, *is assigned to each lattice site. A cluster of invaders grows by occupying the lattice site of its perimeter that has the smallest random number. By perimeter of a cluster we mean all the sites that are nearest neighbors of cluster sites. Assuming that clusters of invaders grow from a single site localized at the origin, write and run a program to generate invader clusters.*

Exercise 6.10 *On page 256 we mentioned that Boyd and Lorberbaum questioned the evolutionary stability of the tit-for-tat strategy. In their discussion, they considered the following two strategies:*

1. *the* tit for two tats *which allows two consecutive defections before retaliating, and*
2. *the* suspicious tit for tat, *which defects on the first encounter but thereafter plays tit for tat.*

   Compare how tit for tat and tit for two tats behave against suspicious tit for tat, and show that tit for tat is not evolutionarily stable.

Exercise 6.11 *In a society, individuals often adopt prevailing political ideas, beliefs, fashion, etc., of their neighbors. In order to model such a process of conforming, consider* *the following two-dimensional two-state cellular automaton. Each cell of a square lattice with periodic boundary conditions is occupied by an individual who at each time step can adopt only one of two possible behaviors (being Democrat or Republican, going or not going to church, buying an American or a foreign car, etc.). In order to fit into his neighborhood, at each time step, each individual adopts the prevailing norm of his neighborhood, i.e., if two or more of his four neighbors share his own idea, he keeps it; otherwise he changes. Run a simulation to exhibit the formation of clusters of individuals having identical behaviors.*

Solutions

Solution 6.1 *The annihilation of two defects for a cellular automaton evolving according to rule 18 is illustrated in Figure 6.27.*

Fig. 6.27. *Rule 18: annihilation of two defects. The defects (two consecutive 1s) at positions 19 and 28 in the initial configuration meet and annihilate at $t = 32$. Time is oriented downwards.*

Solution 6.2 *(i) The process defining the transform Φ of F by the surjective mapping M may be conveniently represented by the following commutative diagram:*

$$
\begin{array}{ccc}
\mathcal{S} & \xrightarrow{\ M\ } & \mathcal{S} \\
{\scriptstyle F}\downarrow & & \downarrow{\scriptstyle \Phi} \\
F(\mathcal{S}) & \xrightarrow{\ M\ } & \Phi(\mathcal{S})
\end{array}
$$

Since M is surjective, each configuration $c \in \mathcal{S}$ has a preimage. Let c_1 and c_2 be two different preimages of a given c (M not being invertible in general, the set $M^{-1}(c)$ of preimages of a given c contains more than one element). If the transform of Rule F exists, then

$$(\Phi \circ M)(c_1) = \Phi(c) = (M \circ F)(c_1),$$
$$(\Phi \circ M)(c_2) = \Phi(c) = (M \circ F)(c_2);$$

that is,

$$(M \circ F)(c_1) = (M \circ F)(c_2).$$

Hence, the transform Φ of a given rule F by a surjective mapping exists if, and only if, for any configuration $c \in S$, the set $(M \circ F)(M^{-1}(c))$ contains one, and only one, element. In other words, all the images by $M \circ F$ of the configurations in $M^{-1}(c)$ are identical. If Φ_1 and Φ_2 are such that, for all $c \in S$,

$$(\Phi_1 \circ M)(c) = (\Phi_2 \circ M)(c) = (M \circ F)(c),$$

then, for all $c \in S$,

$$\big((\Phi_1 - \Phi_2) \circ M\big)(c) = 0,$$

which implies $\Phi_1 = \Phi_2$ and proves that, when Φ exists, it is unique. Since M and F are local, Φ is local, and it is readily verified that its radius cannot be greater than the radius r of F.

(ii) The look-up table of the 3-input mapping 30 is

111	110	101	100	011	010	001	000
0	0	0	1	1	1	1	0

The following table, which lists all the preimages by M_{106} of each 3-block, shows that each 3-block has an equal number of preimages and proves that M_{106} is surjective.

111	110	101	100	011	010	001	000
00100	00101	01010	01101	00010	10101	00001	00000
01001	00110	01011	01110	00011	10110	11010	11101
10010	00111	01100	01111	10100	10111	11011	11110
10011	01000	10001	10000	11001	11000	11100	11111

(iii) To determine all radius 1 rules F that have a transform Φ under the 3-input surjective mapping 30, we have to systematically check when Relation (R) is satisfied for a pair of radius 1 rules (F, Φ). We find that only seven rules F have a transform Φ. The results are summarized in the table below.[47]

F	0 30 170 204 225 240 255
Φ	0 30 170 204 120 240 0

Solution 6.3 *As illustrated in Figure 6.1, the limit set of the cellular automaton rule 184 consists of alternating 0s and 1s if the density of 1s (active sites) is exactly equal to $\frac{1}{2}$. If the density is not equal to $\frac{1}{2}$, as shown in Figure 6.2, the limit set consists of finite sequences of alternating 0s and 1s separated by finite sequences of either 0s, if the density of active sites is less than $\frac{1}{2}$, or 1s, if the density of active sites is greater than $\frac{1}{2}$. Hence, after a number of time steps of the order of magnitude of half the lattice size, any initially disordered configuration becomes ordered as much as possible; that is, according to whether the density of active sites is less than or greater than $\frac{1}{2}$, the 1s or the 0s are isolated.*

---

[47] More results on transformations of cellular automaton rules can be found in Boccara [51].

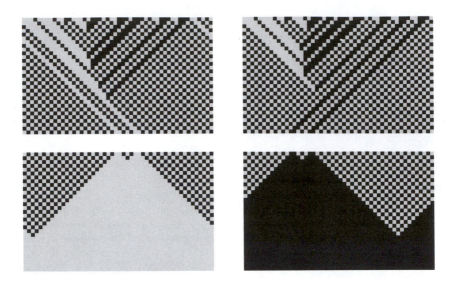

Fig. 6.28. *Successive applications of rules 184 and 232 to determine whether the density of an initial configuration is either less than or greater than $\frac{1}{2}$. Empty (resp. active) sites are light (resp. dark) grey. The density is equal to 0.48 in the left two figures and to 0.52 in the right two ones. The lattice size is 50 and the number of iterations is in all cases equal to 30. Time is oriented downwards.*

If such a configuration evolves according to the 3-input majority rule, that is, the rule f such that

$$f(x_1, x_2, x_3) = \begin{cases} 0 & \text{if } x_1 + x_2 + x_3 = 0 \text{ or } 1, \\ 1 & \text{if } x_1 + x_2 + x_3 = 2 \text{ or } 3, \end{cases}$$

then, after a number of time steps of the order of magnitude of half the lattice size, the configuration consists either of 0s only, if the density of active sites is less than $\frac{1}{2}$, or of 1s only, if the density of active sites is greater than $\frac{1}{2}$. Radius 1 rule 232 is precisely the majority rule above. Figure 6.28 illustrates these considerations. If $\rho = \frac{1}{2}$, we obtain a string of alternating 0s and 1s.

The number of necessary time steps is equal to the number of time steps necessary to completely eliminate one type of defect. Since the two types of defects, for both rules, travel in opposite directions with the same speed $v = 1$ (see Figure 6.28), we find that for each rule the system has to evolve for a number of time steps of the order of magnitude of half the string length.

The density classification problem can also be solved of any rational value of ρ [85, 86].

Solution 6.4 *The 3-input traffic rule* **f** *reads*

$$\mathbf{f}\big((\bullet,\bullet),(1,1),(a,b)\big) =(a,b), \quad \mathbf{f}\big((\bullet,0),(1,0),(0,\bullet)\big) =(0,0),$$
$$\mathbf{f}\big((\bullet,0),(1,0),(1,\bullet)\big) =(0,1), \quad \mathbf{f}\big((\bullet,1),(1,0),(0,\bullet)\big) =(0,1),$$
$$\mathbf{f}\big((\bullet,1),(1,0),(1,\bullet)\big) =(1,1), \quad \mathbf{f}\big((0,\bullet),(0,1),(\bullet,0)\big) =(0,0),$$
$$\mathbf{f}\big((0,\bullet),(0,1),(\bullet,1)\big) =(1,0), \quad \mathbf{f}\big((1,\bullet),(0,1),(\bullet,0)\big) =(1,0),$$
$$\mathbf{f}\big((1,\bullet),(0,1),(\bullet,1)\big) =(1,1), \quad \mathbf{f}\big((a,b),(0,0),(\bullet,\bullet)\big) =(a,b),$$

where $\bullet$ stands for either 0 or 1, and a and b are equal to either 0 or 1. Note that the two lanes play a symmetric role.

Figure 6.29 represents the flow diagram of the two-lane one-way car traffic flow with $v_{\max} = 1$. The average car flow has been determined numerically for the following car densities: 0.1, 0.2, 0.3, 0.4, 0.405, 0.406, 0.407, 0.408, 0.41, 0.43, 0.45, 0.47, 0.49, 0.51, 0.55, 0.6, 0.7, 0.8, 0.9, 0.95.

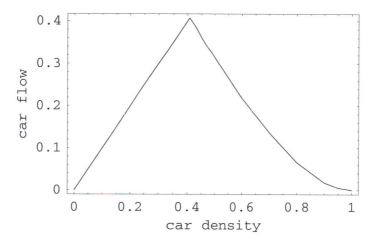

Fig. 6.29. *Fundamental diagram of a two-lane one-way road with* $v_{\max} = 1$. *Lattice size: 1000.*

All simulations were done on a 1000-site lattice. Starting from a random initial road configuration with an exact car density value, the car flow is measured after 2000 iterations. Each point of the fundamental diagram is an average over 10 different simulations. The critical car density ρ_c is approximately equal to 0.406.

Remark 19. Since the speed limit is equal to 1, as for traffic rule 184, one could expect that the critical density should be $\rho_c = \frac{1}{2}$. This would be the case if, for $\rho = \frac{1}{2}$, cars were equally distributed between the two lanes. This is rarely the case, and, due to the existence of defects in the steady state, the average velocity for $\rho = \frac{1}{2}$ is less than 1. This feature is illustrated in Figure 6.30.

Solution 6.5 *We shall prove that an n-input one-dimensional cellular automaton ℓ-vectorial rule $\mathbf{f}$ is number-conserving if, and only if, for all $(\mathbf{x}_1, \mathbf{x}_1, \dots, \mathbf{x}_1) \in \mathbf{Q}^n$, it satisfies*

Fig. 6.30. *Car traffic on a two-lane one-way road for a car density exactly equal to 0.5 on a 60-site lattice. The top figure represents the initial configuration and the bottom one the configuration after 120 iterations.*

$$\sum_{j=1}^{\ell} f_j(\mathbf{x}_1, \mathbf{x}_2, \ldots, \mathbf{x}_n) = \sum_{j=1}^{\ell} x_{1,j} + \sum_{j=1}^{\ell} \left(\sum_{k=1}^{n-1} \Big(f_j(\underbrace{\mathbf{0}, \mathbf{0}, \ldots, \mathbf{0}}_{k}, \mathbf{x}_2, \mathbf{x}_3, \ldots, \mathbf{x}_{n-k+1}) \right.$$
$$\left. - f_j(\underbrace{\mathbf{0}, \mathbf{0}, \ldots, \mathbf{0}}_{k}, \mathbf{x}_1, \mathbf{x}_2, \ldots, \mathbf{x}_{n-k}) \Big) \right). \qquad (6.48)$$

To prove this result, we proceed exactly as for the proof of Theorem 13.

First, if we write that a cyclic configuration of length $L \geq n$, which consists of 0s only, is number-conserving, we verify that

$$\mathbf{f}(\mathbf{0}, \mathbf{0}, \cdots, \mathbf{0}) = \mathbf{0},$$

where $\mathbf{0} = \underbrace{(0, 0, \ldots, 0)}_{\ell}$.

Then, if we consider a cyclic configuration of length $L \geq 2n - 1$ which is the concatenation of a sequence $(\mathbf{x}_1, \mathbf{x}_2, \ldots, \mathbf{x}_n)$ and a sequence of $L - n$ 0s, and express that the n-input rule $\mathbf{f}$ is number-conserving, we obtain

$$\sum_{j=1}^{\ell} f_j(\mathbf{0}, \mathbf{0}, \ldots, \mathbf{0}, \mathbf{x}_1) + \sum_{j=1}^{\ell} f_j(\mathbf{0}, \mathbf{0}, \ldots, \mathbf{0}, \mathbf{x}_1, \mathbf{x}_2) + \cdots$$
$$+ \sum_{j=1}^{\ell} f_j(\mathbf{x}_1, \mathbf{x}_2, \ldots, \mathbf{x}_n) + \sum_{j=1}^{\ell} f_j(\mathbf{x}_2, \mathbf{x}_3, \ldots, \mathbf{x}_n, \mathbf{0}) + \cdots$$
$$+ \sum_{j=1}^{\ell} f_j(\mathbf{x}_n, \mathbf{0}, \ldots, \mathbf{0})$$
$$= \sum_{j=1}^{\ell} (x_{1,j} + x_{2,j} + \cdots + x_{n,j}),$$

where, for $j = 1, 2, \ldots, \ell$, all the terms of the form $f_j(\mathbf{0}, \mathbf{0}, \ldots, \mathbf{0})$ that are equal to zero have not been written. Replacing $x_{1,j}$ by 0 for all $j \in \{1, 2, \ldots, \ell\}$ in the relation above and subtracting the relation so obtained shows that Condition (6.48) is necessary.

Condition (6.48) is obviously sufficient since, when summed over a cyclic configuration, all the left-hand-side terms except the first cancel.

One easily verifies that the two-lane traffic rule of the preceding exercise satisfies Condition (6.48).

Solution 6.6 *Figure 6.31 shows, for the timorous and amiable multiple random walkers models, plots of the reduced Hamming distance between the initial configuration and the*

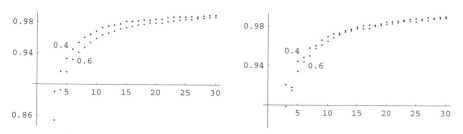

Fig. 6.31. *Reduced Hamming distance between a random initial configuration and the configuration obtained after t time steps for t varying from 1 to 30. Each point represents an average reduced Hamming distance over 40 simulations. Lattice size is 200 × 200. The left (resp. right) figure corresponds to timorous (resp. amiable) random walkers.*

configuration obtained after t time steps for two different densities of random walkers. Increasing the density ρ of random walkers decreases the reduced Hamming distance δ_H. For timorous random walkers, starting from a random initial configuration, δ_H jumps to 0.729 for $\rho = 0.4$ and to 0.615 for $\rho = 0.6$. When the number of time steps increases, as expected, δ_H approaches 1. After 30 time steps, δ_H is equal to 0.989 for $\rho = 0.4$ and to 0.987 for $\rho = 0.6$. For amiable random walkers, the mixing process is, for a small number of iterations, slightly more efficient. δ_H jumps to 0.861 or 0.797 for ρ equal to 0.4 or 0.6. After 30 time steps, δ_H is equal to 0.989 and 0.991. Both models seem to have the same asymptotic behavior. As for the sequential rule, the asymptotic reduced Hamming distances of these synchronous rules probably do not depend upon ρ.

Solution 6.7 *To generate an Eden cluster, we have to manipulate two lists of sites: a list of t sites that, at time $t-1$, form the Eden cluster and the list of perimeter sites (i.e., the list of sites that are first neighbors of sites belonging to the Eden cluster). Then at time t a randomly selected site from the second list is added to the first. Implementing this algorithm, we obtain clusters as illustrated in Figure 6.32.*

Solution 6.8 *Let us first label the sites of the Bethe lattice as shown in Figure 6.33. For $z = 3$, the lattice consists of three binary trees whose root is site 0. Note that, in the case of an infinite lattice, our labelling is purely conventional: all sites play exactly the same role and any site could have been chosen as root 0.*

(i) If p is the probability for a site to be occupied, let $Q(p)$ be the probability that a path starting from an occupied site is connected to infinity along a given branch. The probability that site 0 is not connected to infinity along a specific branch, say branch $(0, 1)$, is then

$$Q(p) = 1 - p + pQ^2(p)$$

since either site 1 is empty (probability $1 - p$) or occupied (probability p) and not connected to infinity along either branch $(1, 11)$ or branch $(1, 12)$ (probability $Q^2(p)$). For an arbitrary value of $z > 2$, the equation above is to be replaced by

$$Q(p) = 1 - p + pQ^{z-1}(p). \tag{6.49}$$

For all values of the positive integer z, $Q(p) = 1$ is always a solution to Equation (6.49). This solution corresponds to $P_\infty^{\mathrm{site}}(p) = 0$ (i.e., for $p < p_c^{\mathrm{site}}$). When p is increasing, it

Fig. 6.32. *An example of Eden cluster obtained on a square lattice after 4000 iterations of the growth rule.*

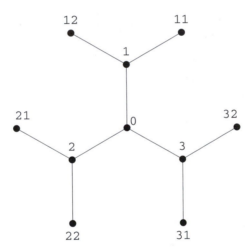

Fig. 6.33. *Labelled Bethe lattice sites for $z = 3$.*

reaches its critical value p_c^{site}; $Q(p_c^{\text{site}}) = 1$ becomes a double root.[48] Thus, p_c^{site} satisfies the equation

$$\frac{\partial}{\partial Q}(1 - p + pQ^{z-1})\bigg|_{Q=1} = p(z-1) = 1;$$

that is,

$$p_c^{\text{site}} = \frac{1}{z-1}.$$

---

[48] As a function of p, the map $Q \mapsto 1 - p + pQ^{z-1}$ exhibits a transcritical bifurcation corresponding to the second-order phase transition characterized by the incipient infinite cluster.

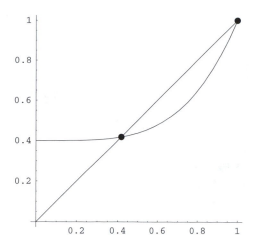

Fig. 6.34. *Graphical solution of the equation* $Q = 1 - p + pQ^{z-1}$. *For* $z = 5$ *and* $p = 0.6$, *we find* $Q = 0.418385\ldots$

As shown in Figure 6.34, for $p > p_c^{\text{site}}$, Equation (6.49) has two real roots, $Q(p) = 1$ and $0 < Q(p) < 1$. For $z = 3$, this second root is

$$Q(p) = \frac{1-p}{p}.$$

The percolation probability $P_\infty^{\text{site}}(p)$ being the probability for an occupied site to be connected to infinity, we have

$$P_\infty^{\text{site}}(p) = p\big(1 - Q^z(p)\big) \tag{6.50}$$

since the site has to be occupied (probability p) and connected to infinity through at least one of the z outgoing edges (probability $1 - Q^z(p)$). For $z = 3$, we have

$$P_\infty^{\text{site}}(p) = p\left(1 - \left(\frac{1-p}{p}\right)^3\right),$$

and in the vicinity of $p_c^{\text{site}} = \frac{1}{2}$, we find

$$P_\infty^{\text{site}}(p) = 6\big(p - \tfrac{1}{2}\big) + O\big((p - \tfrac{1}{2})^2\big),$$

which shows that $\beta = 1$.

In order to verify that $\beta = 1$ for all values of z, we first find the expansion in powers of $p - 1/(z-1)$ of the solution $Q(p) < 1$ to Equation (6.49). Let

$$Q(p) = 1 + a_1\left(p - \frac{1}{z-1}\right) + a_2\left(p - \frac{1}{z-1}\right)^2 + O\left(\left(p - \frac{1}{z-1}\right)^3\right).$$

Then, substituting this expression in (6.49) yields

$$a_1 = \frac{2(z-1)}{2-z}.$$

Replacing

$$Q(p) = 1 + \frac{2(z-1)}{2-z}\left(p - \frac{1}{z-1}\right) + O\left(\left(p - \frac{1}{z-1}\right)^2\right)$$

in the expression of $P_\infty^{\text{site}}(p)$ given by Equation (6.50), we obtain

$$P_\infty^{\text{site}}(p) = \frac{2z}{z-2}\left(p - \frac{1}{z-1}\right) + O\left(\left(p - \frac{1}{z-1}\right)^2\right),$$

which proves that $\beta = 1$ for all values of z.

(ii) In the case of the bond percolation model, p is the probability for a bond to be open. Let $Q(p)$ denote the probability that an outgoing path from a given site along a specific branch is closed. For $z = 3$, the probability that site 0 is not connected to infinity along a specific branch, say branch $(0,1)$, is then

$$Q(p) = \left(1 - p + pQ(p)\right)^2$$

since the two branches $(1,11)$ and $(1,12)$ are either closed (probability $1 - p$) or open but not connected to infinity (probability $pQ(p)$). For an arbitrary value of $z > 2$, the equation above is to be replaced by

$$Q(p) = \left(1 - p + pQ(p)\right)^{z-1}. \tag{6.51}$$

For all values of the positive integer z, $Q(p) = 1$ is always a solution to Equation (6.51). This solution corresponds to $P_\infty^{\text{bond}}(p) = 0$ (i.e., for $p < p_c^{\text{bond}}$). When p is increasing, it reaches its critical value p_c^{bond}; $Q(p_c^{\text{bond}}) = 1$ becomes a double root. Thus, p_c^{bond} satisfies the equation

$$\frac{\partial}{\partial Q}(1 - p + pQ)^{z-1}\bigg|_{Q=1} = p(z-1) = 1,$$

and we find, as for the site percolation model,

$$p_c^{\text{bond}} = \frac{1}{z-1}.$$

For $p > p_c^{\text{site}}$, Equation (6.51) has two real roots, $Q(p) = 1$ and $0 < Q(p) < 1$. For $z = 3$, this second root is

$$Q(p) = \left(\frac{1-p}{p}\right)^2.$$

The percolation probability $P_\infty^{\text{bond}}(p)$ being the probability for a site, say site 0, to be connected to infinity, we have

$$P_\infty^{\text{bond}}(p) = 1 - Q^z(p) \tag{6.52}$$

since site 0 has to be connected to infinity through at least one of z outgoing edges. For $z = 3$, we have

$$P_\infty^{\text{bond}}(p) = 1 - \left(\frac{1-p}{p}\right)^6,$$

and in the vicinity of $p_c^{\text{bond}} = \frac{1}{2}$, we find

$$P_\infty^{\text{bond}}(p) = 24\left(p - \tfrac{1}{2}\right) + O\left((p - \tfrac{1}{2})^2\right),$$

which shows that $\beta = 1$.

To prove that $\beta = 1$ for all values of z, we proceed as for the site model by first finding the expansion in powers of $p - 1/(z - 1)$ of the solution $Q(p) < 1$ to Equation (6.51). Let

$$Q(p) = 1 + a_1\left(p - \frac{1}{z-1}\right) + a_2\left(p - \frac{1}{z-1}\right)^2 + O\left(\left(p - \frac{1}{z-1}\right)^3\right).$$

Then, substituting this expression in (6.51) yields

$$a_1 = \frac{2(z-1)^2}{2-z}.$$

Replacing

$$Q(p) = 1 + \frac{2(z-1)^2}{2-z}\left(p - \frac{1}{z-1}\right) + O\left(\left(p - \frac{1}{z-1}\right)^2\right)$$

in the expression of $P_\infty^{\text{bond}}(p)$ given by Equation (6.52), we obtain

$$P_\infty^{\text{bond}}(p) = \frac{2z(z-1)^2}{z-2}\left(p - \frac{1}{z-1}\right) + O\left(\left(p - \frac{1}{z-1}\right)^2\right),$$

which proves that $\beta = 1$ for all values of z.

Solution 6.9 *To generate a cluster of invaders, we manipulate two lists of sites: a list of t sites that, at time $t - 1$, form the cluster of invaders and the list of perimeter sites. Then, at time t, the perimeter site with the smallest random number is added to the cluster of invaders. Implementing this algorithm, we obtain a typical cluster, as shown in Figure 6.35.*

Solution 6.10 *Tit-for-tat and tit-for-two-tats strategists always cooperate when playing against each other or against a fellow strategist. They cannot, therefore, be distinguished in the absence of a third strategy.*

Playing against each other, a suspicious tit-for-tat and a tit-for-tat strategist lock into a pattern of alternating cooperations and defections, the first one playing the sequence defection, cooperation, defection, cooperation, etc., and the second one cooperation, defection, cooperation, defection, etc. Then, their respective total payoffs are

$$T + wS + w^2T + w^3S + \cdots = \frac{T + wS}{1 - w^2},$$
$$S + wT + w^2S + w^3T + \cdots = \frac{S + wT}{1 - w^2}.$$

Playing against each other, a suspicious tit-for-tat and a tit-for-two-tats strategist, after the first encounter, lock into a pattern of mutual cooperation, and their respective total payoffs are

Fig. 6.35. *Example of a cluster of invaders obtained on a square lattice after 2000 iterations of the invasion rule.*

$$T + wR + w^2R + w^3R + \cdots = T + \frac{wR}{1 - w},$$
$$S + wR + w^2R + w^3R + \cdots = S + \frac{wR}{1 - w}.$$

Hence, in the presence of mutant suspicious tit-for-tat strategists, tit-for-two-tats strategists do better than tit-for-tat strategists showing that the tit-for-tat strategy is not evolutionarily stable.

Remark 20. While this example may cast doubts on the general result of Axelrod and Hamilton [15] as stressed by May [232] it may turn out that tit for tat is more robust than what Boyd and Lorberbaum [70] think since it depends on these strategies being rigidly adhered to. In the real world, cooperation is not always rewarded and defection not always punished. If, for example, a tit-for-tat strategist punishes defection with a probability slightly less than 1, tit-for-tat and suspicious tit-for-tat strategists would eventually settle on mutual cooperation.

Solution 6.11 *If the set of states is $\{0, 1\}$, where state 0 corresponds to say opinion A and 1 to opinion B, the evolution rule is totalistic (i.e., it depends only upon the sum of the states of the nearest neighbors) and can be written*

$$s_{i,j}^{t+1} = f(S_{i,j}^t) = \begin{cases} 0, & \text{if } 0 \leq S_{i,j}^t < 3, \\ 1, & \text{if } 3 \leq S_{i,j}^t \leq 5, \end{cases}$$

where

$$S_{i,j}^t = s_{i,j-1}^t + s_{i-1,j}^t + s_{i,j}^t + s_{i+1,j}^t + s_{i,j+1}^t.$$

Figure 6.36 shows the initial random configuration followed by three other configurations exhibiting cluster formation after 50, 100, and 200 iterations. Note that the patterns after 100 and 200 iterations are identical, showing that clusters do not evolve after a rather small number of time steps. While many individuals change their behavior during the evolution, the number of individuals having a given behavior does not change so much. In our simulation, initially the individuals were divided into two groups of 1214 (light grey) and 1286 (dark grey) individuals. In the steady state these groups consisted of 1145 (light grey) and 1355 (dark grey) individuals, respectively.

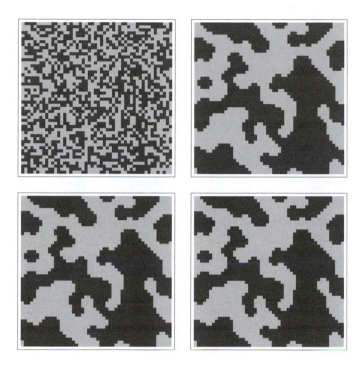

Fig. 6.36. *Formation of clusters of individuals having identical behavior. Initially the individuals have either behavior A or B with equal probability. Lattice size: 50 × 50. From left to right and top to bottom, the figures represent configurations after 0, 50, 100, and 200 iterations.*

7

Networks

7.1 The small-world phenomenon

In the late 1980s, Stanley Milgram [243] performed a simple experiment to show that, despite the very large number of people living in the United States and the relatively small number of a person's acquaintances, two persons chosen at random are very closely connected to one another. Here is how he says his study was carried out:

> The general idea was to obtain a sample of men and women from all walks of life. Each of these persons would be given the name and address of the same target person, a person chosen at random, who lives somewhere in the United States. Each of the participants would be asked to move a message toward the target person, using only a chain of friends and acquaintances. Each person would be asked to transmit the message to the friend or acquaintance who he thought would be most likely to know the target person. Messages could move only to persons who knew each other on a first-name basis.

In a first study, the letters, given to people living in Wichita, Kansas, were to reach the wife of a divinity student living in Cambridge, Massachusetts, and in a second study, the letters, given to people living in Omaha, Nebraska, had to reach a stockbroker working in Boston and living in Sharon, Massachusetts. Out of 160 chains that started in Nebraska, 44 were completed. The number of intermediaries needed to reach the target person in the Nebraska study is distributed as shown below:

$$(2,2)\ (3,4)\ (4,9)\ (5,8)\ (6,11)\ (7,5)\ (8,1)\ (9,2)\ (10,2)\,.$$

The first number of the ordered pairs is the chain length (number of intermediaries) and the second one the number of such completed chains.

The median of this distribution is equal to 5 and its average is 5.43. These results are certainly not very accurate, the statistics being too poor to make serious claims. However, the fact that two randomly chosen persons are connected by only a short chain of acquaintances, referred to as the *small-world phenomenon*, has been verified for many different social networks. One striking example is the collaboration graph of movie actors. Consider the graph whose vertices are the actors listed in the Internet Movie Database[1] in which two actors are connected by an edge if they acted in a film together. In April 1997,[2] this graph had 225,226 vertices, with an average number of edges per vertex equal to 61. It is found that the average chain length (in the sense of Milgram) or *degree of separation* between two actors selected at random from the giant connected component of this graph, which includes about 90% of the actors listed in the database, is equal to 3.65.[3] Networks, like Milgram's network of personal acquaintances, are everywhere, and recently a substantial number of papers dealing with their basic properties have been published. Examples include neural networks, food webs, metabolic networks, power grids, collaboration networks, distribution networks, highway systems, airline routes, the Internet, and the World Wide Web.

7.2 Graphs

From a mathematical point of view, a network is a graph G that is an ordered pair of disjoint sets (V, E), where V is a nonempty set of elements called *vertices*, *nodes*, or *points*, and a subset E of ordered pairs of distinct elements of V, called *directed edges*, *arcs*, or *links*. An edge may join a vertex to itself. If we need to specify that V (resp. E) is the set of vertices (resp. edges) of a specific graph G, we use the notation $V(G)$ (resp. $E(G)$).[4] Edges can also be *un*directed. A network of personal acquaintances is an undirected graph, while the routes of an airline company are a directed graph.

A graph H is a *subgraph* of G if $V(H) \subset V(G)$ and $E(H) \subset E(G)$.

In a directed graph (also called a *digraph*), if (x, y) is an arc joining vertex x to vertex y, vertex y is said to be *adjacent* to x or a *neighbor* of x. The set of all vertices adjacent to a vertex x is the *neighborhood* $N(x)$ of x. In an

---

[1] Web site: http://us.imdb.com.

[2] These data are taken from Watts and Strogatz [342].

[3] On the Web site: http://www.cs.virginia.edu/oracle/ one can select two actors and find their degree of separation. In the world of mathematics, there exists a famous collaboration graph centered on the Hungarian mathematician Paul Erdös, who traveled constantly, collaborated with hundreds of mathematicians, and wrote more than 1400 papers. Mathematicians who coauthored a paper with Erdös have an *Erdös number* equal to 1, mathematicians who wrote a paper with a colleague who wrote a paper with Erdös have an Erdös number equal to 2, and so forth. It appears that among mathematicians who coauthored papers, less than 2% have an Erdös number larger than 8 (http://www.oakland.edu/ grossman/erdoshp.html).

[4] On graph theory, see Bollobás [65].

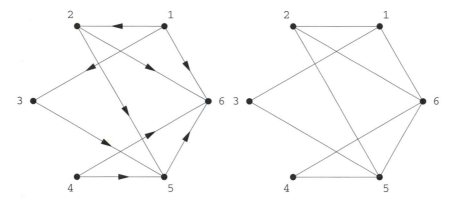

Fig. 7.1. *Examples of directed and undirected graphs with six vertices and nine edges.*

undirected graph, the edge joining x and y is denoted $\{x, y\}$, and, in this case, x and y are said to be *adjacent*. Figure 7.1 shows both types of graphs.

The number of vertices $|V(G)|$ of a graph G is the *order* of G, and the number of edges $|E(G)|$ of G is the *size* of G. The notation $G(N, M)$ represents an arbitrary graph of order N and size M. A graph of order N is *empty* if it has no edges (*i.e.*, if its size $M = 0$) and *complete* if every two vertices are adjacent $\left(i.e.,\ \text{if its size}\ M = \binom{N}{2} \right)$.

Graphs $G_1 = (V_1, E_1)$ and $G_2 = (V_2, E_2)$ are *isomorphic* if there exists a bijection $\varphi : V_1 \to V_2$ such that $\{x, y\} \in E_1$ if, and only if, $\{\varphi(x), \varphi(y)\} \in E_2$. In other words, G_1 and G_2 are isomorphic if the bijection φ between their vertex sets $V_1(G_1)$ and $V_2(G_2)$ preserves adjacency. Isomorphic graphs have, of course, the same order and same size.

The *degree* $d(x)$ of vertex x is the number $|N(x)|$ of vertices adjacent to x. In the case of a directed graph, we have to distinguish the *in-degree* $d_{\text{in}}(x)$ and the *out-degree* $d_{\text{out}}(x)$ of a vertex x, which represent, respectively, the number of incoming and outgoing edges. A graph is said to be *regular* if all vertices have the same degree. A vertex of degree zero is said to be *isolated*.

An alternating sequence

$$(x_0, e_1, x_1, e_2, \ldots, e_\ell, x_\ell)$$

of vertices x_i $(i = 0, 1, 2, \ldots, \ell)$ and edges $e_j = (x_{j-1}, x_j)$ $(j = 1, 2, \ldots, \ell)$ is a *path* or a *chain* of *length* ℓ joining vertex x_0 to vertex x_ℓ. Without ambiguity, a path can also be represented by a sequence of vertices $(x_0, x_1, \ldots, x_\ell)$. This notation assumes that, for all $j = 1, 2, \ldots, \ell$, all edges $e_j = (x_{j-1}, x_j)$ exist. The *chemical distance* $d(x, y)$, or simply *distance*, between vertices x and y is the smallest length of a path joining x to y.

If a path $(x_0, x_1, \ldots, x_\ell)$ is such that $\ell \geq 3$, $x_0 = x_\ell$ and, for all $j = 1, 2, \ldots, \ell$, all vertices x_j are distinct from each other, the path is said to be a *cycle*.

A graph is *connected* if, for every pair of distinct vertices $\{x, y\}$, there is a path joining them. A connected graph of order greater than 1 does not contain isolated vertices. If a graph G is not connected, then it is the union of connected subgraphs. These connected subgraphs are the *components* of G. A connected graph without any cycle is called a *tree*.

Graphs considered here contain neither multiple edges (*i.e.*, more than one edge joining a given pair of vertices) nor loops (*i.e.*, edges joining a vertex to itself).

A graph G of order N can be represented by its $N \times N$ *adjacency matrix* $A(G)$, whose elements a_{ij} are given by

$$a_{ij} = a_{ji} = \begin{cases} 1, & \text{if there exists an edge between vertices } i \text{ and } j, \\ 0, & \text{otherwise.} \end{cases}$$

$$A(G(10, 20)) = \begin{bmatrix} 0 & 1 & 1 & 0 & 1 & 0 & 0 & 1 & 0 & 1 \\ 1 & 0 & 1 & 0 & 1 & 1 & 0 & 0 & 0 & 1 \\ 1 & 1 & 0 & 1 & 0 & 1 & 1 & 0 & 0 & 0 \\ 0 & 0 & 1 & 0 & 0 & 0 & 0 & 1 & 1 & 0 \\ 1 & 1 & 0 & 0 & 0 & 0 & 0 & 1 & 0 & 0 \\ 0 & 1 & 1 & 0 & 0 & 0 & 0 & 0 & 0 & 1 \\ 0 & 0 & 1 & 0 & 0 & 0 & 0 & 1 & 1 & 0 \\ 1 & 0 & 0 & 1 & 1 & 0 & 1 & 0 & 0 & 1 \\ 0 & 0 & 0 & 1 & 0 & 0 & 1 & 0 & 0 & 1 \\ 1 & 1 & 0 & 0 & 0 & 1 & 0 & 1 & 1 & 0 \end{bmatrix}. \tag{7.1}$$

Figure 7.2 shows the graph of order 10 and size 20 represented by the adjacency matrix (7.1).

The ij-element of the square of the adjacency matrix is given by

$$\sum_{k=1}^{N} a_{ik} a_{kj};$$

it is nonzero if, for some $k = 1, 2, \ldots, N$, both a_{ik} and a_{kj} are equal to 1. If, for a specific value of k, $a_{ik} a_{kj} = 1$, this implies that there exists a path of length 2 joining vertex i to vertex j, and the value of $\sum_{k=1}^{N} a_{ik} a_{kj}$ represents, therefore, the number of different paths of length 2 joining vertex i to vertex j. More generally, it is straightforward to verify that the value of the element ij of the nth power of the adjacency matrix is the number of different paths of length n joining vertex i to vertex j. Note that a specific edge may belong to a path more than once. In particular, in the case of an undirected graph, the value of the diagonal element ii of the square of the adjacency matrix is equal to the

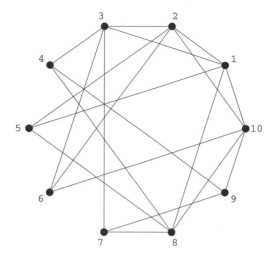

Fig. 7.2. *Undirected graph of order 10 and size 20 represented by the adjacency matrix (7.1).*

number of paths of length 2 joining vertex i to itself. It represents, therefore, the degree of vertex i. These considerations are illustrated by matrix (7.2) which represents the square of the adjacency matrix (7.1).

$$A^2(G(10,20)) = \begin{bmatrix} 5 & 3 & 1 & 2 & 2 & 3 & 2 & 2 & 1 & 2 \\ 3 & 5 & 2 & 1 & 1 & 2 & 1 & 3 & 1 & 2 \\ 1 & 2 & 5 & 0 & 2 & 1 & 0 & 3 & 2 & 3 \\ 2 & 1 & 0 & 3 & 1 & 1 & 3 & 0 & 0 & 2 \\ 2 & 1 & 2 & 1 & 3 & 1 & 1 & 1 & 0 & 3 \\ 3 & 2 & 1 & 1 & 1 & 3 & 1 & 1 & 1 & 1 \\ 2 & 1 & 0 & 3 & 1 & 1 & 3 & 0 & 0 & 2 \\ 2 & 3 & 3 & 0 & 1 & 1 & 0 & 5 & 3 & 1 \\ 1 & 1 & 2 & 0 & 0 & 1 & 0 & 3 & 3 & 0 \\ 2 & 2 & 3 & 2 & 3 & 1 & 2 & 1 & 0 & 5 \end{bmatrix}. \tag{7.2}$$

Sometimes it might be useful to consider that the vertices of a graph are not of the same type. The collaboration graph of movie actors can be viewed as a set of actors (vertices of type 1) and a set of films (vertices of type 2) with links between a film and an actor when the actor appeared in the film. Such graphs are said to be *bipartite*. In a bipartite graph, there exist no edges connecting vertices of the same type.

7.3 Random networks

Except when mentioned otherwise, in what follows we shall deal with undirected graphs. A *random graph* is a graph in which the edges are randomly distributed. Networks of intricate structure could be tentatively modeled by random graphs. But, in order to verify if they are acceptable models, we have to study some of their basic properties.

To give a precise definition of a random graph, we have to define a space whose elements are graphs and a probability measure on that space.[5] For graphs with a given number N of vertices, there exist two closely related models.

1. For $0 \leq M \leq \binom{N}{2}$, there are $\binom{\binom{N}{2}}{M}$ graphs with exactly M edges. If the probability of selecting any of them is $\binom{\binom{N}{2}}{M}^{-1}$, we define the probability space $\mathcal{G}(N, M)$ of *uniform random graphs*.

2. Let $0 \leq p \leq 1$. If the probability of selecting a graph with exactly m edges in the set of all graphs of order N is $p^k(1-p)^{\binom{N}{2}-m}$—that is, if an edge between a pair of vertices exists independently of the other edges with a probability p and does not with a probability $1 - p$—the resulting probability space is denoted $\mathcal{G}(N, p)$, and graphs of this type are called *binomial random graphs*.

In a uniform random graph, the probability that there exists an edge between a given pair of vertices is $M/\binom{N}{2}$, while the average number of edges of a binomial random graph is $p\binom{N}{2}$.

If p is the probability that there exists an edge between two specific vertices, the probability that a specific vertex x has a degree $d(x) = m$ is

$$P\big(d(x) = m\big) = \binom{N-1}{m} p^m (1-p)^{N-1-m}.$$

In most applications, p is small and N is large, so, in the limit $p \to 0$, $N \to \infty$, and $pN = \lambda$, we find that for both uniform and binomial random graphs, the vertex degree probability distribution is Poisson[6]:

$$(\forall x) \qquad P\big(d(x) = m\big) = e^{-\lambda} \frac{\lambda^m}{m!}.$$

For a given number of vertices N, if the number M of edges is small, a random graph tends to have only tree components of orders depending upon the value of M.[7] As the number of edges grows, there is an important qualitative change first studied by Erdös and Rényi and referred to as *the phase transition*, which is the sudden formation of a *giant component* (*i.e.*, a component whose order is a fraction of the order N of the graph). This component, formed from smaller ones, appears when $M = N/2 + O(N^{2/3})$.[8]

---

[5] On random graphs, see Janson, Łuczak, and Ruciński [179].

[6] This classical result, which is very simple to establish using either characteristic functions or generating functions, can be found in almost any textbook on probability theory. See Exercise 7.2.

[7] The interested reader will find precise results in Bollobás [65], p. 240.

[8] For details on this "sudden jump" of the largest component, refer to Janson, Łuczak, and Ruciński [179], Chapter 5.

The phase transition in random graphs is clearly similar to the percolation transition described on page 221. In a random graph of order N, each of the $\binom{N}{2}$ edges exists with a probability $p = 2M/N(N-1)$. The formation of the giant component may, therefore, be viewed as a percolation phenomenon in N dimensions. Since, in most real networks, N is very large, the percolation threshold is extremely small (see Exercise 6.8). This agrees with the result mentioned in the previous paragraph (*i.e.*, the giant component appears for a number of edges $M \sim N$; that is, for a probability $p = O(N^{-1})$).

When studying the properties of a large network, we shall only consider the giant component.

The structural properties of a network are usually quantified by its *characteristic path length*, its *clustering coefficient*, and its *vertex degree probability distribution*.

The characteristic path length L of a network is the average shortest path length between two randomly selected vertices.

The clustering coefficient C of a network is the conditional probability that two randomly selected vertices are connected given that they are both connected to a common vertex. That is, if x is a given vertex and $d(x)$ the number of other vertices linked to x (*i.e.*, the degree of x), since these $d(x)$ vertices may be connected together by at most $\frac{1}{2} d(x)(d(x) - 1)$ edges, the clustering coefficient C_x of vertex x is the fraction of this maximum number of edges present in the actual network, and the clustering coefficient C of the network is the average of the clustering coefficients of all vertices.

The characteristic length takes its minimum value of $L = 1$ and the clustering coefficient its maximum value $C = 1$ in the case of a *complete graph*.

If N is the order of a random graph, $\langle d \rangle = \lambda$ the average vertex degree, and $p \approx \langle d \rangle / N$ the probability that there exists an edge between two vertices, in the case of the random graph, the conditional probability that two randomly selected vertices are connected given that they are both connected to a common vertex coincides with the probability that two randomly selected vertices are connected, which is $\langle d \rangle / N$. Thus

$$C_{\text{random}} \approx \frac{\langle d \rangle}{N}.$$

The shortest path length between any two vertices of a graph is called the *diameter* D of the graph. Since a given vertex has, on average, $\langle d \rangle$ first neighbors, $\langle d \rangle^2$ second neighbors, *etc.*, the diameter D_{random} of a random graph is such that $\langle d \rangle^{D_{\text{random}}} \approx N$; hence

$$D_{\text{random}} \approx \frac{\log N}{\log \langle d \rangle}.$$

This result shows that for large random networks the characteristic path length L_{random} satisfies

$$L_{\text{random}} \sim \frac{\log N}{\log \langle d \rangle}.$$

Let us compare these results with data obtained from real networks.

The *collaboration network of movie actors* mentioned on page 276 has been studied by different authors [342, 37]. The vertices of this collaboration graph are the actors listed in the Internet Movie Database, and there is an edge linking two actors if they acted in a film together. This network is continuously growing; it had 225,226 vertices in April 1997 and almost half a million in May 2000 [267]. The table below indicates a few structural properties of the collaboration graph of movie actors, such as the order N_{actors}, the average vertex degree $\langle d_{\text{actors}} \rangle$, the characteristic path length L_{actors}, and the clustering coefficient C_{actors}. For a comparison, the characteristic path length L_{random} and the clustering coefficient C_{random} of a random network of the same order and same average vertex degree are also given.[9]

Movie actors

N_{actors}	$\langle d_{\text{actors}} \rangle$	L_{actors}	L_{random}	C_{actors}	C_{random}
225,226	61	3.65	2.99	0.79	0.00027

While the characteristic path lengths L_{actors} and L_{random} have the same order of magnitude, the clustering coefficients C_{actors} and C_{random} are very different, suggesting that a random graph is not a satisfactory model for the collaboration network of movie actors. This is the case for many other real networks. Here we shall mention only one more example[10]: *scientific collaboration networks*. These networks have been extensively studied by Newman [263, 264, 265]. The vertices are authors of scientific papers and two authors are joined by an edge if they have written an article together. Bibliographic data from the years 1995–1999 were drawn from the following publicly available databases:

1. The Los Alamos e-Print Archive, a database of unrefereed preprints in physics, self-submitted by their authors.
2. MEDLINE, a database of papers on biomedical research published in refereed journals.
3. SPIRES, a database of preprints and published papers in high-energy physics.
4. NCSTRL, a database of preprints in computer science submitted by participating institutions.

Except in theoretical high-energy physics and computer science where the average number of authors per paper is small (about 2), the size of the giant component is greater than 80%. Since authors may identify themselves in different ways on different papers (*e.g.*, using first initials only, using all initials,

---

[9] The data are taken from Watts and Strogatz [342].

[10] Data on many more networks can be found in Albert and Barabási [5].

or using full names), the determination of the true number of distinct authors in a database is problematic. In the table below, all initials of each author were used. This rarely confuses two distinct authors for the same person but could misidentify the same person as two different people.

The order N, the average vertex degree $\langle d \rangle$, the characteristic path length L, and the clustering coefficient C are given for the various networks. As above, the characteristic path length L_{random} and the clustering coefficient C_{random} of a random network of the same order and same average vertex degree are given for comparison.

Scientific collaboration

Network	N	$\langle d \rangle$	L	L_{random}	C	C_{random}
Los Alamos archive	52,909	9.7	5.9	4.79	0.43	0.00018
MEDLINE	1,520,251	18.1	4.6	4.91	0.066	0.000011
SPIRES	56,627	173	4.0	2.12	0.726	0.003
NCSTRL	11,994	3.59	9.7	7.34	0.496	0.0003

7.4 Small-world networks

7.4.1 Watts-Strogatz model

The collaboration network of movie actors and the various scientific collaboration networks have short characteristic path lengths and high clustering coefficients. In what follows, any large network possessing these two properties will be called a *small-world network*, where by large we mean that the order N of the corresponding graph is much larger than the average vertex degree $\langle d \rangle$.

Random graphs, which have short characteristic path lengths but very small clustering coefficients, are not acceptable models of small-world networks. On the other hand, regular lattices, which have obviously high clustering coefficients but large characteristic path lengths, cannot be used to model small-world networks.[11]

A small-world model with a high clustering coefficient, such as a regular lattice, and a short characteristic path length, such as a random network, has been suggested by Watts and Strogatz [342]. Starting from a ring lattice

---

[11] For a regular one-dimensional lattice with periodic boundary conditions in which each site is linked to its r nearest neighbors (*i.e.*, the degree of each vertex is equal to $2r$), it can be shown (see Exercise 7.1) that the characteristic path length L_{lattice} and the clustering coefficient C_{lattice} are given exactly by

$$L_{\text{lattice}} = \left(1 - \frac{r}{N-1}\left\lfloor \frac{N-1}{2r}\right\rfloor\right)\left(\left\lfloor \frac{N-1}{2r}\right\rfloor + 1\right), \quad C_{\text{lattice}} = \frac{3(r-1)}{2(2r-1)}.$$

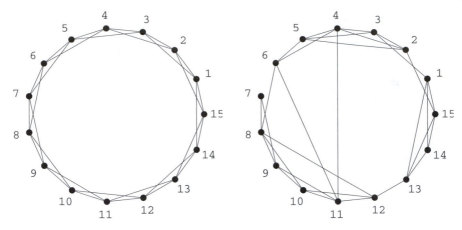

Fig. 7.3. *Constructing the Watts-Strogatz small-world network model. The graph is of order $N = 15$, the average vertex degree $\langle d \rangle = 2r = 4$, and the probability $p = 0.2$.*

with N vertices and $d = 2r$ edges per vertex, they rewire each edge at random with probability p. More precisely, the rewiring is done as follows. The vertices along the ring are visited one after the other; each edge connecting a vertex to one of its r right neighbors is reconnected to a randomly chosen vertex other than itself with a probability p and left unchanged with a probability $1 - p$. Two vertices are not allowed to be connected by more than one edge. This construction is illustrated in Figure 7.3. Note that this rewiring process keeps the average vertex degree unchanged. Varying p from 0 to 1, the perfectly regular network becomes more and more disordered. However, for $p = 1$, the network is not locally equivalent to a random network since the degree of any vertex is larger than or equal to r, as shown in Figure 7.4. However, for $p = 1$, the characteristic path length and the clustering coefficient have the same order of magnitude as the characteristic path length and the clustering coefficient of a random network.

What is particularly interesting in this model is that, starting from $p = 0$, a small increase of p causes a sharp drop in the value of the characteristic path length but almost no change in the value of the clustering coefficient. That is, for small values of p, the Watts-Strogatz model has a clustering coefficient $C_{\mathrm{WS}}(p)$ of the same order of magnitude as C_{lattice}, while its characteristic path length $L_{\mathrm{WS}}(p)$ is of the order of magnitude of L_{random}.

To have an idea of the structural properties of the Watts-Strogatz small-world network model compared to those of regular lattices and random networks, it suffices to consider a rather small network with $N = 500$ vertices, $\langle d \rangle = 8$, and $p = 0.05$. We find:

1. For a regular lattice, $L_{\mathrm{lattice}} = 31.6894$, $C_{\mathrm{lattice}} = 0.642857$.
2. For a random network, $L_{\mathrm{random}} = 3.21479$, $C_{\mathrm{random}} = 0.00930834$.

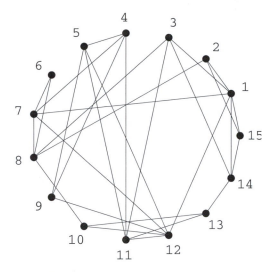

Fig. 7.4. *For* $p = 1$, *the Watts-Strogatz small-world network model is NOT a random graph.* $N = 15$ *and* $\langle d \rangle = 2r = 4$.

3. For a Watts-Strogatz small-world network, $L_{\mathrm{WS}} = 5.50771$, $C_{\mathrm{WS}} = 0.555884$.

Hence, for a small value of p, the clustering coefficient remains of the same order of magnitude, while the characteristic path length is much shorter.

It is possible to modify the rewiring process of the Watts-Strogatz small-world network model in order to obtain a random network for $p = 1$. Starting from a regular ring lattice, we visit each edge and replace it with a probability p by an edge between two randomly selected vertices or leave it unchanged with a probability $1 - p$, subject to the condition that no more than one edge may exist between two vertices. Small-world graphs of this type may have disconnected subgraphs, as shown in Figure 7.5.

For the same values of N, $\langle d \rangle$, and p as above, we find, in this case, $L_{\mathrm{modifiedWS}} = 5.73836$, $C_{\mathrm{modifiedWS}} = 0.556398$, which are very close to the values obtained for the Watts-Strogatz original rewiring process.

If we say that a network in which the characteristic path length behaves as the number of vertices is a "large-world" network as opposed to a small-world network in which the characteristic path length behaves as the logarithm of the number of vertices, on the basis of numerical simulations, it has been shown that, for the Watts-Strogatz model, as p increases from zero, the appearance of the small-world behavior is not a phase transition but a crossover phenomenon [36, 262, 8, 37]. As a function of N and p, the characteristic path length satisfies the scaling relation

$$ L_{\mathrm{WS}}(N, p) = N f \left(\frac{N}{\xi(p)} \right), $$

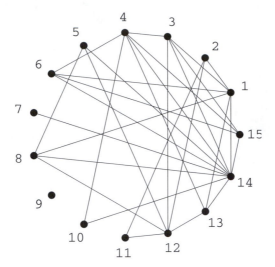

Fig. 7.5. *For the modified Watts-Strogatz small-world network model, the graph may become disconnected, and some vertices may have a degree equal to 0. $N = 15$, $\langle d \rangle = 2r = 4$, and $p = 1$.*

where $\xi(p) \sim p^{-1}$ and

$$f(x) \sim \begin{cases} \text{constant}, & \text{if } x \ll 1, \\ \dfrac{\log x}{x}, & \text{if } x \gg 1. \end{cases}$$

7.4.2 Newman-Watts model

While the rewiring process of the original Watts-Strogatz small-world network model cannot yield a disconnected graph, some rewiring process may disconnect the graph. Although, for small values of p, there always exists a giant connected subgraph, the existence of small disconnected subgraphs makes the characteristic path length poorly defined since we may find pairs of vertices without a path joining them. And assuming that, in such a case, the path length is infinite does not solve the problem. What is usually done is to forget about small detached subgraphs and define the characteristic path length for the giant connected part. While such a procedure may be acceptable for numerical studies, for analytic work, having to deal with poorly defined quantities is not very satisfactory.

To avoid disconnecting the network, Newman and Watts [261] suggested adding new edges between randomly selected pairs of vertices instead of rewiring existing edges. That is, in order to randomize a regular lattice, on average, pN new links, referred to as *shortcuts*, between pairs of randomly selected vertices are added, where, as usual, p is a probability and N the number

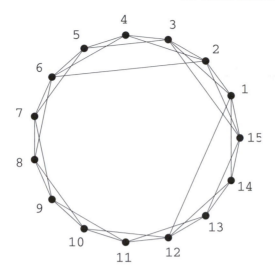

Fig. 7.6. *Newman-Watts small-world network model with shortcuts. The graph is of order $N = 15$ and the probability $p = 0.2$.*

of vertices of the lattice (see Figure 7.6). In this model, it is allowed to have more than one edge between a pair of vertices and to have edges connecting a vertex to itself.

If we take as above $N = 500$, $\langle d \rangle = 8$, and $p = 0.05$, for the Newman-Watts small-world model, we find $L_{\mathrm{NW}} = 7.03808$ and $C_{\mathrm{NW}} = 0.636151$.

7.4.3 Highly connected extra vertex model

Instead of creating direct shortcuts between sites as in the Newman-Watts model, we can create shortcuts connecting distant sites going through an extra site connected to an average pN sites, as illustrated in Figure 7.7. In the case of a large network, this extra site would represent the existence of a well-connected individual. More realistically, for a large network, we should consider the existence of a small group of well-connected individuals. In the case of the collaboration network of movie actors, these extra vertices would correspond to the small group of very popular actors, while for the collaboration network of scientists, they would correspond to particularly productive authors.

If we take, as above, $N = 500$, $\langle d \rangle = 8$, and $p = 0.05$, for the model with only one highly connected extra vertex, we find $L_{\mathrm{HCEV}} = 6.07453$ and $C_{\mathrm{HCEV}} = 0.633522$.

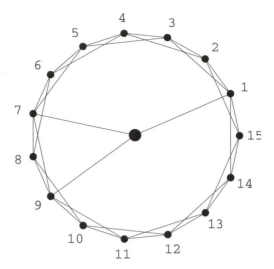

Fig. 7.7. *Extra vertex connected on average to pN sites. The graph is of order*
$N = 15$ *and the probability* $p = 0.2$.

7.5 Scale-free networks

7.5.1 Empirical results

At the end of Section 7.3, we reported results on the existence of a short characteristic path length and a high clustering coefficient for the collaboration
network of movie actors and different scientific collaboration networks showing that these networks are therefore better modeled by small-world network
models than by random graphs. There is, however, a third important structural property of real networks that we have not yet examined, namely the
vertex degree probability distribution.

World Wide Web

The World Wide Web, which has become the most useful source of information, is a huge network whose order, larger than one billion, is continuously
growing. This growth is totally unregulated: any individual or institution is
free to create Web sites with an unlimited number of documents and links.
Moreover, existing Web sites keep evolving with the addition and deletion of
documents and links. The vertices of this network are HTML documents called
Web pages, and the edges are the hyperlinks pointing from one document to
another. Since the Web has directed links, as mentioned on page 277, we have
to distinguish the *in-degree* $d_{\text{in}}(x)$ and the *out-degree* $d_{\text{out}}(x)$ of a vertex x,
which represent, respectively, the number of incoming and outgoing edges. In
1999, Albert, Jeong, and Barabási [2] studied a subset of the Web (the nd.edu

domain) containing 325,729 documents and 1,469,680 links.[12] They found that both vertex in-degree and out-degree probability distributions have power-law tails extending over about 4 orders of magnitude; *i.e.*,

$$P(d_{\text{out}} = k) \sim k^{-\gamma_{\text{out}}}, \quad P(d_{\text{in}} = k) \sim k^{-\gamma_{\text{in}}},$$

with $\gamma_{\text{out}} = 2.45$ and $\gamma_{\text{in}} = 2.1$. As pointed out by the authors, this result shows that, while the owner of a Web page has complete freedom in choosing the number of links and the addresses to which they point, the overall system obeys scaling laws characteristic of self-organized systems.

Despite its huge size, Barabási, Albert, and Jeong [34] found that the World Wide Web is a highly connected graph: two randomly chosen documents are, on average, 19 clicks away from each other. More precisely, they showed that the average distance between any two documents is given by $0.35 + 2.06 \log N$, where N is the total number of documents. This logarithmic dependence shows that an "intelligent" agent should be able to find in a short time the information he is looking for by navigating the Web.

Scientific collaborations

Like the World Wide Web, networks of scientific collaboration are evolving networks.[13] An extensive investigation of the dynamical properties of these networks has been recently carried out [35]. To reveal the dynamics of these networks, it is important to know the time at which the vertices (authors) and edges (coauthored papers) have been added to the network.

The databases analyzed by the authors contain titles and authors of papers in important journals in mathematics and neuroscience published over an 8-year period from 1991 to 1998. In mathematics, the number of different authors is 70,975 and the number of published papers is 70,901. In neuroscience, these numbers are, respectively, 209,293 and 210,750. Both networks are scale-free. The exponents characterizing the power-law behavior of the vertex degree probability distributions are $\gamma_{\text{M}} = 2.4$ for mathematics and $\gamma_{\text{NS}} = 2.1$ for neuroscience.

An important part of this study is devoted to the time evolution of various properties of these networks.

1. While the order of these networks increases with time, the characteristic path length is found to decrease with time. This apparently surprising result could be due to new edges being created between already existing vertices: two authors, who had not yet written a paper together, coauthor a new article. Since the average number of coauthors in neuroscience is

---

[12] In Barabási, Albert, and Jeong [34], the authors also studied the domains whitehouse.gov, yahoo.com, and snu.ac.kr.

[13] The collaboration network of movie actors is also an evolving network, but the decision to collaborate is usually the responsibility of the casting director. In contrast, the decision for two scientists to collaborate is made at the author level.

larger than in mathematics, the characteristic path length is smaller in neuroscience than in mathematics.

2. Clustering coefficients are found to decay as a function of time. Convergence to an asymptotic value, if it exists, appears to be very slow.
3. The size of the largest component, greater for neuroscience than for mathematics, grows with time. For neuroscience, this size seems to be very close to an asymptotic value, while this is, apparently, not the case for mathematics.
4. The average vertex degree increases with time, much faster for neuroscience than for mathematics.

These time-dependent properties suggest that one has to be very careful when interpreting certain results. For instance, numerical simulations of a model based on the empirical results show that the vertex degree probability distribution exhibits a crossover between two different power-law regimes characterized by the exponents $\gamma_{small} = 1.5$ for small degrees and $\gamma_{large} = 3$ for large degrees.

First papers by new authors (*i.e.*, authors who did not belong to the network), are found to be preferentially coauthored with old authors who already have large numbers of coauthors than with those less connected. This phenomenon is called *preferential attachment*. More exactly, it is found that the probability that an old author with d links is selected by—or selects—a new author is proportional to d^{α} with $\alpha \approx 0.8$ for mathematics and $\alpha \approx 0.75$ for neuroscience. Moreover, old authors also write papers together. If d_1 and d_2 are the number of coauthors of author 1 and author 2, respectively, it is found that the probability that two old authors write a new paper together is proportional to $d_1 d_2$, exhibiting another aspect of preferential attachment.

Citations

Another interesting study of the vertex connectivity within a network has been carried out by Redner [295], who investigated two large data sets:

1. The citation distribution of 783,339 papers published in 1981 and cited between 1981 and June 1997 that have been cataloged by the Institute for Scientific Information with a number of 6,716,198 citations.
2. The citation distribution, as of June 1997, of the 24,296 papers cited at least once that where published in volumes 11 through 50 of Physical Review D from 1975 to 1994 with a number of 351,872 citations.

In this network, the vertices are the papers and the directed edges are the links from citing to cited papers.

The main result of the study is that the citation distribution is, for large citation numbers, reasonably well-approximated by a decreasing power law, characterized by an exponent $\gamma_{cite} \approx 3$, while, for small citation numbers, a stretched exponential (*i.e.*, $\exp\left(-(x/x_0)^{\beta}\right)$ with $\beta \approx 0.4$) provides a better fit. The fact that two different functions are needed to fit the data is, according to Redner, a consequence of different mechanisms generating the citation distributions of rarely cited and frequently cited papers. The former are essentially cited by their authors and close associates and usually forgotten a short time after their publication, while the latter become known through collective effects and their impact extends over a much longer time.

In contrast with Redner's suggestion, Tsallis and de Albuquerque [333] showed that the whole distribution is fairly well fit by a single function, namely $x \mapsto N_0\left(1 + \lambda(1-q)x\right)^{q/(1-q)}$ with $N_0 = 2332$, $\lambda = 0.13$, and $q = 1.64$ for the Physical Review D papers, and $N_0 = 46604$, $\lambda = 0.11$, and $q = 1.53$ for the Institute for Scientific Information papers. In both cases, the distribution has a decreasing power-law tail with an exponent $\gamma_{\text{cite}} = q/(q-1)$ equal to 2.56 for the first data set and to 2.89 for the second one. The fit by a single function might suggest that the underlying mechanisms responsible for the citation distribution of rarely and frequently cited papers are not necessarily different, but the authors do not put forward any theoretical argument indicating why the parameter q should be approximately equal to 1.5. It is worth mentioning that, in 1966, Mandelbrot [223] suggested a modification of Zipf's law (see page 314) concerning the asymptotic behavior of the frequency f of occurrence of a word as a function of its rank r of the form

$$f(r) \sim \frac{A}{(1 + ar)^{\gamma}},$$

which coincides with the form used by Tsallis and de Albuquerque.

The empirical results above concerning the citation network, in which the vertices are the papers and the directed edges are links from citing to cited papers, are related to the vertex *in-degree* distribution. Vázquez [336] investigated the *out-degree* distribution (*i.e.*, the distribution of the number of papers cited). Analyzing the data collected from the Science Citation Index report concerning 12 important journals for the period 1991–1999, he found that, in all cases, the out-degree probability distribution exhibits a maximum for an out-degree value d_m. The values of d_m and $f(d_m)$ depend upon the journal. d_m varies from 13 to 39 and $f(d_m)$ from 0.024 to 0.079. A plot of the scaled probability distribution $f(d)/f(d_m)$ as a function of d/d_m reveals that, for large out-degrees, the probability distribution decreases exponentially, with a rate equal to either 0.4 or 1.6 according to whether there is or there is not a restriction on the number of pages per article.

Human sexual contacts

The network of human sexual contacts has been recently investigated [206]. The authors analyzed data gathered in a 1996 Swedish survey of sexual behavior. They found that the cumulative distribution functions of number of sexual partners, either during the 12-month period preceding the survey or during the entire life, have power-law behaviors for males and females. For the 12-month period, the characteristic exponents are equal to 2.54 ± 0.2 for females with more than four partners and to 2.31 ± 0.2 for males with more than five partners. Over the entire life, these exponents are 2.1 ± 0.3 for females, and 1.6 ± 0.3 for males when the number of partners is greater than 20. Since the number of different partners cannot be very large over a one-year period, power-law behaviors are observed over only one order of magnitude. Over an entire life, the number of partners of some promiscuous individuals could be quite large, but the survey apparently did not gather enough data on older individuals. The authors did not mention whether the survey they analyzed took into consideration only heterosexual contacts or also homosexual ones. If only heterosexual contacts are considered, the network of human sexual contacts is an example of a bipartite graph (see page 279).

E-mails

E-mail networks, which consist of e-mail addresses connected by e-mails, appear to have a scale-free structure. A study of the log file of the e-mail server at the University of Kiel recording the source and destination of all e-mails exchanged between students over a 3-month period has shown that the vertex degree distribution has a power-law behavior characterized by an exponent $\gamma_{e-mail} = 1.82 \pm 0.10$ over about 2 orders of magnitude [113]. Since the study was restricted to one e-mail server, only the accounts at this server are known exactly. When only e-mails exchanged between students having an account at this server are taken into account, the exponent characterizing the power-law behavior of the "internal" vertex degree distribution, equal to 1.47 ± 0.12, is smaller than the exponent for the whole network.

The authors also determined the characteristic path length and clustering coefficient of this e-mail network. They found $L_{e-mail} = 5.33 \pm 0.03$ and $C_{e-mail} = 0.113$, showing that the e-mail network is a small world.

7.5.2 A few models

A common feature of both the random network model and the Watts-Strogatz small-world network model is that the probability of finding a vertex with a large degree decreases exponentially. In contrast, power-law vertex degree probability distributions imply that highly connected vertices have a large chance of being present.

Example 57. Barabási-Albert models. According to Barabási and Albert [37, 5], there are two generic aspects of real networks that are not incorporated in the two types of models above.

First, most real networks continuously expand. The collaboration network of movie actors grows by the addition of new actors, and the World Wide Web grows by the addition of new Web pages.

Second, when constructing a random network, the addition, with a probability p, of an edge is done by selecting the end vertices with equal probability among all pairs of vertices. In growing real networks, a new vertex has a higher probability of being connected to a vertex that already has a large degree. For instance, a new actor is more likely to be cast with a well-known actor than with a less well-established one; a new paper has a higher probability of citing an already much-cited article.

In order to incorporate these two features, Barabási and Albert suggested the following model. Start at $t = 0$ with an empty graph of small order N_0. At $t = 1$, add a new vertex and a number $m \leq N_0$ of edges linking that vertex to m randomly selected different vertices. Then, at all times $t > 1$, add a new vertex and m new edges connecting this vertex to m existing vertices. In order to take into account preferential attachment of a new vertex to highly connected vertices, the probability of connecting the new vertex to an existing vertex x_i is

$$\frac{d(x_i)}{\sum_{j=1}^{N_{t-1}} d(x_j)},$$

where $d(x_i)$ is the degree of vertex x_i and N_{t-1} is the number of already existing vertices.[14] At time t, the network is of order $N_t = N_0 + t$ and of size $M_t = mt$. These linear behaviors imply that the average vertex degree is constant in time.

Numerical simulations by the authors showed that their model evolves into a stationary scale-free network with a power-law probability distribution for the vertex degree $P(d = k) \sim k^{-\gamma_{\mathrm{BA}}}$, where $\gamma_{\mathrm{BA}} = 2.9 \pm 0.1$ over about 3 orders of magnitude.[15] In their simulations, the authors considered different values of $m = N_0 = 1, 3, 5$, and 7 for $t = 300{,}000$, and different values of $t = 100{,}000$, $150{,}000$, and $200{,}000$ for $m = N_0 = 5$. To verify that continuous growth and preferential attachment are necessary ingredients to build up a model of a scale-free network, the authors studied two other models. In the first one, they consider a growing network in which each new added vertex has an equal probability of connecting to any of the already existing vertices. In the second model, they consider a network with a fixed number N of vertices. Initially the network consists of N vertices and no edges; then, at each time step a randomly selected vertex is connected to another vertex with preferential attachment. They found that neither of these two models developed a stationary power-law probability distribution.

Because of preferential attachment, a vertex that acquires more connections than another will increase its connectivity at a higher rate (*i.e.*, a sort of "rich get richer" phenomenon). Barabási and Albert showed that, as a function of time t, the degree of a vertex increases as $t^{\beta_{\mathrm{BA}}}$, where $\beta_{\mathrm{BA}} = \frac{1}{2}$.

---

[14] The linear dependence upon $d(x_i)$ of the probability to connect is an essential assumption. See Remark 22.

[15] See Exercise 7.3.

The Barabási–Albert exponent $\gamma_{BA} \approx 3$ is close to the value found by Redner [295] for the network of citations, but other real growing networks such as the World Wide Web or scientific collaboration networks have power-law vertex degree probability distributions characterized by different exponent values.

Remark 21. Error and attack tolerance. In a scale-free network, the power-law vertex degree probability distribution implies that the majority of vertices have only a few links. Thus, if a small fraction of randomly selected vertices are removed, the characteristic path length remains essentially the same since highly connected vertices have a very small probability of being selected. A scale-free network should, therefore, exhibit a high degree of tolerance against errors. On the other hand, the intentional removal of a few highly connected vertices could destroy the small-world properties of the network, making scale-free networks highly vulnerable to deliberate attacks. These properties have been verified for the Barabási-Albert model [4]. In contrast, random networks, in which all vertices have approximately the same number of links, are much less vulnerable to attack, but their characteristic path length, which increases monotonically with the fraction of removed vertices, shows they are less tolerant against errors.

While the Barabási-Albert model predicts the emergence of power-law scaling, as pointed out by the authors, the agreement between the measured and predicted exponents is less than satisfactory. This led them to propose an extended model [3]. The dynamical growth process of this new model, which incorporates addition of new vertices, addition of new edges, and rewiring of edges, starts like the first model, but, instead of just adding new vertices linked to existing ones with preferential attachment, at each time step, one of the following operations is performed.

1. With probability p, add m new edges. That is, randomly select a vertex as one endpoint of an edge and attach that edge to another existing vertex with preferential attachment. Repeat this m times.
2. With a probability q, rewire m edges. That is, randomly select a vertex x_i and an edge $\{x_i, x_j\}$. Replace that edge by a new edge $\{x_i, x_{j'}\}$ with preferential attachment. Repeat this m times.
3. With a probability $1 - p - q$, add a new vertex and m new edges linking the new vertex to m existing vertices with preferential attachment.

In the (p, q)-plane, the system exhibits two regimes:

1. an exponential regime as for the uniform or binomial random network models and the Watts-Strogatz small-world network model;
2. a scale-free regime, but with an exponent γ characterizing the power-law decay depending continuously on the parameters p and q, showing a lack of universality, and accounting for the variations observed for real networks.

Using this model to reproduce the properties of the collaboration network of movie actors, Albert and Barabási found an excellent fit taking $p = 0.937$, $q = 0$ (rewiring is absent), and $m = 1$.

Example 58. Dorogovtsev-Mendes-Samukhin model. Starting from the assumption that preferential attachment is the only known mechanism of self-organization of growing a network into a scale-free structure, Dorogovtsev, Mendes, and Samukhin [104] proposed a model similar to the first Barabási-Albert model but in which a new parameter, representing the site *initial attractiveness*, leads to a more general power-law behavior.

In this model, at each time step, a new vertex and m new edges are added to the network. Here, the edges are *directed* and coming out from *nonspecified* vertices; that is, they may come from the new vertex, from already existing vertices, or even from outside of the network. In the Barabási-Albert model, the new links were all coming out from the new vertex. The probability that a new edge points to a given vertex x_t, added at time t, is proportional to the *attractiveness* of this site, defined as

$$A_{x_t} = A + d(x_t),$$

where $A \geq 0$ is the *initial attractiveness* of a site (the same for all sites) and $d(x_t)$ denotes, as usual, the degree of vertex x_t. In the Barabási-Albert model, all sites have an initial attractiveness A equal to m. The Dorogovtsev-Mendes-Samukhin model is equivalent to a system of increasing number of particles distributed in a growing number of boxes, such that, at each time step, m new particles (incoming links) have to be distributed between an increasing number (one per time step) of boxes (sites) according to the rule above.

Writing down the master equation for the evolution of the distribution of the connectivity of a given site as a function of time, the authors were able to derive the exact expression of the vertex degree probability distribution. In particular, for the Barabási-Albert model, they obtained (see Exercise 7.4)

$$P(d = k) = \frac{2m(m + 1)}{(k + m)(k + m + 1)(k + m + 2)},$$

which, for large k, behaves as k^{-3}. Another interesting result is the existence of a simple *universal relation* between the exponent β characterizing the connectivity growth rate of a vertex and the exponent γ characterizing the power-law behavior of the vertex degree probability distribution. They found that

$$\beta(\gamma - 1) = 1,$$

which is, in particular, satisfied by the Barabási-Albert exponents $\beta_{\mathrm{BA}} = \frac{1}{2}$ and $\gamma_{\mathrm{BA}} = 3$.

Example 59. Krapivsky-Rodgers-Redner model. Using a simple growing network model, Krapivsky, Rodgers, and Redner were able to reproduce the observed in-degree and out-degree probability distributions of the World Wide Web as well as find correlations between in-degrees and out-degrees of each vertex [196]. In their model, the network grows according to one of the following two processes:

1. With probability p, a new vertex is added to the network and linked to an existing target vertex, the target vertex being selected with a probability depending only upon its in-degree.
2. With probability $1-p$, a new edge is created between two already existing vertices. The originating vertex is selected with a probability depending upon its out-degree and the target vertex with a probability depending upon its in-degree.

If N is the order of the network, and D_{in} and D_{out} denote, respectively, the total in-degree and out-degree, according to the growth processes, the evolution equations of these quantities are

$$N(t+1) = \begin{cases} N(t)+1, & \text{with probability } p, \\ N(t), & \text{with probability } 1-p, \end{cases}$$

$$D_{in}(t+1) = D_{in}(t)+1,$$
$$D_{out}(t+1) = D_{out}(t)+1.$$

Hence,
$$N(t) = pt, \quad D_{in}(t) = D_{out} = t.$$

These results show that the average in-degree and out-degree are both time-independent and equal to $1/p$.

In order to determine the joint degree distributions, the authors assume that the *attachment rate* $A(d_{in}, d_{out})$, defined as the probability that a newly introduced vertex connects with an existing vertex of in-degree d_{in} and out-degree d_{out}, and the *creation rate* $C\big((d_{in}(x_1), d_{out}(x_1)), (d_{in}(x_2), d_{out}(x_2))\big)$, defined as the probability of adding a new edge pointing from vertex x_1, whose in-degree and out-degree are, respectively, $d_{in}(x_1)$ and $d_{out}(x_1)$, to vertex x_2, whose in-degree and out-degree are, respectively, $d_{in}(x_2)$ and $d_{out}(x_2)$, are simply given by

$$A(d_{in}, d_{out}) = d_{in} + \lambda$$
$$C\big((d_{in}(x_1), d_{out}(x_1)), (d_{in}(x_2), d_{out}(x_2))\big) = \big(d_{in}(x_2) + \lambda\big)\big(d_{out}(x_1) + \mu\big),$$

where $\lambda > 0$ and $\mu > -1$ to ensure that both rates are positive for all permissible values of in-degrees $d_{in} \geq 0$ and out-degrees $d_{out} \geq 1$.

Using the assumptions above regarding the attachment and creation rates to write down and solve the rate equation for the evolution of the joint degree distribution $N(d_{in}, d_{out}, t)$, defined as the average number of vertices with d_{in} incoming and d_{out} outgoing links, Krapivsky, Rodgers, and Redner obtained

$$N(d_{in}, d_{out}, t) = n(d_{in}, d_{out})t.$$

The expression of $n(d_{in}, d_{out})$ reveals interesting features. Power-law probability distributions for the in-degrees and out-degrees of vertices are dynamically generated. Choosing the values [5]

$$p = 0.1333, \quad \lambda = 0.75, \quad \mu = 3.55,$$

it is possible to match the empirical values regarding the World Wide Web:

$$\langle d_{\text{in}} \rangle = \langle d_{\text{out}} \rangle = 7.5, \quad \gamma_{\text{in}} = 2.1, \quad \gamma_{\text{out}} = 2.7.$$

Correlations between the in-degree and out-degree of a vertex develop spontaneously. For instance, the model predicts that popular Web sites (*i.e.*, those with large in-degree) tend to list many hyperlinks (*i.e.*, have large out-degree), and sites with many links tend to be popular. Finally, the model also predicts power-law behavior when, *e.g.*, the in-degree is fixed and the out-degree varies.

Remark 22. The *linear dependence* of the attachment rate upon d_{in} appears to be a *necessary* condition for a growing network to display a scale-free behavior [195, 197, 104]. If the attachment rate varies as d_{in}^{α}, then,

1. for $0 < \alpha < 1$, the degree distribution exhibits a stretched exponential decay;
2. for $\alpha = 1$, the degree distribution has a power-law behavior characterized by the exponent $\gamma_{\text{BA}} = 3$;
3. for an attachment rate asymptotically linear in d_{in}, the degree distribution has a *nonuniversal* power-law behavior characterized by an exponent $\gamma > 2$;
4. for $1 < \alpha < 2$, the number of vertices with a smaller number of links grows slower than linearly in time, while one vertex has the rest of the links;
5. for $\alpha > 2$, all but a finite number of vertices are linked to one particular vertex.

The statistical analysis of the two databases containing paper titles and authors of important journals in mathematics and neuroscience seems to indicate a *non*linear dependence of the attachment rate (see page 290), in contradiction with the reported power-law behavior of the vertex degree probability distribution.

Example 60. Vázquez model. When a scientist enters a new field, he is usually familiar with only some of the already published papers dealing with this field. Then, little by little, he discovers other papers on the subject by following the references included in the papers he knows. When, finally, the new paper is accepted for publication, it includes, in its list of references, papers discovered in this way; *i.e.*, with the addition of a new vertex to the citation network, new outgoing links, discovered recursively, are simultaneously added to the network.

This *walking on the network* mechanism is the basic ingredient of a citation network model proposed by Vázquez [335]. Starting from a single vertex, at each time step the following subrules are applied successively.

1. A new vertex with an outgoing link to a randomly selected existing vertex is added.
2. Then, other outgoing links between the new vertex and each neighbor of the randomly selected existing vertex are added with a probability p. If at least one link has been added in this way, the process is repeated by adding outgoing links between the new vertex and the second neighbors of the randomly selected vertex, and so forth. When no links are added, the time step is completed.

This evolution rule does not include preferential attachment as in Barabási-Albert models, but vertices with large in-degrees being more frequently reachable during the walking process, the basic mechanism of the evolution rule induces preferential attachment.

Numerical simulations show that, if $p < p_c \approx 0.4$, the asymptotic behavior of the vertex in-degree distribution decreases exponentially, while, for $p > p_c$, this distribution decreases as a power law with an exponent $\gamma_V \approx 2$. A value close to 2 has been found for the distribution of in-degrees in the case of the World Wide Web (see page 289).

The small-world network models described in the preceding section were built up adding some randomness to an underlying regular lattice. While these models are certainly interesting toy models, real small-world networks, such as acquaintance networks, form dynamically, starting from some random structure. On the other hand, many real networks, and in particular some acquaintance networks, have been shown to possess a scale-free structure. This important feature is successfully explained in terms of network growth and preferential attachment. But models based on these two ingredients have a small clustering coefficient, which arguably makes them small-world network models.

Example 61. Davidsen-Ebel-Bornholdt model. Davidsen, Ebel, and Bornholdt proposed a small-world network model (*i.e.*, a graph with short characteristic path length and a high clustering coefficient) evolving towards a scale-free network despite the fact that it keeps a fixed number of vertices [100]. The two essential ingredients of this model are a local connection rule based on *transitive linking* and a *finite vertex life span*.

In order to model an acquaintance network, the authors start from a random graph of order N in which each vertex represents a person and each undirected edge between two vertices indicates that these two persons know each other. The network evolves as the result of the application at each time step of the following subrules:

1. Assuming that people are usually introduced to each other by a common acquaintance, randomly select one vertex and two of its neighbors, and, if these two vertices are not connected, add an edge between them, otherwise do nothing. If the selected vertex has less than two neighbors, add an edge between this vertex and another one selected at random.
2. Since people have a finite life span, with probability $p \ll 1$, one randomly selected vertex and all the edges connected to this vertex are removed from the network and replaced by a new vertex and an edge connecting this vertex to a randomly selected existing vertex.

The condition $p \ll 1$ implies that the average life span of a person is much larger than the rate at which people make new acquaintances. Due to this finite life span, the average vertex degree does not grow forever but tends to a finite value that depends upon the value of p.

Numerical simulations of a model with $N = 7000$ vertices show that, for $p = 0.0025$, the vertex degree probability distribution behaves as a decreasing power law characterized by an exponent $\gamma_{\mathrm{DEB}} = 1.35$, the average vertex degree $\langle d \rangle = 149.2$, the clustering coefficient $C_{\mathrm{DEB}} = 0.63$, and the characteristic path length $L_{\mathrm{DEB}} = 2.38$. The vertex degree distribution exhibits a cutoff at high degree values resulting from the finite vertex life span.

Exercises

Exercise 7.1 *Consider a regular finite one-dimensional lattice having N sites and periodic boundary conditions (i.e., a ring), in which each site is linked to its r nearest neighbors.*

(i) Find the exact expression of the characteristic path length L_{lattice}.

(ii) Find the exact expression of the clustering coefficient C_{lattice}.

(iii) What is the asymptotic expression of L_{lattice} for large N and fixed r? What is the maximum value of C_{lattice}?

Exercise 7.2 *Consider a rather large random graph of a few thousand vertices, and check numerically that the probability distribution of the vertex degrees of a random graph are a Poisson distribution.*

Exercise 7.3 *In order to study the evolution as a function of time t of the degree $d(x_i)$ of a given vertex x_i of the Barabási-Albert scale-free network model, Barabási, Albert, and Jeong [33] assume that the degree is a continuous function of time whose growth rate is proportional to the probability $d(x_i)/2mt$ that the new vertex, added at time t, will connect to vertex x_i. Taking into account that, when adding a new vertex to the network, m new edges are added at the same time, the authors show that $d(x_i)$ satisfies the equation*

$$\frac{\partial d(x_i)}{\partial t} = \frac{d(x_i)}{2t}.$$

(i) Find the solution to this equation.

(ii) Assuming that, at time t, the random variable t_i at which vertex x_i has been added to the network is uniformly distributed in the interval $[0, N_0 + t[$, use the result of (i) to determine in the infinite-time limit the vertex degree probability distribution.

(iii) In order to verify that continuous growth and preferential attachment are necessary ingredients to build up a model of a scale-free network, Barabási and Albert studied two other models (see page 293). In the first one, they consider a growing network in which each new added vertex has an equal probability of connecting to any of the already existing vertices. In the second model, they consider a network with a fixed number N of vertices. Initially the network consists of N vertices and no edges. Then, at each time step, a randomly selected vertex is connected to another vertex with preferential attachment. Use the method of (i) and (ii) to find the time evolution of the degree of a specific vertex and the vertex degree probability distribution for these two models.

Exercise 7.4 *Following a method due to Krapivsky, Redner, and Leyvraz [195, 296], it can be shown that, for the Barabási-Albert network model, the average number $N_d(t)$ of vertices with d outgoing edges at time t is a solution of the differential equation*

$$\frac{dN_d}{dt} = \frac{(d-1)N_{d-1} - dN_d}{2t} + \delta_{dm},$$

where $\delta_{dm} = 1$ if $d = m$ and 0 otherwise.

(i) Justify this equation.

(ii) Show that the vertex degree probability distribution

$$P(d) = \lim_{t \to \infty} \frac{N_d(t)}{t}$$

satisfies a linear recurrence equation, and obtain its expression.

Exercise 7.5 *When dealing with discrete random variables, the generating function technique is a very effective problem-solving tool,[16] and it can be used successfully when dealing with random graphs [267]. Let X be a discrete random variable defined on the set of nonnegative integers $\mathbb{N}_0 = \{0, 1, 2, \ldots\}$. If the probability $P(X = k)$ is equal to p_k, the generating function G_X of this probability distribution is defined by*

$$G_X(x) = \sum_{k=0}^{\infty} p_k x^k.$$

The relation

$$(\forall k \in \mathbb{N}_0) \qquad 0 \le p_k \le 1$$

implies that the radius of convergence of the series defining G_X is at least equal to 1; that is, for $|x| < 1$ the series is normally convergent.

In this exercise, G_1 denotes the generating function of the vertex degree probability distribution (i.e., p_k is the probability that a randomly chosen vertex has k first neighbors).

(i) What is the generating function of the sum of the degrees of n randomly chosen vertices?

(ii) What is the generating function of the vertex degree probability distribution of an end vertex of a randomly chosen edge?

(iii) What is the generating function of the number of outgoing edges of an end vertex of a randomly chosen edge?

(iv) We have seen that, for uniform and binomial random graphs, the vertex degree probability distribution is the Poisson distribution

$$p_k = e^{-\lambda} \frac{\lambda^k}{k!}$$

when $N \to \infty$, $p \to 0$, and $Np = \lambda$. In this case, what is the generating function of the probability distribution of the number of outgoing edges of an end vertex of a randomly chosen edge?

---

[16] See, for example, Boccara [56], pp. 42–49.

(v) What is the generating function of the probability distribution of the number of second neighbors of a randomly selected vertex?

(vi) What is the average number of second neighbors of a randomly selected vertex?

Exercise 7.6 *(i) Let S_t and I_t denote the respective densities of susceptible and infective individuals occupying the sites of a random network at time t. If p_i is the probability for a susceptible to be infected by a neighboring infective, show that the average value of the probability for a susceptible to be infected, at time t, by one of his neighbors is*

$$1 - \exp(-\langle d \rangle p_i I_t),$$

where $\langle d \rangle$ is the average vertex degree.

(ii) If p_r is the probability for a susceptible to recover and become susceptible again, within the mean-field approximation, show that the system undergoes a transcritical bifurcation from a disease-free to an endemic state. Denote by ρ the total constant density of individuals.

(iii) The previous model neglects births and deaths. In order to model the spread of a more serious disease, assume that infected individuals do not recover and denote by d_s and d_i, respectively, the probabilities of dying for a susceptible and an infective. Dead individuals are permanently removed from the network. If b_s and b_i are the respective probabilities for a susceptible and an infective to give birth to a susceptible at a vacant vertex of the network during one time unit, show that, within the mean-field approximation, this model exhibits various bifurcations. Discuss numerically the existence and stability of the different steady states.

Solutions

Solution 7.1 *(i) Let the sites of the regular lattice be numbered from 0 to $N-1$. If $\ell < N/2$, we have*

$$d(0, \ell) = d(0, N - \ell) = \begin{cases} \dfrac{\ell}{r}, & \text{if } \ell = 0 \mod r, \\[2mm] \left\lfloor \dfrac{\ell}{r} \right\rfloor + 1, & \text{otherwise,} \end{cases}$$

where $d(x, y)$ denotes the distance between sites x and y. Hence, grouping sites in r-blocks on both sides of site 0, we are left with $N - 1 - 2r \lfloor (N-1)/2r \rfloor$ sites that cannot be grouped into two r-blocks, and their distance from 0 is $\lfloor (N-1)/2r \rfloor + 1$ (if $N - 1 = 0 \mod 2r$, there are no sites left, but the final result is always valid). The sum of the distances from site 0 to all the other $N-1$ sites of the lattice is therefore given by

$$2r \left(1 + 2 + \cdots + \left\lfloor \frac{N-1}{2r} \right\rfloor \right) + \left(N - 1 - 2r \left\lfloor \frac{N-1}{2r} \right\rfloor \right) \left(\left\lfloor \frac{N-1}{2r} \right\rfloor + 1 \right)$$

$$= r \left\lfloor \frac{N-1}{2r} \right\rfloor \left(\left\lfloor \frac{N-1}{2r} \right\rfloor + 1 \right) + \left(N - 1 - 2r \left\lfloor \frac{N-1}{2r} \right\rfloor \right) \left(\left\lfloor \frac{N-1}{2r} \right\rfloor + 1 \right)$$

$$= \left(N - 1 - r \left\lfloor \frac{N-1}{2r} \right\rfloor \right) \left(\left\lfloor \frac{N-1}{2r} \right\rfloor + 1 \right)$$

and, since site 0 does not play any particular role, dividing this result by $N - 1$ yields the characteristic path length:

$$L_{\text{lattice}} = \left(1 - \frac{r}{N-1} \left\lfloor \frac{N-1}{2r} \right\rfloor\right)\left(\left\lfloor \frac{N-1}{2r} \right\rfloor + 1\right).$$

(ii) Let $-r, -r+1, -r+2, \ldots, -1, 1, 2, \ldots, r-1, r$ be the $2r$ nearest neighbors of site 0. To avoid counting edges twice, starting from site $-r$, we only count edges whose endpoints are on the right of each site of the list above. For $0 \le j < r$, site $-r+j$ is linked to $r-1$ sites on its right, and for $0 \le k < r$, site $r - k$ is linked to k sites on its right. The total number of edges is thus

$$r(r-1) + (r-1) + (r-2) + \cdots + 2 + 1 + 0 = \frac{3}{2}r(r-1).$$

The number of edges connecting all pairs of nearest-neighboring sites of site 0 is $r(2r-1)$. Thus, the clustering coefficient of the lattice s given by

$$C_{\text{lattice}}(r) = \frac{3(r-1)}{2(2r-1)}.$$

(iii) For large N and fixed r, we find

$$L_{\text{lattice}} \sim \frac{N}{4r}.$$

This result can be obtained without knowing the exact expression of L_{lattice}. It suffices to note that site $\lfloor N/2 \rfloor$, which is one of the most distant sites from site 0, is at distance $\lfloor N/2r \rfloor$. The characteristic path length is, therefore, approximately equal to $N/4r$.
For $r \ge 1$, C_{lattice} is an increasing function of r that tends to $\frac{3}{4}$ when $r \to \infty$. $C_{\text{lattice}}(r)$ is, therefore, bounded by $\frac{3}{4}$.

Solution 7.2 *Writing a small program to build up a random graph with $N = 5,000$ vertices and a total number of edges $E = 20,000$, for a specific example, we find that the degrees of the vertices vary between 0 and 13. Degrees equal to 0 imply the existence of disconnected subgraphs. Actually, the graph connectivity, which is defined as the fraction of vertices belonging to the giant component, is found to be equal to 0.964242. The numerical frequencies of the different degrees compared to the theoretical frequencies predicted by the Poisson distribution are listed in the table below.*

degree	0	1	2	3	4	5	6
numerical	89	363	715	996	986	787	520
theoretical	91.58	366.31	732.63	976.83	976.83	781.47	520.98
degree	7	8	9	10	11	12	13
numerical	297	143	65	24	9	5	1
theoretical	297.70	148.85	66.16	26.46	9.62	3.21	0.99

The agreement is quite good. The plots of the numerical and theoretical frequencies shown in Figure 7.8 almost coincide. The standard deviation of the numerical frequencies is found equal to 1.98544 compared to the theoretical value $\sqrt{\langle d \rangle} = 2$.

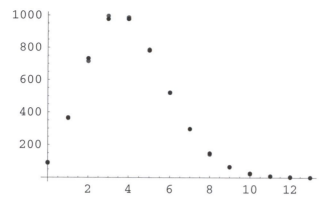

Fig. 7.8. *Poisson test. Plots of the degree's frequencies for z varying from 0 to 13 obtained from a numerical simulation of a random graph with 5000 vertices and an average degree $\langle d \rangle = 4$ (light grey) compared to theoretical frequency values predicted by a Poisson distribution (darker grey).*

Solution 7.3 *(i) The solution $d(x_i, t)$ to the equation*

$$\frac{\partial d(x_i)}{\partial t} = \frac{d(x_i)}{2t},$$

which, for $t = t_i$, is equal to m, is

$$d(x_i, t) = m\sqrt{\frac{t}{t_i}}.$$

(ii) Taking into account the expression of $d(x_i, t)$, the probability $P(d(x_i, t) \leq d)$ that the degree of a given vertex is less than or equal to d can be written

$$P(d(x_i, t) \leq d) = P\left(t_i \geq \frac{m^2 t}{d^2}\right).$$

But

$$P\left(t_i \geq \frac{m^2 t}{d^2}\right) = 1 - P\left(t_i < \frac{m^2 t}{d^2}\right),$$

and since, at time t, the random variable t_i is uniformly distributed in the interval $[0, N_0 + t[$, the probability $P(t_i < T)$ that, at time t, t_i is less than T is $T/(N_0 + t)$. Hence,

$$P(d(x_i, t) \leq d) = 1 - \frac{m^2 t}{d^2 (N_0 + t)}.$$

Thus, in the infinite-time limit, the vertex degree cumulative distribution function F_{vertex} is such that

$$F_{\text{vertex}}(d) = \lim_{t \to \infty} \left(1 - \frac{m^2 t}{d^2 (N_0 + t)}\right)$$

$$= 1 - \frac{m^2}{d^2},$$

and the vertex degree probability density is given by

$$f_{\text{vertex}}(d) = \frac{\partial F(d)}{\partial d}$$
$$= \frac{2m^2}{d^3}.$$

As it should be, the function F_{vertex}, defined on the interval $[m, \infty[$, is monotonically increasing from 0 to 1.

In agreement with numerical simulations, the exponents β_{BA} and γ_{BA} are, respectively, equal to $\frac{1}{2}$ and 3 (see page 293).

(iii) For the first model, with growth but no preferential attachment, the differential equation satisfied by $d(x_i)$ is

$$\frac{\partial d(x_i)}{\partial t} = \frac{m}{N_0 + t - 1}$$

since the probability that, at time t, one of the m new edges added to the network connects to an already existing vertex is the same for all the $N_0 + t - 1$ vertices of the network. The solution to this equation, which, for $t = t_i$, is equal to m, is

$$d(x_i, t) = m \left(\log \frac{N_0 + t - 1}{N_0 + t_i - 1} + 1 \right).$$

As above, the probability $P(d(x_i, t) \le d)$ that the degree of a given vertex is less than or equal to d can be written

$$P(d(x_i, t) \le d) = P\left(t_i \ge (N_0 + t - 1) \exp\left(1 - \frac{d}{m}\right) - N_0 + 1 \right).$$

Assuming that, at time t, the random variable t_i is uniformly distributed in the interval $[0, N_0 + t[$, we can write

$$P\left(t_i \ge (N_0 + t - 1) \exp\left(1 - \frac{d}{m}\right) - N_0 + 1 \right)$$
$$= 1 - P\left(t_i < (N_0 + t - 1) \exp\left(1 - \frac{d}{m}\right) - N_0 + 1 \right)$$
$$= 1 - \frac{(N_0 + t - 1) \exp\left(1 - \frac{d}{m}\right) - N_0 + 1}{N_0 + t}.$$

For this first model, in the infinite-time limit, the cumulative distribution function F^1_{vertex} of the vertex degree is given by

$$F^1_{\text{vertex}}(d) = 1 - \exp\left(1 - \frac{d}{m}\right),$$

and the corresponding vertex degree probability density f^1_{vertex} is such that

$$f^1_{\text{vertex}}(d) = \frac{e}{m} e^{-d/m}.$$

For the second model with preferential attachment but no growth (i.e., a fixed number of vertices N). The degree growth rate is given by the probability $1/N$ of first selecting the vertex from which the edge originates plus the probability of preferentially attaching that edge to vertex x_i. This second probability is of the form

$$A\frac{d(x_i)}{\displaystyle\sum_{j=1}^{N} d(x_j)},$$

where $\sum_{j=1}^{N} d(x_j) = 2t$ and $A = N/(N-1)$ to exclude from the summation edges originating and terminating at the same vertex. Hence, we have

$$\frac{\partial d(x_i)}{\partial t} = \frac{N}{N-1}\frac{d_{x_i}}{2t} + \frac{1}{N}.$$

The solution that, for $t = 0$, satisfies $d(x_i, 0) = 0$ (the initial network consists of N vertices and no edges) is

$$d(x_i, t) = \frac{2(N-1)}{N(N-2)}t + Ct^{N/2(N-1)},$$

where C is a constant. Since $N \gg 1$, the expression above can be written

$$d(x_i, t) \approx \frac{2}{N}t + Ct^{1/2}.$$

Taking into account the relation $\sum_{j=1}^{N} d(x_j) = 2t$, the constant C is equal to zero, and we finally obtain

$$d(x_i, t) \approx \frac{2}{N}t.$$

That is, for large values of N and t all vertices have approximately the same degree. This implies a normal vertex degree distribution centered around its mean value.

In their paper, Barabási, Albert, and Jeong [33] give the results of numerical simulations that agree with all the analytical results above.

Solution 7.4 (i) When a new vertex with m outgoing edges is added to the network, the probability that a new edge connects to a vertex with degree d is given by

$$\frac{md}{\displaystyle\sum_{k\geq m} kN_k} = \frac{md}{2mt} = \frac{d}{2t}.$$

Hence, the first and second terms of the right-hand side of the Krapivsky-Redner-Leyvraz rate equation represent, respectively, the increase and decrease of N_d when new edges are connected either to vertices of degree $d-1$ or to vertices of degree d. The last term accounts for new vertices of degree m.

(ii) The term $\sum_{k\geq m} kN_k = 2mt$ represents the sum of the degrees of all the vertices, clearly equal to twice the sum of all edges. On average, in the infinite-time limit, $N_d(t)$ is therefore linear in t. Thus, replacing $N_d(t)$ by $tP(d)$ in the differential equation, we obtain

$$P(d) = \tfrac{1}{2}\left((d-1)P(d-1) - dP(d)\right) + \delta_{dm}$$

or

$$P(d) = \begin{cases} \dfrac{d-1}{d+2} P(d-1), & \text{if } d > m, \\[3mm] \dfrac{2}{m+2}, & \text{if } d = m. \end{cases}$$

Hence, for $d > m$,

$$P(d) = \frac{2m(m+1)}{d(d+1)(d+2)}.$$

For large d, $P(d) \sim d^{-\gamma_{\mathrm{BA}}}$ with $\gamma_{\mathrm{BA}} = 3$.

Solution 7.5 *(i) This is a classical result: If G_X and G_Y are the respective generating functions of the discrete random variables X and Y, then the generating function of the random variable $X + Y$ is $G_{X+Y} = G_X G_Y$. Hence, the generating function of the sum of the degrees of n randomly chosen vertices is the nth power of the generating function G_1 of the vertex degree probability distribution.*

(ii) The probability that an edge reaches a vertex of degree k is proportional to $k p_k$. Since the sum of all probabilities has to be equal to 1, the generating function of the vertex degree probability distribution of an end vertex of a randomly chosen edge is defined by

$$\frac{\displaystyle\sum_{k=0}^{\infty} k p_k x^k}{\displaystyle\sum_{k=0}^{\infty} k p_k} = \frac{x G_1'(x)}{G_1'(1)}.$$

This result assumes that the value of the average vertex degree $G_1'(1) = \sum_{k=1}^{\infty} k p_k$ exists (i.e., is finite).

(iii) The previous result gave us the generating function of the vertex degree probability distribution of an end vertex of a randomly chosen edge. The number of outgoing edges of one end vertex is equal to its degree minus 1, since one edge was selected to reach this vertex. Thus, the generating function G_{out} of the number of outgoing edges is equal to the generating function of the vertex degree probability distribution of an end vertex of a randomly chosen edge divided by x; that is,

$$G_{\mathrm{out}}(x) = \frac{G_1'(x)}{G_1'(1)}.$$

(iv) The expression of the generating function of the Poisson probability distribution is

$$\begin{aligned} G_1(x) &= \sum_{k=0}^{\infty} e^{-\lambda} \frac{\lambda^k}{k!} x^k \\ &= \sum_{k=0}^{\infty} e^{-\lambda} \frac{(\lambda x)^k}{k!} \\ &= e^{\lambda(x-1)}, \end{aligned}$$

where $\lambda = G_1'(0)$ is the average vertex degree. Hence, the generating function of the probability distribution of the number of outgoing edges of an end vertex of a randomly selected edge is, in this case,

$$G_{\mathrm{out}}(x) = \frac{G_1'(x)}{G_1'(1)} = e^{\lambda(x-1)} = G_1(x);$$

that is, the generating function of the vertex degree probability distribution. Whether a vertex is randomly selected or it is an end vertex of a randomly selected edge, the probability distribution of the number of outgoing edges of this vertex is the same!

(v) A randomly selected vertex has a probability p_k of having k first neighbors, and each of these neighbors has a probability q_ℓ of having a number ℓ of outgoing edges, where q_ℓ is the coefficient of x^ℓ in the series expansion of $G_{\text{out}}(x)$; that is,

$$q_\ell = \frac{1}{\ell!} \frac{d^\ell}{dx^\ell} G_{\text{out}}(x)\bigg|_{x=0}.$$

The generating function of the number of second neighbors of a randomly selected vertex is therefore defined by

$$G_2(x) = G_1\left(G_{\text{out}}(x)\right) = G_1\left(\frac{G_1'(x)}{G_1'(1)}\right).$$

(vi) If G_X is the generating distribution of a discrete random variable X, the average value of X, if it exists, is given by

$$\langle X \rangle = \sum_{k=0}^{\infty} k p_k = G_X'(1).$$

The average number of second neighbors of a randomly selected vertex is then

$$\frac{d}{dx} G_1\left(\frac{G_1'(x)}{G_1'(1)}\right)\bigg|_{x=1} = G_1''(1).$$

Solution 7.6 *(i) Let S_t and I_t denote the respective densities of susceptible and infective individuals occupying the network vertices at time t. If p_i is the probability for a susceptible to be infected by a neighboring infective, the probability for a susceptible to be infected, at time t, by one of his k neighbors is*

$$P(p_i, I_t, k) = 1 - (1 - p_i I_t)^k.$$

Since, in a random network, the vertex degree probability distribution is Poisson, the average value of $P(p_i, I_t, k)$ as a function of the random variable k is given by

$$\langle P(p_i, I_t, k) \rangle_{\text{Poisson}} = \sum_{k=0}^{\infty} \left(1 - (1 - p_i I_t)^k\right) e^{-\langle d \rangle} \frac{\langle d \rangle^k}{k!}$$

$$= 1 - \exp(-\langle d \rangle p_i I_t).$$

(ii) If p_r is the probability for an infective to recover and become susceptible again, the mean-field equations governing the time evolution of this SIS epidemic model are

$$S_{t+1} = \rho - I_{t+1},$$
$$I_{t+1} = I_t + S_t\left(1 - \exp(-\langle d \rangle p_i I_t)\right) - p_r I_t,$$

where ρ is the total constant density of individuals. Eliminating S_t in the second equation yields

$$I_{t+1} = (1 - p_r) I_t + (\rho - I_t)\left(1 - \exp(-\langle d \rangle p_i I_t)\right).$$

In the infinite-time limit, the stationary density of infective individuals I_∞ satisfies the equation

$$I_\infty = (1 - p_r)I_\infty + (\rho - I_\infty)\big(1 - \exp(-\langle d\rangle p_i I_\infty)\big).$$

$I_\infty = 0$ *is always a solution to this equation. This value characterizes the disease-free state. It is a stable stationary state if, and only if, $\langle d\rangle \rho p_i - p_r < 0$.*

If $\langle d\rangle \rho p_i - p_r > 0$, the stable stationary state is given by the unique positive solution of the equation above. In this case, a nonzero fraction of the population is infected. The system is in the endemic state.

For $\langle d\rangle \rho p_i - p_r = 0$, within the framework of the mean-field approximation, the system undergoes a transcritical bifurcation similar to a second-order phase transition characterized by a nonnegative order parameter, whose role is played, in this model, by the stationary density of infected individuals I_∞.

(iii) If b_s and b_i are, respectively, the probabilities for a susceptible and an infective to give birth to a susceptible at a vacant site during one time unit, reasoning as in (i), the probability that either a susceptible or an infective will give birth to a susceptible at a neighboring vacant site is

$$(1 - S_t - I_t)\big(1 - \exp\big(-\langle d\rangle(b_s S_t + b_i I_t)\big)\big).$$

Taking into account this result, the mean-field equations governing the time evolution of the model are

$$S_{t+1} = S_t + (1 - S_t - I_t)\big(1 - \exp\big(-\langle d\rangle(b_s S_t + b_i I_t)\big)\big) - d_s S_t,$$
$$- (1 - d_s)S_t\big(1 - \exp(-\langle d\rangle p_i I_t)\big)$$
$$I_{t+1} = I_t + (1 - d_s)S_t\big(1 - \exp(-\langle d\rangle p_i I_t)\big) - d_i I_t,$$

where d_s and d_i are, respectively, the probabilities for a susceptible and an infective to die during one time unit.

In the infinite-time limit, the stationary densities of susceptible and infective individuals S_∞ and I_∞ satisfy the equations

$$S_\infty = S_\infty + (1 - S_\infty - I_\infty)\big(1 - \exp\big(-\langle d\rangle(b_s S_\infty + b_i I_\infty)\big)\big) - d_s S_\infty$$
$$- (1 - d_s)S_\infty\big(1 - \exp(-\langle d\rangle p_i I_\infty)\big)$$
$$I_\infty = I_\infty + (1 - d_s)S_\infty\big(1 - \exp(-\langle d\rangle p_i I_\infty)\big) - d_i I_\infty.$$

There exist different solutions.

1. *$(S_\infty, I_\infty) = (0, 0)$ is always a solution of the system of equations above. The eigenvalues of the Jacobian at $(0,0)$ being $1 - d_i$ and $1 - \langle d\rangle b_s - d_s$, this point, which corresponds to extinction of all individuals, is asymptotically stable if $\langle d\rangle b_s < d_s$ (i.e., when the average birth rate of susceptible individuals is less than their death rate).*
2. *$I_\infty = 0$ is always a solution to the second recurrence equation. There exists, therefore, a disease-free state $(S_\heartsuit, 0)$, where the infinite-time limit of the susceptible density $S_\heartsuit$ is the nonzero solution to the equation*

$$(1 - S_\heartsuit)\big(1 - \exp(-\langle d\rangle b_s S_\heartsuit)\big) - d_s S_\heartsuit = 0.$$

This solution is stable if

$$\langle d \rangle p_i (1 - d_s) S_\heartsuit - d_i < 0.$$

It can be verified that, during the evolution towards the disease-free steady state,

- *if $S_0 > di/(\langle d \rangle p_i(1 - d_s))$, $I_1 > I_0$, there is an epidemic,*
- *if $S_0 < di/(\langle d \rangle p_i(1 - d_s))$, $I_1 < I_0$, there is no epidemic,*

where S_0 and I_0 denote the susceptible and infective initial densities.[17] For example, if $\langle d \rangle = 4$, $p_i = 0.34$, $b_s = 0.2$, $b_i = 0.1$, $d_s = 0.355$, and $d_i = 0.5$, the condition $\langle d \rangle b_s > d_s$ is verified. In the infinite-time limit, the system is in a disease-free state, and since $d_i/(\langle d \rangle p_i(1 - d_s)) = 0.57$ there is no epidemic for $S_0 = 0.54$ and an epidemic for $S_0 = 0.6$. For these parameter values $S_\heartsuit = 0.468$, and the eigenvalues of the Jacobian at the point $(S_\heartsuit, 0)$ are 0.911 and 0.625, confirming the asymptotic stability of this point.

3. *There also exists a solution $(S_\clubsuit, I_\clubsuit) \neq (0, 0)$ that satisfies the system of equations*

$$0 = (1 - S_\clubsuit - I_\clubsuit)\left(1 - \exp\left(-\langle d \rangle(b_s S_\clubsuit + b_i I_\clubsuit)\right)\right) - d_s S_\clubsuit$$
$$- (1 - d_s)S_\clubsuit\left(1 - \exp(-\langle d \rangle p_i I_\clubsuit)\right),$$
$$0 = (1 - d_s)S_\clubsuit\left(1 - \exp(-\langle d \rangle p_i I_\clubsuit)\right) - d_i I_\clubsuit.$$

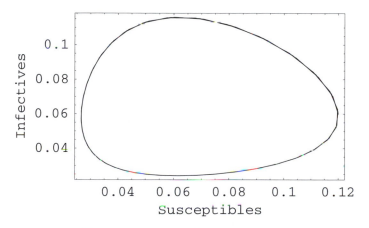

Fig. 7.9. *Limit cycle. Parameter values: $\langle d \rangle = 4$, $p_i = 0.8$, $b_s = 0.049$, $b_i = 0.0001$, $d_s = 0.001$, and $d_i = 0.18$.*

For $\langle d \rangle = 4$, $p_i = 0.8$, $b_s = 0.6$, $b_i = 0.1$, $d_s = 0.355$, and $d_i = 0.5$, the nontrivial fixed point $(S_\clubsuit, I_\clubsuit) = (0.3495, 0.2450)$ and the eigenvalues of the Jacobian at the point are complex conjugate and equal to $0.448522 \pm i\,0.401627$. This indicates the existence of damped oscillations in the neighborhood of this asymptotically stable point representing an endemic state.

4. *Playing with the numerical values of the parameters, we can modify the location of the nontrivial fixed point $(S_\clubsuit, I_\clubsuit)$ and increase the modulus of the eigenvalues of*

---

[17] This result is another illustration of the Kermack-McKendrick threshold theorem; see pages 45 and 137.

the Jacobian at this point. When the modulus of these eigenvalues becomes equal to 1, the system exhibits a Hopf bifurcation. For $\langle d \rangle = 4$, $p_i = 0.8$, $b_s = 0.054$, $b_i = 0.0001$, $d_s = 0.001$, and $d_i = 0.18$, the convergence to the fixed point $(S_\clubsuit, I_\clubsuit) = (0.0623303, 0.0646393)$ is extremely slow since the eigenvalues of the Jacobian at this point, equal to $0.983503 \pm i\,0.180775$, have a modulus equal to 0.999979, indicating the proximity of a Hopf bifurcation.

5. Decreasing the value of the parameter b_s, the system exhibits a limit cycle as illustrated in Figure 7.9.

8
Power-Law Distributions

8.1 Classical examples

Many empirical analyses suggest that power-law behaviors in the distribution of some quantity are quite frequent in nature.[1] However, we have to keep in mind that these behaviors may be purely apparent (see pages 324 and 353) or simply transient or spurious (see page 345). Here are two classical examples completed with some recent developments.

Example 62. Pareto law. Italian economist and sociologist Vilfredo Pareto (1848–1923) is known, in particular, for his pioneering work on the distribution of income. Pareto [277] considered data for England, a few Italian towns, some German states, Paris, and Peru. A log-log plot of the number $N(x)$ of individuals with income greater than x, showed[2] that $N(x)$ behaved as $x^{-\alpha}$, where, for all the data, the exponent α, referred to as the *Pareto exponent*, is in all cases close to 1.5.

x	N	x	N	x	N
150	400,648	600	42,072	2000	9880
200	234,185	700	34,269	3000	6069
300	121,996	800	29,311	4000	4161
400	74,041	900	25,033	5000	3081
500	54,419	1000	22,899	10,000	1104

---

[1] In Subsection 7.5.1, we presented a few scale-free networks characterized by power-law vertex degree probability distributions.

[2] If N_{total} is the total number of individuals, the quantity

$$F(x) = \frac{N_{\text{total}} - N(x)}{N_{\text{total}}}$$

is the probability that an individual has an income less than or equal to x. F is known in probability theory as the cumulative distribution function; see Subsection 8.2.1.

The table above gives the number $N(x)$ as a function of x in pounds sterling of a category of British taxpayers (businessmen and professionals) for the fiscal year 1893–1894. The data are taken from Pareto's *Cours d'Économie Politique*, volume 2, page 305. Figure 8.1 shows a log-log plot of these data (dots) and the least-squares fit (straight line). Actually, from our fit of Pareto's data, we find $\alpha = 1.34$.

Many papers have been published on the probability density of income distribution in various countries.[3] It appears that this probability density follows a power-law behavior in the high-income range, but, according to Gibrat [142],[4] it is *lognormal* in the low–middle range. This probability distribution, which depends upon two parameters is discussed in Subsection 8.2.3. It can can be mistaken for a power-law distribution as shown in Figure 8.5.

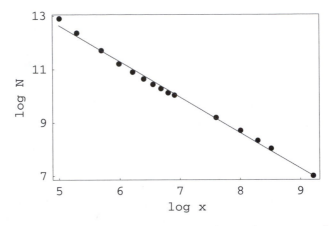

Fig. 8.1. *Distribution of income of a category of British taxpayers (businessmen and professionals) for the year 1893–1894. x is the annual income and $N(x)$ the number of taxpayers whose annual income is greater than x.*

In a recent study of the overall profile of Japan's personal income distribution, Souma [322][5] found that only the top 1% of the Japanese income data for the year 1998 follows a power-law distribution with a Pareto exponent equal to $\alpha = 2.06$; the remaining 99% of the data fit a lognormal distribution with $m = 4$ million yen and $\beta = 1/\sqrt{2\pi\sigma^2} = 2.68$.[6] Souma also studied the time dependence from 1955 to 1998 of Pareto's exponent α and the index β and

---

[3] Pareto law, which states $N(x) \sim x^{-\alpha}$, implies the probability density $f(x) \sim x^{-(1+\alpha)}$.

[4] See also Badger [22] and Montroll and Shlesinger [248].

[5] See also analyses of large and precise data sets by Fujiwara, Souma, Aoyama, Kaizoji, and Aoki [132].

[6] β is called the *Gibrat index*.

found that the values of the Pareto exponent for Japan and the US during that period were strongly correlated.

Pointing out that many recent papers dealing with income probability distributions "do little or no comparison at all with real statistical data," Drăgulescu and Yakovenko [107, 108] analyzed the data on income distribution in the United States from the Bureau of the Census and the Internal Revenue Service. They found that the individual income of about 95% of the population is not lognormal but described by an exponential distribution of the form $\exp(-r/R)/R$.[7] The parameter R, which represents the average income, changes every year. For example, from data collected from the Internal Revenue Service[8] for the year 1997, $R = 35,200$ US\$ from a database size equal to 1.22×10^8. Above 100,000 US\$ the income distribution has a power-law behavior with a Pareto exponent $\alpha = 1.7 \pm 0.1$.

The exponential distribution is valid for individual incomes. For households with two persons filing jointly, the total income r is the sum $r_1 + r_2$ of the two individual incomes, and the income probability distribution p_2 of the household is

$$p_2(r) = \int_0^r p(r')p(r - r')\, dr' = \frac{r}{R^2}\mathrm{e}^{-r/R}.$$

Data obtained from the Web site of the bureau of the Census[9] are in good agreement with this distribution.

The individual and family income distributions differ qualitatively. The former monotonically increases towards low incomes and reaches its maximum at zero. The latter has a maximum at a finite income $r_{\max} = R$ and vanishes at zero. That is, for individuals, the most probable income is zero, while, for a family with two earners, it is equal to R.

According to Drăgulescu and Yakovenko, hierarchy could explain the existence of two regimes, exponential and power-law, for the income distribution. People have leaders, which have leaders of higher order, and so forth. The number of people decreases exponentially with the hierarchical level. So, if the income increases linearly with the hierarchical level, an exponential distribution follows; but if the income increases exponentially with the hierarchical level, a power law is obtained. The linear increase is probably more realistic for moderate incomes, while the exponential increase should be the case for very high incomes.

---

[7] This discrepancy with Souma's result may be due, according to the authors, to the fact that, in Japan, below a relatively high threshold, people are not required to file a tax form.

[8] Information on tax statistics can be found on the Internal Revenue Service Web site: http://www.irs.gov/taxstats/.

[9] http://ferret.bls.census.gov/. FERRET, which stands for Federal Electronic Research and Review Electronic Tool, is a tool developed and supported by the U.S. Bureau of the Census.

Example 63. Zipf law. Analyzing word frequency in different natural languages, George Kingsley Zipf (1902–1950) discovered that the frequency f of a word is inversely proportional to its rank r [355, 356], where a word of rank r is the rth word in the list of all words ordered with decreasing frequency.

For example, James Joyce's novel *Ulysses* contains 260,430 words. If words such as *give, gives, gave, given, giving, giver*, and *gift* are considered to be different, there are, in *Ulysses*, 29,899 different words. Zipf data, taken from *Human Behavior and the Principle of Least Effort*, p. 24, are reproduced in the table below.

r	f	r	f	r	f
10	2653	200	133	3000	8
20	1311	300	84	4000	6
30	926	400	62	5000	5
40	717	500	50	10,000	2
50	556	1000	20	20,000	1
100	265	2000	12	29,899	1

A log-log plot of the word frequency f as a function of the word rank r is shown in Figure 8.2 with a straight line representing the least-squares fit. The slope of this line is equal to 1.02.

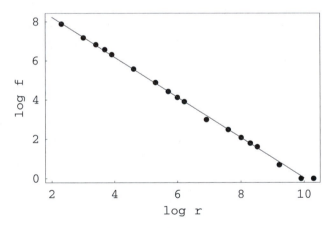

Fig. 8.2. *Log-log plot of word frequency f as a function of word rank r (dots) and the least-squares fit (straight line). Zipf data taken from Human Behavior and the Principle of Least Effort are reproduced in the table above.*

In its original form, Zipf's law, observed on a rather small corpus, accounts for the statistical behavior of word frequencies for ranks between a few hundred and a few thousand. To test its validity for higher rank values, much larger corpora have to be analyzed. Recently, Montemurro [246] studied a very large corpus made up of 2606 books with a vocabulary size of 448,359 different words. He found that Zipf's law is verified with an exponent equal to 1.05 up to $r \approx 6000$. Above this value, there is a crossover to a new regime characterized by an exponent equal to 2.3. These two regimes have also been found by Ferrer i Cancho and Solé [125], who processed the British National Corpus, which is a collection of text samples both spoken and written and comprising about 9×10^7 words. The values of the two exponents found by these authors are 1.01 ± 0.02 and 1.92 ± 0.07. Words seem, therefore, to belong to two different sets: a kernel vocabulary used by most speakers and a much larger vocabulary used for specific communications. For words belonging to the second set, being much less usual, it is not so surprising that their frequency decreases much faster as a function of their rank than the more usual words belonging to the first set.

It is clear that the frequency-ordered list carries only a small part of the information on the role of words. If, for instance, we reorder at random the different words of a text, the word frequency as a function of word rank is unchanged, while the text meaning is totally lost. It is, therefore, not sufficient to determine how often a word is used, we need also to assess where a word is used. A step in this direction has been taken by Montemurro and Zanette [247]. Consider a text corpus (36 plays by Shakespeare containing a total of 885,535 words divided into 23,150 different words) as made up of the concatenation of P parts (the different plays). Determine the frequency f_i of a particular word in part i ($i = 1, 2, \ldots, P$). The quantity

$$p_i = \frac{f_i}{\sum\limits_{i=1}^{P} f_i}$$

is the probability of finding the word in part i and

$$S = -\frac{1}{\log P} \sum_{i=1}^{P} p_i \log p_i$$

the entropy associated with this word. Note that $0 \leq S \leq 1$, where the minimum value 0 corresponds to a word that appears only in one part and the maximum value to a word equally distributed among all parts. It is clear that very frequent words tend to be more uniformly distributed, which implies that the entropy S of a word increases with its global frequency.

To explain the behavior of the entropy, the authors generated a random version of Shakespeare's 36 plays. This was done by shuffling the list of all the 885,535 words and then dividing this list in 36 parts, part j containing exactly the same number of words as play j ($j = 1, 2, \ldots, 36$). Comparing the entropy of the actual plays with the entropy of the randomized plays, while the tendency of the entropy to increase with the global word frequency is preserved, it is found that large fluctuations of the value of the entropy are absent in the randomized version. On average, words have higher entropies in the random version than in the actual corpus.

Language can also be described as a network of linked words. That is, a language can be described as a graph whose vertices are words, and undirected edges represent connections between words. Word interactions are not easy to define precisely, but it seems that different reasonable definitions provide very similar network structures. Following Ferrer i Cancho and Solé [126], an edge may connect an adjective and a noun (*e.g.*, red flowers), an article and a noun (*e.g.*, the house), a verb and an adverb (*e.g.*, stay here), *etc.* This means that two words are at distance 1, *i.e.*, they are connected by an edge if they are nearest neighbors in at least one sentence. The existence of a link between two words could be given a more precise definition saying that words i and j are at distance 1 if the probability p_{ij} of the co-occurrence of the pair (i, j) is larger than the product $p_i p_j$ of their probabilities of occurring in the language. Taking into account this restrictive condition, Ferrer i Cancho and Solé found that in a network of 460,902 vertices (words) and 1.61×10^7 edges, the average degree $\langle d \rangle = 70.13$. The vertex degree probability distribution has two different power-law regimes characterized by the exponents $\gamma_1 = 1.50$ for vertex degrees less than 10^3 and $\gamma_2 = 2.70$ for vertex degrees larger than 10^3. Furthermore, the characteristic path length of the word network $L_{wn} = 2.67$, and the clustering coefficient $C_{wn} = 0.437$, which shows that the word network is a small world.

Considering the language as a self-organizing growing network, and using the Barabási-Albert idea of preferential attachment (see page 292) (*i.e.*, a new word has a probability of being connected to an old word x proportional to the degree $d(x)$ of that word), Dorogovtsev and Mendes [105] found that the vertex degree probability distribution exhibits a crossover between two power-law behaviors with exponents $\gamma_1 = 1.5$ for vertex degrees $d \lesssim 5000$ and $\gamma_2 = 3$ for vertex degrees $d \gtrsim 5000$.

Another application of statistical linguistics is the idea of extracting keywords from interword spacing [275]. The interword spacing is defined as the word count between a word and the next occurrence of the same word in a text. For each word, all interword spacings are counted up and the standard deviation is computed. It is found that high standard deviation is a better criterion as a search engine keyword than high frequency.

Standard deviations of interword spacings seem also to offer the possibility of identifying texts due to the same author. If, for a given text, words are ranked according to the value of the standard deviation of their interword spacing and standard deviation is plotted versus the logarithm of the rank, it appears that plots corresponding to different texts by the same author almost coincide. This result, reported by Berryman, Allison, and Abbot [41], is in favor of the hypothesis that the biblical books of *Luke* and *The Acts of the Apostles* have a common author.

8.2 A few notions of probability theory

8.2.1 Basic definitions

Although ideas about probability are very old, the development of probability theory, as a scientific discipline, really started in 1654 when Pascal (1623–1662) and Fermat (1601–1665) analyzed simple games of chance in a correspondence published for the first time in 1679 [124]. In the meantime, the basic concepts of probability theory were clearly defined and used by Huygens (1629–1695) in his treatise *De Ratiociniis in AleæLudo* published in 1657. In the early 18th century, Jakob Bernoulli (1654–1705) proved the law of large numbers, and de Moivre (1667–1754) established a first version of the central limit theorem. In the early 19th century, important contributions, essentially on limit theorems, were made by Laplace (1749–1827), Gauss (1777–1855) and Poisson (1781–1840). These results were extended in the late 19th and early 20th centuries by Chebyshev (1821–1894), Markov (1856–1922), and Lyapunov (1857–1918). The axiomatization of the theory started with the works of Bernstein (1878–1956), von Mises (1883–1953), and Borel (1871–1956), but the formulation based on the notion of measure, which became generally accepted, was given by Kolmogorov (1903–1987) in 1933.

According to Kolmogorov [192], a *probability space* is an ordered triplet $(\Omega, \mathcal{A}, P)$, where Ω is the *sample space* or *space of elementary events*, $\mathcal{A}$ is the set of all *events*, and P is a *probability measure* defined on $\mathcal{A}$.

The set $\mathcal{A}$ is such that:

1. The empty set $\emptyset$ is an event.
2. If $\{A_n \mid n \in \mathbb{N}\}$ is a countable family of events, then $\bigcup_{n \in \mathbb{N}} A_n$ and $\bigcap_{n \in \mathbb{N}} A_n$ are events.
3. If A is an event, the complement A_c of A is an event.

The probability measure P defined on $\mathcal{A}$ has the following properties:

1. $P(\emptyset) = 0$.
2. For all countable sequences (A_n) of pairwise disjoint events

$$P\left(\bigcup_{n=1}^{\infty} A_n\right) = \sum_{n=1}^{\infty} P(A_n).$$

3. $P(\Omega) = 1$.

A *random variable* X is a real function defined on Ω such that, for all $x \in \mathbb{R}$, the set $\{\omega \in \Omega \mid X(\omega) \leq x\}$ is an event; *i.e.*, an element of $\mathcal{A}$. The *cumulative distribution function* of the random variable X is the function $F_X : \mathbb{R} \to [0, 1]$ defined by

$$F_X(x) = P(X \leq x),$$

where the notation $P(X \leq x)$ represents the probability of the event $\{\omega \in \Omega \mid X(\omega) \leq x\}$.

The cumulative distribution function is a measure. This measure is *absolutely continuous* if there exists a nonnegative function f_X, called the *probability density* of the random variable X, such that

$$F_X(x) = \int_{-\infty}^{x} f_X(u) \, du.$$

In this case, X is said to be an *absolutely continuous random variable*.

If the cumulative distribution function is piecewise constant with discontinuities at points x_k ($k \in \mathbb{N}$) such that, for all k, $\Delta F_X(x_k) = F(x_k) - F(x_k - 0) > 0$, the random variable is said to be *discrete*, and the sequence of numbers (p_k), where $p_k = P(\{x_k\})$, is called a *discrete probability distribution* (see an example on page 280).

There exist continuous cumulative distribution functions that are constant almost everywhere; *i.e.*, they increase on a set of zero Lebesgue measure. Such distributions are said to be *singular* (see an example on page 219).

In most applications, when dealing with a random variable X, we never refer to the sample space Ω of elementary events. We always refer to the cumulative distribution function F_X, or, when X is absolutely continuous, to its probability density f_X. Both functions are often abusively called probability distributions. It is clear that any nonnegative integrable function f such that

$$\int_{-\infty}^{\infty} f(x) \, dx = 1$$

defines a probability density.

The following characteristics are of frequent use. They do not, however, always exist.

The *mean value* or *average value* of a random variable X is

$$m_1(X) = \langle X \rangle = \int_{-\infty}^{\infty} x \, dF_X(x),$$

$$= \begin{cases} \sum_{k=1}^{\infty} x_k p_k, & \text{if } X \text{ is discrete,} \\ \int_{-\infty}^{\infty} x f_X(x) \, dx, & \text{if } X \text{ is absolutely continuous.} \end{cases}$$

More generally, the moment of order r of X is defined by

$$m_r(X) = \langle X^r \rangle = \int_{-\infty}^{\infty} x^r \, dF_X(x).$$

The variance of X is

$$\sigma^2(X) = \langle (X - \langle X \rangle)^2 \rangle$$
$$= \langle X^2 \rangle - \langle X \rangle^2 = m_2(X) - m_1^2(X).$$

The square root $\sigma(X)$ of the variance is called the *standard deviation*. If there is no risk of ambiguity, we shall omit X and simply write m_1, m_r, and σ^2.

An important function, which is always defined, is the *characteristic function* φ_X of the probability distribution of X defined by

$$\varphi_X(t) = \langle e^{itX} \rangle = \int_{-\infty}^{\infty} e^{itx} \, dF_X(x).$$

In the case of absolutely continuous random variables, the characteristic function is the Fourier transform of the probability density: $\varphi_X = \widehat{f_X}$.[10]

If the moment $m_r(X)$ of the random variable X exists, then

$$\varphi_X(t) = \sum_{k=0}^{r} \frac{m_k(X)}{k!} (it)^k + o(t^r);$$

that is, if $m_r(X)$ exists, then φ_X is at least of class C^r and, for $k = 0, 1, \ldots, r$, $\varphi^{(k)}(0) = i^k m_k(X)$.

There exist various modes of convergence for sequences of random variables. Let us just mention one mode that usually leads to short and elegant proofs of limit theorems:

Definition 24. *The sequence of random variables (X_n) converges in distribution to the random variable X if for every bounded continuous function f*

$$\lim_{n \to \infty} \langle f(X_n) \rangle = \langle f(X) \rangle.$$

This mode of convergence is denoted

$$X_n \xrightarrow{\ d\ } X.$$

That is, to prove that (X_n) converges to X in distribution, we may just check if either

$$\lim_{n \to \infty} F_{X_n}(x) = F_X(x)$$

at every point of continuity of F_X or

$$\lim_{n \to \infty} \varphi_{X_n}(t) = \varphi_X(t).$$

---

[10] The Fourier transform of the function f is denoted $\widehat{f}$.

8.2.2 Central limit theorem

The most important absolutely continuous probability distribution is the *normal* or *Gauss* distribution, whose density, defined on $\mathbb{R}$, is

$$f_{\text{Gauss}}(x) = \frac{1}{\sqrt{2\pi}\sigma} \exp{-\frac{(x-m)^2}{2\sigma^2}},$$

where m is real and σ positive. It is easily verified that, for a Gaussian random variable,

$$\langle X \rangle = m, \quad \text{and} \quad \langle X^2 \rangle - \langle X \rangle^2 = \sigma^2.$$

A Gaussian random variable of mean m and variance σ^2 is denoted $N(m, \sigma^2)$.

As its name emphasizes, the central limit theorem plays a central role in probability theory. In its simplest form it can be stated:

Theorem 20. *Let (X_j) be a sequence of independent identically distributed random variables with mean m and standard deviation σ; if $S_n = X_1 + X_2 + \cdots + X_n$, then*

$$\frac{S_n - nm}{\sigma\sqrt{n}} \xrightarrow{\text{d}} N(0,1).$$

That is, the scaled sum of n independent identically distributed random variables with mean m and standard deviation σ converges in distribution to a normal random variable of zero mean and unit variance.

Following Lyapunov, the simplest method for proving the central limit theorem is based on characteristic functions of probability distributions. Since the random variables X_j are identically distributed, they have the same characteristic function φ_X. For $j \in \mathbb{N}$, let $Y_j = (X_j - m)/\sigma$ and denote by φ_Y their common characteristic function. Then, the characteristic function φ_{U_n} of the random variable

$$U_n = \frac{1}{\sqrt{n}} \sum_{j=1}^{n} Y_j = \frac{S_n - nm}{\sigma\sqrt{n}}$$

is

$$\varphi_{U_n}(t) = \left(\varphi_Y(t/\sqrt{n})\right)^n.$$

For a fixed t and a large n,

$$\varphi_Y\left(\frac{t}{\sqrt{n}}\right) = 1 - \frac{t^2}{2n} + o\left(\frac{t^2}{n}\right).$$

Thus

$$\varphi_{U_n}(t) = \left(1 - \frac{t^2}{2n} + o\left(\frac{t^2}{n}\right)\right)^n$$

and

$$\lim_{n\to\infty} \varphi_{U_n}(t) = \exp\left(-\tfrac{1}{2}t^2\right),$$

which proves the theorem since $t \mapsto e^{-\frac{1}{2}t^2}$ is the characteristic function of a Gaussian random variable $N(0,1)$ (see page 325). $\qquad\square$

Example 64. The probability distribution of a Poisson random variable is

$$P_{\text{Poisson}}(X = k) = e^{-\lambda k}\frac{\lambda^k}{k!}.$$

It is easy to verify that the sum of n independent identically distributed Poisson variables is a Poisson variable. Thus, for $\lambda = 1$, the central limit theorem implies

$$\lim_{n\to\infty} \sum_{k\leq n+x\sqrt{n}} e^{-nk}\frac{n^k}{k!} = \frac{1}{\sqrt{2\pi}} \int_{-\infty}^{x} e^{-u^2/2}\, du.$$

Figure 8.3 shows the approach to a normal distribution for the scaled sum of 50 independent Poisson random variables with a unit parameter value; see also Exercise 8.2.

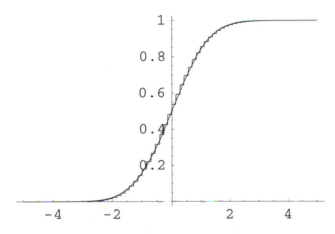

Fig. 8.3. *Convergence to the normal distribution of the scaled sum of 50 independent Poisson random variables with unit parameter value.*

The central limit theorem holds under more general conditions. Here is an example[11]:

Theorem 21. *Let (X_j) be a sequence of independent random variables such that, for all j,*

$$\langle X_j\rangle = m_j, \quad \langle (X_j - m_j)^2\rangle = \sigma_j^2, \quad \langle |X_j - m_j|^3\rangle < \infty.$$

If

---

[11] See, for example, Loève [210].

$$\lim_{n \to \infty} \frac{\sum_{j=1}^{n} \langle |X_j - m_j|^3 \rangle}{\left(\sum_{j=1}^{n} \sigma_j^2 \right)^{3/2}} = 0,$$

then

$$\frac{\sum_{j=1}^{n} (X_j - m_j)}{\sqrt{\sum_{j=1}^{n} \sigma_j^2}} \xrightarrow{\text{d}} N(0,1).$$

Remark 23. According to the central limit theorem, the normal distribution is *universal* in the sense that the distribution of the scaled sum of a large number of random variables subject to some restrictive conditions is a Gaussian random variable. In the averaging process, we are thus losing information concerning the specific random variables. This implies that, for given average value and variance, the normal distribution is the distribution containing minimal information.[12]

Random variables observed in nature can often be interpreted as sums of many individual mutually independent random variables, which, as a consequence of the central limit theorem, should be normally distributed. Coefficients characterizing the shape of a probability distribution as *skewness* and *kurtosis excess* are often used to check possible deviations from normality. These coefficients are, respectively, defined by the scale-invariant expressions[13]

$$\alpha_3 = \frac{\langle (X - m_1)^3 \rangle}{\sigma^3},$$

$$\alpha_4 = \frac{\langle (X - m_1)^4 \rangle}{\sigma^4} - 3.$$

They are both equal to zero for normally distributed random variables. The coefficient α_3 is a measure of the asymmetry of the probability distribution, and α_4 can be, more specifically, used as a measure of departure from Gaussian behavior.

In the case of a *unimodal* probability distribution (*i.e.*, probability distribution with only one relative maximum—the *mode*), the *median* $\tilde{m}$, defined by $F_X(\tilde{m}) = \frac{1}{2}$, may give, when it differs from the average value m_1, a measure of skewness (see page 332).[14]

Example 65. A discrete random variable X that can take the value $k = 0, 1, 2, \ldots, n$ with the probability

$$P_{\text{binomial}}(X = k) = \binom{n}{k} p^k (1 - p)^{n-k}$$

is a *binomial* random variable. Its skewness and kurtosis excess are given by

---

[12] This result can, in fact, be established directly, as shown in Exercise 8.1.

[13] The coefficient $\langle (X - m_1)^4 \rangle / \sigma^4$, which is also used in statistics, is called the *kurtosis*.

[14] On the technical aspects of statistics, the interested reader may consult Sachs [305]. On the mathematical aspects, see Wilks [346].

$$\alpha_3 = \frac{1 - 2p}{\sqrt{np(1-p)}} \quad \text{and} \quad \alpha_4 = \frac{1 - 6p(1-p)}{np(1-p)}.$$

As $n \to \infty$, both coefficients go to zero. This result follows from the central limit theorem. Since the binomial distribution of parameters n and p is the probability distribution of the sum of n independent identically distributed Bernoulli random variables X_j $(j = 1, 2, \ldots, n)$, where, for all j,

$$P_{\text{Bernoulli}}(X_j = 1) = p \quad \text{and} \quad P_{\text{Bernoulli}}(X_j = 0) = 1 - p,$$

the central limit theorem implies[15]

$$\lim_{n \to \infty} \sum_{k \le np + x\sqrt{np(1-p)}} \binom{n}{k} p^k (1-p)^{n-k} = \frac{1}{\sqrt{2\pi}} \int_{-\infty}^{x} e^{-u^2/2}\, du.$$

The central limit theorem tells us that the sum S_n of n independent identically distributed random variables $X_1, X_2, \ldots, X_n$ with finite variance is well-approximated by a Gaussian distribution with mean nm_1 and variance $n\sigma^2$ when n is large enough. Typically, for a large finite value of n, deviations of S_n from nm_1 are of the order of $n^{1/2}$, which is small compared to nm_1. Deviations of S_n from nm_1 may be much greater. The theory of *large deviations* studies, in particular, the asymptotic behavior of $P(|S_n - nm_1| > an)$, where $a > 0$. The following result is often useful[16]:

Theorem 22. *Let (X_j) be a sequence of independent identically distributed random variables; if the function[17] $M : t \mapsto \langle \exp(t(X_1 - m_1)) \rangle$ is finite in some neighborhood of the origin, then, if $P(X_1 - m_1 > a) > 0$ for a positive a,*

$$\lim_{n \to \infty} \left(P(S_n - nm_1 > na) \right)^{1/n} = e^{-\psi(a)},$$

where $S_n = X_1 + X_2 + \cdots + X_n$ and

$$\psi(a) = -\log \left(\inf_{t > 0} \left(e^{-at} M(t) \right) \right)$$

is positive.

That is, $P(S_n - nm_1 > na)$ decays exponentially. For an application of this result, see Exercise 8.3.

---

[15] This result was first established by de Moivre for $p = \frac{1}{2}$ and extended by Laplace to any value of p.

[16] See Grimmett and Stirzaker [158], p. 184.

[17] M is the *moment-generating function*.

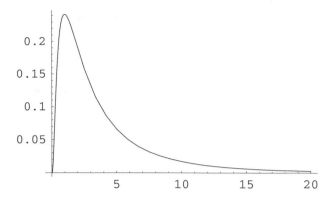

Fig. 8.4. *Probability density function of the lognormal distribution for $x_0 = e$ and $\sigma = 1$.*

8.2.3 Lognormal distribution

The lognormal probability density function, which depends upon two parameters x_0 and σ, is given by

$$f_{\text{lognormal}}(x) = \frac{1}{x\sqrt{2\pi\sigma^2}} \exp\left(-\frac{\left(\log(x/x_0)\right)^2}{2\sigma^2}\right).$$

That is, if X is a lognormal random variable, $\log X$ is normally distributed with $\langle \log X \rangle = \log x_0$ and $\langle (\log X)^2 \rangle - \langle \log X \rangle^2 = \sigma^2$. The graph of $f_{\text{lognormal}}$ is represented in Figure 8.4.

A lognormal distribution results when many random variables cooperate multiplicatively. Let X_i $(i = 1, 2, \ldots, n)$ be independent random variables. Since

$$\log \prod_{i=1}^{n} X_i = \sum_{i=1}^{n} \log X_i,$$

if the independent random variables $\log X_i$ $(i = 1, 2, \ldots, n)$ satisfy the conditions of validity of the central limit theorem, the scaled sum of such random variables is a *lognormal* random variable.

The lognormal distribution is used in biology, where, for instance, the sensitivity of an individual to a drug may depend multiplicatively upon many characteristics such as weight, pulse frequency, systolic and diastolic blood pressure, cholesterol level, *etc.* It is also used in finance, where it is assumed that the increments of the logarithm of the price rather than the change of prices are independent random variables (see page 336). In data analysis, power-law behavior of certain observables is detected using log-log plots. When adopting such a procedure, a lognormal distribution can be mistaken for a power-law distribution over a rather large interval, as illustrated in Figure 8.5. (See Exercise 8.6 for another example of apparent power-law behavior.)

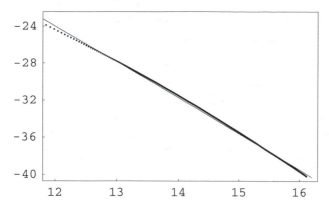

Fig. 8.5. *Apparent power-law behavior of the lognormal distribution. The dotted line represents the sequence* $\big(\log(10,000\,n), \log\big(f_{\text{normal}}(10,000\,n)\big)\big)_{n=1,2,\ldots,1000}$. *The parameters of the lognormal distribution are* $x_0 = \mathrm{e}$ *and* $\sigma = -3.873$. *The slope of the straight line giving the least-squares fit is equal to* -3.873.

8.2.4 Lévy distributions

Let X_1 and X_2 be two absolutely continuous random variables and f_{X_1} and f_{X_2} their respective probability densities. If X_1 and X_2 are defined on the same space and are independent, the probability density $f_{X_1+X_2}$ of their sum is given by

$$f_{X_1+X_2}(x) = \int_{-\infty}^{x} f_{X_1}(x_1) f_{X_2}(x - x_1)\, dx_1.$$

That is, the probability density $f_{X_1+X_2}$ of the sum of the two independent random variables X_1 and X_2 is the convolution $f_{X_1} * f_{X_2}$ of their probability densities.

Since the characteristic function φ_X of an absolutely continuous random variable X is the Fourier transform of its probability density, we have

$$\varphi_{X_1+X_2}(t) = \varphi_{X_1}(t)\varphi_{X_2}(t).$$

It is straightforward to verify that the characteristic function of a normal (Gaussian) random variable with probability density function

$$x \mapsto \frac{1}{2\pi\sigma} \exp\left(\frac{x-m)^2}{2\sigma^2}\right),$$

where m is real and σ positive, is

$$t \mapsto \exp\left(\mathrm{i}tm - \tfrac{1}{2}t^2\sigma^2\right).$$

Using this result, we find that the characteristic function of the sum of two independent normal random variables is given by

$$\exp\left(itm_1 - \tfrac{1}{2}t^2\sigma_1^2\right)\exp\left(itm_2 - \tfrac{1}{2}t^2\sigma_2^2\right)$$
$$= \exp\left(it(m_1 + m_2) - \tfrac{1}{2}t^2(\sigma_1^2 + \sigma_2^2)\right),$$

which proves that the sum of two independent normal random variables is also a normal random variable. Probability distributions having this property are said to be *stable under addition*.[18]

Paul Lévy [202, 203, 145] addressed the problem of determining the class of all stable distributions. He found that the general form of the characteristic function $\widehat{L}_{\alpha,\beta}$ of a stable probability distribution is such that

$$\log \widehat{L}_{\alpha,\beta}(t) = \begin{cases} imt - a|t|^\alpha\left(1 - i\beta\dfrac{t}{|t|}\tan\left(\dfrac{\pi}{2}\alpha\right)\right), & \text{if } \alpha \neq 1, \\[2ex] imt - a|t|\left(1 - i\beta\dfrac{k}{|t|}\dfrac{2}{\pi}\log|t|\right), & \text{if } \alpha = 1, \end{cases} \qquad (8.1)$$

where $0 < \alpha \leq 2$,[19] $a > 0$, m is real, and $-1 \leq \beta \leq 1$. The indices α and β refer, respectively, to the *peakedness* and *skewness* of Lévy probability density functions. If $\beta = 0$, the Lévy probability density function is symmetric about $x = m$ (see Figure 8.6).

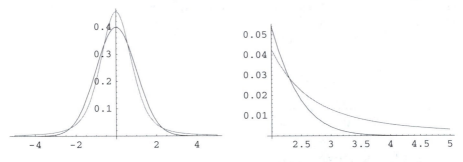

Fig. 8.6. *Right figure: symmetric Lévy stable probability density functions for $\alpha = 1.5$ (lighter grey) and $\alpha = 2$ (normal distribution). Left figure: tails of both probability density functions.*

The particular case $\alpha = 2$, $a = \tfrac{1}{2}\sigma^2$, and $\beta = 0$ corresponds to the normal distribution. Another simple example of a stable distribution is the Cauchy distribution (also called the Lorentzian distribution). It corresponds to $\alpha = 1$ and $m = \beta = 0$, and its probability density is

$$L_{1,0}(x) = \frac{a}{\pi(x^2 + a^2)}, \qquad (a > 0).$$

---

[18] Using the characteristic function method, one can easily verify that binomial and Poisson probability distributions are stable under addition.

[19] For $\alpha > 2$, the Fourier transform of $e^{-a|t|^\alpha}$ may become negative for some values of x and cannot be an acceptable probability density function.

If $m = \beta = 0$, we have

$$L_{\alpha,0}(x) = \frac{1}{2\pi} \int_{-\infty}^{\infty} \exp(-itx - a|t|^\alpha)\, dt$$

$$= \frac{1}{\pi} \int_0^{\infty} e^{-a|t|^\alpha} \cos tx\, dt.$$

Except for α equal to either 1 or 2, there is no simple expression for $L_{\alpha,0}(x)$. It is, however, possible to find its asymptotic expansions for large values of x. For $0 < \alpha < 2$, one finds

$$L_{\alpha,0}(x) \sim \frac{a\Gamma(1+\alpha)}{\pi|x|^{1+\alpha}} \sin \tfrac{1}{2}\pi\alpha,$$

where Γ is the Euler function.[20] The case $\alpha = 2$, which corresponds to the normal distribution, is singular with its exponential tail.

For $0 < \alpha \leq 1$, the mean value of a Lévy random variable is not defined, while for $1 < \alpha < 2$, the mean value is defined but not the variance.

The probability that the sum S_n of n independent identically distributed symmetric Lévy random variables lies in the interval $]s_n, s_n + ds_n]$ is

$$\left(\frac{1}{\pi} \int_0^{\infty} e^{-an|t|^\alpha} \cos t s_n\, dt \right) ds_n.$$

Changing $n^{1/\alpha}t$ in τ yields

$$\left(\frac{1}{\pi} \int_0^{\infty} e^{-a|\tau|^\alpha} \cos \tau(s_n n^{-1/\alpha})\, d\tau \right) d(s_n n^{-1/\alpha}),$$

showing that the correct scaled sum is the random variable $U_n = S_n n^{-1/\alpha}$. Note that for $\alpha = 2$, which corresponds to the Gaussian case, we obtain the right scaling.

The central limit theorem may be restated saying that the Gaussian distribution with zero mean and unit variance is, in the space of probability distributions, the attractor of scaled sums of large numbers of independent identically distributed random variables with finite variance; or, equivalently, scaled sums of large numbers of independent identically distributed random variables with finite variance are in the basin of attraction of the Gaussian distribution with zero mean and unit variance.

Is there an equivalent result for scaled sums of independent identically distributed random variables with infinite variance? The answer is yes.

---

[20] For $\text{Re}\, z > 0$, the Γ function is defined by

$$\Gamma(z) = \int_0^{\infty} t^{z-1} e^{-t}\, dt.$$

Consider the random variable

$$S_n = \sum_{j=1}^{n} X_j,$$

where the X_j $(j = 1, 2, \ldots, n)$ are independent identically distributed random variables. Suppose that their common probability density function f behaves for large x as

$$f(x) \sim \begin{cases} C_- |x|^{-(1+\alpha)}, & \text{as } x \to -\infty, \\ C_+ x^{-(1+\alpha)}, & \text{as } x \to \infty, \end{cases}$$

where $0 < \alpha < 2$, and let

$$\beta = \frac{C_+ - C_-}{C_+ + C_-}.$$

Then, the probability density function of the scaled sum tends to a Lévy probability density of index α and asymmetry parameter β.

The asymptotic power-law behavior of stable Lévy probability distributions makes them good candidates when modeling systems characterized by Pareto's tails with a Pareto exponent equal to α.

If, in a random walk, the jump length probability distribution is a symmetric Lévy distribution, the existence of a long tail implies unfrequent long jumps connecting distant clusters of visited points. This feature is illustrated in Figure 8.7 on Cauchy random walks in one and two dimensions.[21] Random walks with a jump length probability distribution given by a Lévy stable distribution are called *Lévy flights*.

8.2.5 Truncated Lévy distributions

Finite-size effects prohibit the observation of infinitely long tails. If a measured probability distribution looks Lévy-like, since it cannot extend to infinity, there must exist some sort of cutoff. This remark led Mantegna and Stanley [226] to define *truncated Lévy probability densities* by

$$T(x) = \begin{cases} CL_{\alpha,0}(x), & \text{if } |x| \leq \ell, \\ 0, & \text{otherwise}, \end{cases}$$

where C is a normalizing factor and ℓ a cutoff length. An example of such a distribution is represented in Figure 8.8.

For a Lévy random variable X, the probability $P(X > \ell)$ that X is greater than ℓ is a decreasing function of ℓ. The truncation therefore eliminates the

---

[21] The probability density function of the Cauchy distribution considered here is

$$x \mapsto \frac{1}{\pi(1 + x^2)},$$

which is the Lévy distribution $L_{1,0}$ for $a = 1$.

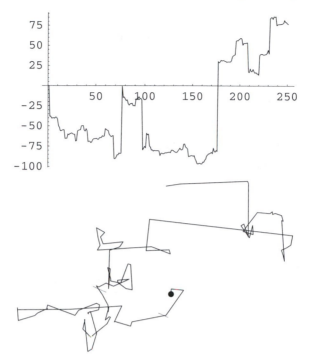

Fig. 8.7. *Cauchy random walks showing occasional long jumps. Top: one-dimensional walk, number of steps: 250. Bottom: two-dimensional walk, dot indicates initial point, number of steps: 100.*

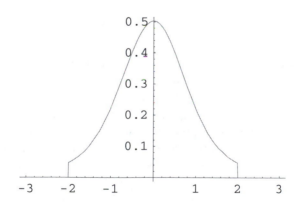

Fig. 8.8. *A truncated symmetric Lévy probability density function for $\alpha = 1.5$ equal to zero for $|x| > 2$.*

possible but rare occurrence of certain events. For very large values of n, the scaled sum of n independent identically distributed truncated Lévy random variables should approach a Gaussian distribution as a consequence of the central limit theorem. But it is clear that the larger ℓ is, the slower the convergence will be, since, for a large ℓ, the truncation eliminates values of X that, even for sums of nontruncated Lévy random variables, would have had a negligible probability of occurring. There should therefore exist a value n^* of n such that, for $n < n^*$, the scaled sum would approach a Lévy random variable, while, for $n > n^*$, the scaled sum would approach a Gaussian random variable. This crossover has indeed been observed by Mantegna and Stanley, who found n^* to be proportional to ℓ^α.

If, instead of the sharp cutoff of Mantegna and Stanley, one considers with Koponen a smooth cutoff [194]—that is, a probability density of the form

$$T_{\mathrm{Koponen}}(x) = \begin{cases} A_- e^{-\lambda x} |x|^{-1-\alpha}, & \text{if } x < 0, \\ A_+ e^{-\lambda x} x^{-1-\alpha}, & \text{if } x \geq 0, \end{cases}$$

where $0 < \alpha < 2$—the expression of its characteristic function can be exactly calculated. This allowed Koponen to prove for sums of independent random variables, whose common probability distribution is $T_{\mathrm{Koponen}}(x)$, the existence of the two regimes observed by Mantegna and Stanley.

8.2.6 Student's t-distribution

The t-distribution was introduced in statistics by William Sealy Gosset (1876–1937), writing under the pseudonym Student [327]. For $x \in \mathbb{R}$, its probability density function is given by

$$f_{\mathrm{t}}(x) = \frac{\Gamma\left(\frac{1}{2}(1+\mu)\right)}{(\pi\mu)^{\frac{1}{2}} \Gamma\left(\frac{1}{2}\mu\right)} \left(1 + \frac{x^2}{\mu}\right)^{-\frac{1}{2}(1+\mu)}.$$

In Figure 8.9 are represented the graphs of two Student's probability density functions for $\mu = 1$ and $\mu = 10$. Since $\Gamma(\frac{1}{2}) = \sqrt{\pi}$, for $\mu = 1$, $f_{\mathrm{t}}(x)$ coincides with the Cauchy distribution $\left(\pi(1+x^2)\right)^{-1}$.

The Student's probability distribution arises in a commonly used statistical test of hypotheses concerning the mean of a small sample of observed values drawn from a normally distributed population when the population standard deviation is unknown (see page 333).

It is simple to show that, as $\mu \to \infty$, the Student's distribution tends to the Gaussian distribution with zero mean and unit variance. We have

$$\frac{\Gamma\left(\frac{1}{2}(1+\mu)\right)}{(\pi\mu)^{\frac{1}{2}} \Gamma\left(\frac{1}{2}\mu\right)} = \frac{1}{\sqrt{2\pi}} \left(1 - \frac{1}{4\mu} + \frac{1}{32\mu^2} - \cdots\right)$$

and

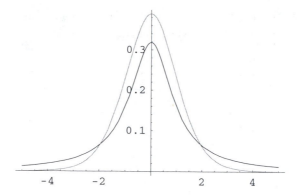

Fig. 8.9. *Probability density function of the Student t-distribution for $\mu = 1$ (black) and $\mu = 10$ (grey).*

$$\log\left(1+\frac{x^2}{\mu}\right)^{-\frac{1}{2}(1+\mu)} = -\frac{1}{2}(1+\mu)\log\left(1+\frac{x^2}{\mu}\right)$$

$$= -\frac{1}{2}(1+\mu)\left(\frac{x^2}{\mu}-\frac{x^4}{2\mu^2}+\cdots\right)$$

$$= -\frac{x^2}{2}+\frac{x^4-2x^2}{4\mu}+\cdots .$$

Thus

$$f_{\mathrm t}(x) = \frac{1}{\sqrt{2\pi}}\,\mathrm e^{-\frac{1}{2}x^2}\left(1-\frac{1}{4\mu}+\cdots\right)\left(1+\frac{x^4-2x^2}{4\mu}+\cdots\right),$$

which proves that

$$\lim_{\mu\to\infty} f_{\mathrm t}(x) = \frac{1}{\sqrt{2\pi}}\,\mathrm e^{-\frac{1}{2}x^2}.$$

8.2.7 A word about statistics

If, for example, we are interested in the distribution of annual incomes of people living in a specific country, we assume that the income is a random variable X taking the observed values $x_1, x_2, \ldots$, where each index refers to a specific individual of the population. In order to estimate the parameters characterizing the probability distribution of X, we have to select a set of observed values $\{x_i \mid i = 1, 2, \ldots, n\}$, which is said to be a *sample of size n*.[22]

---

[22] In mathematical statistics a *random sample* is a sequence of random variables $(X_1, X_2, \ldots, X_n)$ with a common probability distribution. Such a random sample can be viewed as the result of n successive drawings from an infinite population, the population probability distribution remaining unchanged throughout the drawings.

Assuming that the sample is *representative*,[23] numerical values computed from the sample are called *statistics*. Here are some examples:

1. the *arithmetic mean*

$$\overline{x} = \frac{\sum\limits_{i=1}^{n} x_i}{n};$$

2. the *standard deviation*

$$s = \sqrt{\frac{\sum\limits_{i=1}^{n} (x_i - \overline{x})^2}{n - 1}};$$

3. the *skewness*

$$a_3 = \frac{\sum\limits_{i=1}^{n} (x_i - \overline{x})^3}{ns^3};$$

4. the *kurtosis excess*

$$a_4 = \frac{\sum\limits_{i=1}^{n} (x_i - \overline{x})^4}{ns^4} - 3.$$

Remark 24. To estimate the skewness of a distribution, statisticians have other coefficients at their disposal such as,

$$\frac{3(\overline{x} - \widetilde{x})}{s},$$

where $\widetilde{x}$ is the estimate of the median. For instance, in the case of the distribution of incomes, the median, which is not influenced by the extreme values, is a more informative quantity than the arithmetic mean. Smaller than the mean, it reveals the positive skewness of the distribution.

The right-hand side of the relation that defines $\overline{x}$ is said to be the *estimate* of $\overline{x}$. It should be distinguished from the *estimator* $\overline{X}$ defined by

$$\overline{X} = \frac{1}{n} \sum_{i=1}^{n} X_i,$$

where X_i are independent random variables distributed as X. In this specific case, since

$$\langle \overline{X} \rangle = m_1,$$

---

[23] A representative sample of a population consists of elements selected with equal probability (*i.e.*, if the population has a total number N of elements, each element has a probability $1/N$ of being selected.

$\overline{X}$ is said to be an *unbiased estimator* for the mean.[24]
Similarly,

$$S^2 = \frac{1}{n-1} \sum_{i=1}^{n} (X_i - \overline{X})^2$$

is an unbiased estimator for the variance.[25]

Remark 25. The estimator $\overline{X}$ converges to $\langle X \rangle$ in the limit of an infinite sample size (refer to Footnote 24). In practice, the sample size is finite, and, as a consequence, we can only expect to obtain an approximate value of $\langle X \rangle$. But, according to the central limit theorem,

$$\lim_{n \to \infty} \left(|\overline{X} - m_1| < \frac{\sigma}{\sqrt{n}} \right) = \sqrt{\frac{2}{\pi}} \int_0^1 \exp\left(-\tfrac{1}{2} x^2 \right) dx.$$

Hence, if we want $|\overline{X} - m_1| < \varepsilon$, the sample size n should be of the order of ε^{-2}.

Given a sample $(x_1, x_2, \ldots, x_n)$ of size n, Student's test allows acceptance or rejection of a hypothesis concerning its mean $\overline{x}$. Let (X_i) be a sequence of n independent identically distributed Gaussian random variables $N(m, \sigma)$. The random variable

$$T = \frac{(\overline{X} - m)\sqrt{n}}{S},$$

where S is the unbiased estimator of the variance, has a Student's probability distribution; *i.e.*,

$$P(|T| \leq t_*) = \int_{-t_*}^{t_*} \frac{\Gamma\left(\tfrac{1}{2} n\right)}{(\pi(n-1))^{\frac{1}{2}} \Gamma\left(\tfrac{1}{2}(n-1)\right)} \left(1 + \frac{x^2}{n-1}\right)^{-\frac{1}{2} n} dx.$$

Thus, if we have a small sample $(x_1, x_2, \ldots, x_n)$ of size n, Student's test tells us that the mean m belongs to the interval $[\overline{x} - st_* n^{-1/2}, \overline{x} + st_* n^{-1/2}]$ with a probability given by the expression above if

---

[24] As a consequence of the *law of large numbers*, the estimator $\overline{X}$ converges to the mean m_1 as $n \to \infty$. There exist actually two forms of the law of large numbers. The *weak law* states that, for any positive a,

$$\lim_{n \to \infty} P(\overline{X} - m_1) = 0,$$

while the *strong law* states that the event

$$\left\{ \omega \in \Omega \mid \lim_{n \to \infty} \overline{X}(\omega) = m_1(\omega) \right\}$$

has a probability 1. $\overline{X}$ is said to converge to m_1 *in probability* in the first case and *almost surely* in the second case.

[25] For the estimator of the standard deviation, the denominator has to be equal to $n-1$ and not n. For a proof, see Boccara [56], p. 95.

$$\left| \frac{(\overline{x} - m)\sqrt{n}}{s} \right| \leq t_*.$$

For example, if d_n is a sample of size n from a Gaussian distribution with zero mean and unit variance, the table below shows the mean $\overline{x}_n$ and the standard deviation s_n of the sample, the value of $t_n = \overline{x}_n\sqrt{n}/s_n$, and the 90% *confidence interval* for n equal to 10 and 100.

n	$\overline{x}_n$	s_n	t_n	t_*	confidence interval
10	0.0617	0.847	0.231	1.8331	$[-0.429, 0.552]$
100	0.129	0.988	1.304	1.6604	$[-0.0352, 0.293]$

In both cases, the nonzero value of $\overline{x}_n$ is not significant, or, more precisely, the hypothesis of a zero mean value is accepted with 90% confidence. Similarly, the Student's distribution could also be used to test the hypothesis that two independent random samples have the same mean. Since the Student's distribution tends to the normal distribution when μ tends to infinity (see page 330), as the sample size increases, instead of the Student's distribution, the normal distribution is usually applied. For instance, for $n = 100$, using the normal distribution, the t_* value differs by less than 1%.

A sample $(x_1, x_2, \ldots, x_n)$ of size n is supposed to represent successive drawings from the probability distribution of a certain random variable X. All the statistics computed from this sample give a more or less accurate description of the probability distribution of X. It is clear that, compared to the information contained in the sample of size n, the values of a few statistics contain much less information. Due to the finite bin size, even the histogram corresponds to a loss of information. Not to lose information, it is much better to use all the sample points to represent the approximate cumulative distribution function F_X of the random variable X. To do it, we first have to order by increasing values the sequence $(x_1, x_2, \ldots, x_n)$. If $(x_{j_1}, x_{j_2}, \ldots, x_{j_n})$ is the ordered sequence, the sequence of n points of coordinates (ξ_k, η_k) $(k = 1, 2, \ldots, n)$, where $\xi_k = x_{j_k}$ and $\eta_k = j_k/n$, are approximately located on the graph of F_X.

Figure 8.10 represents the approximate cumulative distribution function determined from a random sequence of size 2000 drawn from a normal distribution with zero mean and unit variance.

8.3 Empirical results and tentative models

In order to determine a reliable value for the exponent characterizing the asymptotic power-law behavior of the probability distribution of a certain quantity, it is indispensable to carry out statistical analyses on very large samples. For the time being, this is far from being the case for most examples presented below.

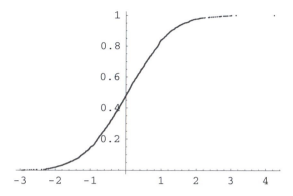

Fig. 8.10. *Approximate cumulative distribution function determined from a random sequence of size 2000 drawn from a normal distribution with zero mean and unit variance.*

8.3.1 Financial markets

Financial markets are systems of many interacting agents that can be divided into two broad categories: *traders* and *assets*. Banks, brokerage firms, mutual funds, and individual investors are traders; stocks, bonds, futures, and options are assets. The interactions between these agents generate different observables such as transaction prices and traded volumes.

The statistical analysis of financial data has a long history, starting with the publication of Louis Bachelier's thesis in 1900 [21]. If $S(t)$ is the price of a stock at time t, Bachelier's model assumes that the successive differences $S(t + T) - S(t)$ are independent normally distributed random variables with zero mean and variance proportional to the time interval T.[26] The abundant data accumulated since Bachelier's pioneering contribution showed, however, a clear departure from normality. This fact led Mandelbrot to suggest a "radically new approach" called the *stable Paretian hypothesis* [222].[27] Mandelbrot's basic assumption is that $\log S(t+T) - \log S(t)$ are independent identically distributed Lévy random variables (see Subsection 8.2.4).

Nowadays, an important number of papers presenting empirical results and theoretical interpretations are regularly submitted for publication.[28] Physicists, in particular, have been increasingly interested in the analysis and modeling of financial markets, giving birth to *econophysics* [227].

An example of a studied quantity is the *return* $G(t)$ defined by

---

[26] $S = \{S(t) \mid t \geq 0\}$ is a *Wiener process* (see Grimmett and Stirzaker [158], Chapter 13). If $S(t + T) - S(t)$ is $N(0, \sigma^2 T)$, for all positive t and T, Bachelier, five years before Einstein, was the first to show that such a process satisfies the diffusion equation.

[27] See also Fama [119].

[28] Refer to the Web site http://www.unifr.ch/econophysics.

$$G_{\Delta t}(t) = \log S(t + \Delta t) - \log S(t),$$

where $S(t)$ may represent the market capitalization of a specific company or a market index at time t, and Δt is the *sampling time interval*. If Δt is small, $G_{\Delta t}(t)$ is approximately equal to the relative price change:

$$G_{\Delta t}(t) \approx \frac{S(t + \Delta t) - S(t)}{S(t)}.$$

In order to compare returns of different companies, it is preferable to study the *normalized return*, defined as

$$g_{\Delta t}(t) = \frac{G_{\Delta t}(t) - \langle G_{\Delta t} \rangle_T}{\sqrt{\langle G_{\Delta t}^2 \rangle_T - \langle G_{\Delta t} \rangle_T^2}},$$

where $\langle X \rangle_T$ represents the time average of X over an interval of time T. The standard deviation $\sqrt{\langle G_{\Delta t}^2 \rangle_T - \langle G_{\Delta t} \rangle_T^2}$ is the *volatility*.[29]

The statistical analysis of the normalized returns of the 1000 largest US companies, with $\Delta t = 5$ min over the 2-year period 1994–1995, and where the time t runs over the working hours of the stock exchange; *i.e.*, removing nights, weekends, and holidays shows that the probability $P(|g| > x)$ has, for both positive and negative tails, a power-law behavior [284, 148]. That is,

$$P(|g| > x) \sim \frac{1}{x^{\alpha}},$$

where, in the region $2 \leq |g| \leq 80$,

$$\alpha = \begin{cases} 3.10 \pm 0.03, & \text{for positive } g, \\ 2.84 \pm 0.12, & \text{for negative } g. \end{cases}$$

The S&P 500 index, defined as the sum of the market capitalizations of 500 companies representative of the US economy, has also been analyzed [149]. The probability $P(|g| > x)$ of the normalized return also exhibits power-law tails characterized by the exponents

$$\alpha = \begin{cases} 2.95 \pm 0.07, & \text{for positive } g, \\ 2.75 \pm 0.13, & \text{for negative } g, \end{cases}$$

for $\Delta t = 1$ min, over the 13-year period 1984–1996, and in the region $3 \leq |g| \leq 50$. This result is surprising since the returns of the S&P 500 are weighted sums

---

[29] There is no unique definition of the volatility. For example, it could also be defined as an average over a time window $T = n\Delta t$, *i.e.*,

$$\frac{1}{n} \sum_{t=\tau}^{\tau+n-1} |G_{\Delta t}(t)|,$$

where n is an integer.

of the returns of 500 companies, and, as a consequence of the central limit theorem, one would expect to find that these returns are normally distributed unless there exist significant dependencies between the returns of the different companies.

Investigating the power-law behavior for sampling time interval Δt, it appears that the exponent $\alpha \approx 3$ for Δt varying from a few minutes up to several days. For larger values of Δt, there is a slow convergence to a Gaussian behavior.

There exist many other stock market indices. A study, by the same authors, of the NIKKEI index of the Tokyo stock exchange (for the 14-year period 1984–1997) and the Hang Seng index of the Hong Kong stock exchange (for the 18-year period 1980–1997) shows that, in both cases, the probability $P(|g| > x)$ has similar power-law tails with exponents close to 3, suggesting the existence of a *universal* behavior for these distributions.

In the recent past, most papers on price dynamics have essentially tried to answer the question: How can one approximately determine with good accuracy the probability distribution function of the logarithm of price differences from market data?[30] While, for small Δt, it is clear that this distribution is not Gaussian, it is less clear to decide whether it is a truncated Lévy distribution with an exponential cutoff or a Student's t-distribution, which, incidentally, are not very easy to distinguish. More recently, the search for plausible mechanisms to explain the observed phenomena and build up models that are not just curve-fitting has been suggested by some authors.[31] To conclude this section, we briefly present an example of a time series model widely used in finance.[32]

One important feature of financial time series is the phenomenon of *conditional heteroskedasticity*; *i.e.*, the time-dependence of the variance of the price increments. For instance, traders may react to certain news by buying or selling many stocks, and a few days later, when the news is valued more properly, a return to the behavior preceding the arrival of the news may be observed. This could induce a large increase or decrease of the returns followed by the opposite change in the following days; and it may be reasonable to assume that these two abrupt changes are correlated.

Let $W = \{W_t \mid t \in \mathbb{N}_0\}$ be a stochastic process, where the W_t are independent identically distributed random variables with zero mean and unit variance, and consider the process $X = \{X_t \mid t \in \mathbb{N}_0\}$ defined by

$$X_t = \sigma_t W_t, \quad \text{with} \quad \sigma_t^2 = a_0 + a_1 X_{t-1}^2 \quad (a_0 > 0, \ 0 < a_1 < 1 \ \sigma_0 = 1). \quad (8.2)$$

---

[30] Mantegna and Stanley [227], Chapter 8.

[31] See Bouchaud and Potters [68], Chapter 2; Lux and Marchesi [218]; Huang and Solomon [175]; and Bouchaud [69].

[32] On time series models, see Franses [129].

Processes such as X are called *ARCH* processes.[33] ARCH processes are specified by the probability density f_W of the random variables W_t and the parameters a_0 and a_1.[34]

Since $\langle W_t \rangle = 0$, the unconditional mean value of X_t is equal to zero, and its unconditional variance $\langle X_t^2 \rangle$ is given by

$$\langle X_t^2 \rangle = a_0 + a_1 \langle X_{t-1}^2 \rangle,$$

which shows the autocorrelation of the squared time series. Solving the equation above for $\langle X_{t-1}^2 \rangle = \langle X_t^2 \rangle$ yields

$$\langle X_t^2 \rangle = \frac{a_0}{1 - a_1}.$$

Relations (8.2) show that large (resp. small) absolute values of X_t are expected to be followed by large (resp. small) absolute values even while $\langle X_{t-\tau} X_t \rangle = 0$; *i.e.*, the time series is uncorrelated. The ARCH model can therefore describe time series with sequences of data points looking like outliers, and as a consequence of the relation $\sigma_t^2 = a_0 + a_1 X_{t-1}^2$, these outliers will appear in clusters (see Exercise 8.7).

Clusters of outliers lead to nonzero kurtosis excess. From (8.2), we have

$$\langle X_t^4 \rangle = 3 \langle (a_0^2 + 2 a_0 a_1 X_t^2 + a_1^2 X_{t-1}^4) \rangle$$
$$= 3 a_0^2 + 6 a_0 a_1 \frac{a_0}{1 - a_1} + 3 a_1^2 \langle X_{t-1}^4 \rangle.$$

Solving for $\langle X_{t-1}^4 \rangle = \langle X_t^4 \rangle$ yields

$$\langle X_t^4 \rangle = \frac{3 a_0^2 (1 + a_1)}{(1 - a_1)(1 - 3 a_1^2)},$$

which shows that the unconditional kurtosis excess is thus given by

$$\frac{\langle X_t^4 \rangle}{\langle X_t^4 \rangle^2} - 3 = \frac{6 a_1^2}{1 - 3 a_1^2}$$

if $0 < a_1 < 3^{-\frac{1}{2}}$. Since $\langle W_t^3 \rangle = 0$, the skewness of X_t is zero.

An interesting property of ARCH processes is that they can dynamically generate power-law tails [285, 286] (see Exercises 8.9 and 8.10). For instance, the probability distribution of the first differences of the S&P 500 index over the 12-year period 01/84–12/95, for $\Delta t = 1$ min, is well-fit by the probability distribution of an ARCH process with f_W given by a truncated Lévy probability density determined by the parameters $\alpha = 1.4$, $a = 0.275$, and $\ell = 8$

---

[33] The acronym ARCH stands for autoregressive conditional heteroskedasticity. Such time series models were initially proposed by Engle [116].

[34] This is actually an ARCH model of order 1. In an ARCH model of order τ, σ_t^2 is given by $a_0 + a_1 X_{t-1}^2 + a_2 X_{t-2}^2 + \cdots + a_\tau X_{t-\tau}^2$.

(see pages 326 and 328 for parameter definitions) and the parameters of the ARCH process $a_1 = 0.4$ and a_0 chosen to fit the empirical standard deviation $\sigma = 0.07$. The probability density f_X appears to exhibit two power-law regimes: a first one characterized by the exponent $\alpha_1 = 1.4$ of the Lévy distribution up to the cutoff length ℓ followed by a second one with a dynamically generated exponent $\alpha_2 = 3.4$ for $x > \ell$.

8.3.2 Demographic and area distribution

The demographic distribution of humans on Earth is extremely heterogeneous. Highly concentrated areas (cities) alternate with large extensions of much lower population densities. If s is the population size of a city (number of inhabitants) and $r(s)$ its rank, where the largest city has rank 1, Zipf [355] found empirically that

$$r(s) \sim \frac{1}{s},$$

which, for the frequency as a function of population size, implies

$$f(s) \sim \frac{1}{s^2}.$$

This result appears to be universal. Statistical studies of 2700 cities of the world having more than 10^5 inhabitants, 2400 cities of the United States having more than 10^4 inhabitants, and 1300 municipalities of Switzerland having more than 10^3 inhabitants give exponents very close to 2. More precisely: 2.03 ± 0.05 for the world (in the range $s < 10^7$), 2.1 ± 0.1 for the United States, and 2.0 ± 0.1 for Switzerland [351]. Moreover, around huge urban centers, the distribution of the areas covered by satellite cities, towns, and villages obeys the same universal law.

Zanette and Manrubia proposed a simple model to account for the exponent -2 characterizing the power-law behavior of the frequency f as a function of population size s [351]. Consider an $L \times L$ square lattice and denote by $s(i,j;t) \geq 0$ the population size at site $(i,j) \in \mathbb{Z}_L^2$ at time t. This system evolves in discrete time steps according to the following two subrules.

1. The one-input rule:

$$s(i,j;t+\tfrac{1}{2}) = \begin{cases} \dfrac{1-q}{p}\, s(i,j;t), & \text{with probability } p, \\[2mm] \dfrac{q}{1-p}\, s(i,j;t), & \text{with probability } 1-p, \end{cases}$$

where $0 < p < 1$ and $0 \leq q \leq 1$.
2. A diffusion process in which a fraction a of the population at site (i,j) at time $t+\tfrac{1}{2}$ is equally divided at time $t+1$ between its nearest-neighboring sites.

Note that for such an evolution rule the total population is conserved in the limit $L \to \infty$.

Zanette and Manrubia claim that numerical simulations on a 200×200 square lattice starting from an initial homogeneous population size distribution $(s(i, j; 0) = 1$ for all $(i, j) \in \mathbb{Z}_{200}^2)$, and for different values of the parameters p, q and a, show that, after a transient of 10^3 time steps, the time average of the size frequency has a well-defined power-law dependence characterized by an exponent $\alpha = -2.01 \pm 0.01$. This result has been questioned by Marsili, Maslov, and Zhang [228], who found that, for the same parameter values, the total population tends to zero as a function of time. In their reply, Zanette and Manrubia [352] argued that extinction is a finite-size effect and that simulations in large systems clearly show a transient in which a power-law distribution builds up before finite-size effects play their role.

8.3.3 Family names

A statistical study of the distribution of family names in Japan shows that, if s denotes the size of a group of individuals bearing the same name, the number $n(s)$ of groups of size s has a decreasing power-law behavior characterized by an exponent equal to 1.75 ± 0.05 in five different Japanese communities [245]. The data were obtained from 1998 telephone directories. Similar results have been obtained from (i) a sample of the United States 1990 census and (ii) all the family names of the Berlin phone book beginning with A. In both these cases, however, the exponent is close to 2 [353]. It has been argued that the discrepancy between these two results could be attributed to transient effects: most Japanese family names having been created in the rather recent past (about 120 years ago) [353]. In order to obtain reliable results, statistical analyses of family name frequency should be carried out on very large population sizes. This is not the case for the two studies above.

8.3.4 Distribution of votes

On October 4, 1998, Brazil held general elections with compulsory vote. The state of São Paulo, with 23,321,034 voters and 1260 local candidates, is the largest electoral group. It has been found [93] that the number of candidates $N(v)$ receiving a fraction v of the votes behaves as $v^{-\alpha}$ with $\alpha = 1.03 \pm 0.03$ over almost two orders of magnitude. The authors claimed that this power-law behavior is also observed for the other Brazilian states having a large enough number of voters and candidates. A similar result ($\alpha = 1.00 \pm 0.02$) holds also for all the candidates for state deputy in the country.[35]

A similar analysis was carried out for the 2002 elections [94]. In spite of a new rule concerning alliances between parties approved by the National

---

[35] Detailed information on Brazilian elections can be found on the Web site of the Tribunal Superior Eleitoral: http://www.tse.gov.br. For the 1998 elections, go to http://www.tse.gov.br/eleitorado/eleitorado98/index.html.

Congress, the same power law was observed for the number of candidates $N(v)$ receiving a fraction v of the votes. It should be noted that, while a scale-free behavior is found for elections to the National Congress and State Houses, there is no evidence of such a behavior for the municipal elections held in 2000. According to the authors, this discrepancy may be due to the fact that, in municipal elections, candidates are much closer to voters while candidates to the National Congress or State Houses are usually only known through the media.

According to the authors, a possible explanation of the power-law behavior is to assume that the success of a candidate is the result of a multiplicative process in which each factor "should be related to attributes and/or abilities of the candidate to persuade voters." Assuming that the corresponding random variables are independent and in large number, the central limit theorem implies that the distribution of votes should be lognormal, which can be mistaken for a power-law distribution over a few orders of magnitudes (see page 324).[36]

8.4 Self-organized criticality

In the late 1980s, Bak, Tang, and Wiesendfeld [24, 25, 27] reported "the discovery of a general organizing principle governing a class of dissipative coupled systems." These systems evolve naturally towards a *critical* state, with no intrinsic time or length scale, in which a minor event starts a chain reaction that can affect any number of elements in the system. This *self-organized criticality* therefore would be an explanation of the occurrence of power-law behaviors. In a recent book [29], Bak even argues that self-organized criticality "is so far the only known general mechanism to generate complexity."

8.4.1 The sandpile model

Consider a finite $L \times L$ square lattice. The state of a cell is a nonnegative integer representing the number of sand grains stacked on top of one another in that cell. In the initial state, each cell contains a random number of sand grains equal to 0, 1, 2, or 3. Then, to cells sequentially chosen at random, we add one sand grain. When the number of sand grains in one cell becomes equal to 4, we stop adding sand grains, remove the four grains, and equally distribute them among the four nearest-neighboring cells. That is, the number of sand grains in the cell that reached the threshold value 4 at time t becomes equal to 0 at time $t + 1$, and the number of sand grains in the four nearest-neighboring cells increases by one unit. This process is repeated as long as

---

[36] Playing with the two parameters of the lognormal distribution, it is not difficult to adjust the value of the exponent of the apparent power-law behavior.

the number of sand grains in a cell reaches the threshold value.[37] When the number of sand grains in a boundary cell reaches the value 4, the same process applies but the grains "falling off" the lattice are discarded.

A sequence of toppling events occurs when nearest-neighboring cells of a cell reaching the threshold value contain three sand grains. Such a sequence is called an *avalanche*. An avalanche is characterized by its *size* (that is, the number of consecutive toppling events) and its *duration*, equal to the number of update steps. An avalanche of size 6 and duration 4 is shown in Figure 8.11.

Fig. 8.11. *An example of an avalanche of size 6 and duration 4. Cells occupied by n sand grains, where n is equal to 0, 1, 2, 3, and 4 are, respectively, white, light grey, medium grey, dark grey, and black. The system evolves from left to right and top to bottom.*

The sandpile model illustrates the basic idea of self-organized criticality. For a large system, adding sand grains leads at the beginning to small avalanches. But as time increases, the system reaches a stationary state where the amount of added sand is balanced by the sand leaving the system along the boundaries. In this stationary state, there are avalanches of all sizes up to the size of the entire system. Numerical simulations show that the number $N(s)$ of avalanches of size s behaves as $s^{-\tau}$, where, in two-dimensional systems, the exponent τ is close to 1.

Careful experiments carried out on piles of rice [130] have shown that self-organized criticality

---

[37] A similar transition rule defined on a one-dimensional lattice was first studied by J. Spencer [324]; for exact results on one-dimensional sandpile models, see E. Goles [146].

is not as 'universal' and insensitive to the details of a system as was initially supposed, but that instead its occurrence depends on the detailed mechanism of energy dissipation.

Power-law behaviors were only observed for rice grains of elongated shape. The authors of this experimental study were then led to reanalyze previous experiments on granular systems. They found that the flow over the rim of the pile had a stretched-exponential distribution that is definitely incompatible with self-organized criticality.

A distinctive observable consequence of self-organized critical dynamics, as illustrated by the sandpile model, is that the distribution of fluctuations or avalanche size has a power-law behavior. But, as we shall discover, the converse is not necessarily true.

8.4.2 Drossel-Schwabl forest fire model

An important number of numerical studies have been dedicated to the analysis of this model. Originally, a forest fire model that should exhibit self-organized criticality had been proposed by Bak, Chen, and Tang [26]. Their model is a three-state cellular automaton. State 0 represents an empty site, state 1 a green tree, and state 2 a burning tree. At each time step, a green tree has a probability p of growing at an empty site, a green tree becomes a burning tree if there is, at least, one burning tree in its von Neumann neighborhood, and sites occupied by burning trees become empty. As stressed by Grassberger and Kantz [155], for finite p, "the model is in the universality class of the epidemic model with recovery which is isomorphic to directed percolation." Numerical simulations on rather small lattices led Bak, Chen, and Tang to conjecture that the system exhibits self-organized criticality in the limit $p \to 0$, *i.e.*, when the time for a tree to burn is much smaller than the average time to wait for a green tree to grow at an empty site.

Simulations on much larger lattices and for much smaller values of p performed by Grassberger and Kantz [155] and Moßner, Drossel, and Schwabl [250] provided evidence that the system does not exhibit self-organized criticality. In two dimensions, starting from a random distribution of trees (green and burning), for small values of p, after a number of iterations of the order of $1/p$, the fire propagates along rather regular fronts, as illustrated in Figure 8.12.

Following Drossel and Schwabl [109], for the model to become critical, a mechanism allowing small forest clusters to burn has to be included. This can be done introducing the following extra subrule: A green tree with no burning tree in its von Neumann neighborhood has a probability $f \ll p$ of being struck by lightning and becoming a burning tree. This model involves three time scales: the fire propagation rate from burning trees to green neighbors, the growth rate of green trees, and the lightning rate $(1 \gg p \gg f)$. As emphasized by Grassberger [157]:

Fig. 8.12. *Forest fire model of Bak, Chen, and Tang. Lattice size:* 200; *initial green tree density:* 0.3; *initial burning tree density:* 0.01; *tree growth rate* $p = 0.05$; *number of iterations:* 40. *Color code: burning trees are black, green trees are grey, and empty sites are light grey.*

> The growth of trees does not lead to a state which is inherently unstable (as does the addition of sand grains to the top of the pile does), but only to a state susceptible to being burnt. Without lightning, the tree density in any patch of forest can go far beyond criticality. When the lighting strikes finally, the surviving trees have a density far above critical.

Actually, Grassberger showed [156] that the "forest" consists of large patches of roughly uniform density, most of which are either far below or far above the critical density for spreading. Since these patches occur with all sizes, fires also have a broad distribution of sizes. Grassberger confirmed the criticality of the Drossel-Schwabl model, but he found critical exponents in disagreement with those obtained by Drossel and Schwabl:

1. If s is the number of trees burnt in one fire, in the limit $p/f \to \infty$, the fire size distribution behaves as $s^{1-\tau}$ with $\tau = 2.15 \pm 0.02$.
2. For a finite value of p/f, the power-law behavior above is cut off at $s_{\max} \sim (p/f)^{\lambda}$ with $\lambda = 1.08 \pm 0.02$.
3. The average cluster radius $\langle R^2 \rangle^{1/2}$ scales as $(p/f)^{\nu}$ with $\nu = 0.584 \pm 0.01$.

In their paper, Drossel and Schwabl [109], using mean-field arguments, found $\tau = 2$, $\lambda = 1$, and $\nu = 0.5$. Their numerical simulations apparently confirmed these results, but they were carried out for rather small values of p/f.

This, however, is not the end of the story. Following recent studies [308, 279, 309], that introduced corrections to the scaling laws above, Grassberger [157] presented numerical simulations on very large systems ($65,536 \times 65,536$) with helical boundary conditions, large p/f values (256,000), long transients (approximately 10^7 lightnings), and a very large number of fires (between 9.3×10^6 for $p/f = 256,000$, and 10^9 for $p/f \leq 250$). While confirming previous results obtained for much smaller lattice sizes and p/f values, the essential conclusion of this study is that, without ambiguity, previous results do not describe the true critical behavior of the Drossel-Schwabl forest fire model. Still worse, according to Grassberger, his own simulations do not reach the true asymptotic regime. Most probably, all proposed scaling laws are just transient and there seems to be no indications of any power laws. These results are to be taken as a warning, especially when dealing with real systems [157]:

> The situation becomes even worse when going to real life phenomena. It does not seem unlikely that many of the observed scaling laws are just artifacts or transients. Problems of spurious scaling in models which can be simulated with very high precision such as the present one should be warnings that not every power law supposedly seen in nature is a real power law.

8.4.3 Punctuated equilibria and Darwinian evolution

In the foreword to the Canto edition of John Maynard Smith's book on the theory of evolution [236], Richard Dawkins tells us that:

> Natural selection is the only workable explanation for the beautiful and compelling illusion of 'design' that pervades every living body and every organ. Knowledge of evolution may not be strictly useful in everyday commerce. You can live some sort of life and die without ever hearing the name of Darwin. But if, before you die, you want to understand why you lived in the first place, Darwinism is the one subject that you must study. This book is the best general introduction to the subject now available.

And, in the introduction of this highly recommended book, John Maynard Smith so presents Eldredge and Gould's theory of *punctuated equilibria* [115, 325]:

> A claim of which perhaps too much has been made concerns 'punctuated equilibria'. Gould and Eldredge suggested that evolution has not proceeded at a uniform rate, but that most species, most of the time, change very little, and that this condition of 'stasis' is occasionally interrupted by a rapid burst of evolution. I do not doubt that this picture is sometimes, perhaps often, true. My difficulty is that I cannot see that it makes a profound difference to our view of evolution.

Punctuated equilibrium behavior, with its intermittent bursts interrupting long periods of stasis, is very similar to self-organized critical behavior.

Bak and Sneppen [28] proposed a model of biological evolution of interacting species that exhibits co-evolutionary avalanches. Consider a one-dimensional lattice of size L, with periodic boundary conditions. Each lattice site represents a species, and its state is a random number between 0 and 1 that measures the *evolutionary barrier* of the species. At each time step, the system—called an ecology—is updated by locating the species with the lowest barrier and mutating it by assigning a new random barrier to that species. At the same time, the left and right neighbors are also assigned new random barriers.

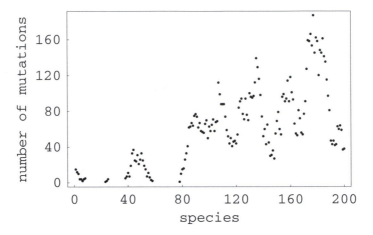

Fig. 8.13. *Number of mutations of each species. The number of species is 200. The system has been updated 20,000 times, and the first 10,000 iterations have been discarded.*

Figure 8.13 shows that some species do not change at all while others change many times.

Flyvberg, Sneppen, and Bak [128] gave a mean-field theory of a slightly different model in which, instead of modifying three random barriers—the barrier of the species with the lowest barrier and the barriers of its two nearest neighbors—in the new model, the random barriers of K species are modified: the barrier of the species with the lowest barrier and the barriers of $K - 1$ other species chosen at random. In the limit $N \to \infty$, the authors found that avalanches come in all sizes, and the larger ones are distributed according to a power law with an exponent equal to $-\frac{3}{2}$.

Actually this model has been solved exactly [63], and the mean-field exponent was found to be exact.

Analyzing the distribution of distances between successive mutations, Bak and Sneppen found a power-law behavior indicating that the ecology has organized into a self-critical state.

8.4.4 Real life phenomena

Here are presented some observed phenomena that may provide illustration of the idea of self-organized criticality.

Example 66. Mass extinctions. Most species that ever lived on Earth are extinct. Recently, several well-documented mass extinctions have been analyzed, and the interpretation of the available data has generated many controversies.

Analyzing the extinction record of the past 250 ma (millions of years), Raup and Sepkoski [294] discovered, using a variety of tests, a periodicity of 26 ma. Discarding the existence of purely biological or earthbound cycles, the authors favor extraterrestrial causes, such as meteorite impacts due to the passage of the solar system through the spiral arms of the Milky Way, as a possible explanation of periodic mass extinctions.

More recently, the existence of power-law functional forms in the distribution of the sizes of extinction events has been put forward in favor of a self-organized criticality mechanism [317, 319, 320]. In particular, it has been reported [320] that many time series are statistically self-similar with $1/f$ power spectra. These findings, according to the authors, support the idea that a nonlinear response of the biosphere to perturbations provides the main mechanism for the distribution of events.

These results have been criticized by Kirchner and Weil [189], who claimed that the apparent self-similarity and $1/f$ scaling are artefacts of interpolation methods—the analyzed time series containing actually much more interpolated (86%) than real data.

On the other hand, in a detailed paper on mass extinction, Newman [260] reviewed the various models based on self-organized criticality and analyzed the available empirical evidence. Using a simple model he previously used in collaboration with Sneppen [259, 318] to study earthquake dynamics, he showed that self-organized criticality is not the only mechanism leading to power laws.

Newman's model assumes that the ultimate cause for the extinction of any species is environmental stress of any kind such as climate change or meteorite impact. His model may be described as follows. Consider a system of N species, and assign to species i a random number x_i from a probability distribution with a given density $f_{\text{th}}(x)$, which, for example, may be uniform on the interval $[0, 1]$. The number x_i, called the *threshold level* of species i, represents the amount of stress above which it becomes extinct. Then, at each time step, the system evolves according to two subrules:

1. Select a random number η, called the *stress level*, from a probability distribution with a given density $f_{\text{st}}(\eta)$. Assuming that small stresses are

more common than large ones, the probability density is chosen to fall off away from zero. All species whose threshold level is less than η become extinct. Extinct species are replaced by an equal number of new species to which are assigned threshold levels distributed according to $f_{th}(x)$. The number of extinct species s is the size of the avalanche taking place at this time step.

2. A small fraction f of all species, chosen at random, are assigned new random threshold levels.

Although this system never becomes critical in the sense of possessing long-range spatial correlations, it does organize into a stationary state characterized by avalanches with a power-law size distribution. If $f_{st}(x)$ is a Gaussian distribution with zero mean and standard deviation $\sigma = 0.1$, it is found that the distribution of the size of extinction events has a power-law behavior, over many orders of magnitude, with an exponent equal to 2.02 ± 0.02. Moreover, for a variety of stress probability distributions—e.g., Cauchy or exponential—the exponent varies slightly but remains close to 2, in agreement with fossil data.

In self-organized critical models, the system is *slowly* driven towards some instability and the relaxation mechanism is *local*. In Newman's model, on the contrary, the system is *not* slowly driven to produce intermittent bursts and there are *no* (local) interactions between the species.

Example 67. Earthquakes. Earthquakes are the result of seismic waves produced when some sort of stored energy is suddenly released. Earthquakes are classified in magnitude according to the Richter scale [298]. Essentially, the *magnitude number* is the logarithm of the maximum seismic wave amplitude measured at a certain distance from the *epicenter*.[38] Earthquakes having a magnitude equal to 8 or more are extremely devastating. For example, the San Francisco earthquake of April 18, 1906, had a magnitude of 8.25, and was felt along the west coast from Los Angeles in the south to Coos Bay in southwestern Oregon in the north.

In the late 1960s, the *plate tectonics* model was developed to explain seismic patterns. According to this theory, the *lithosphere* (*i.e.*, the Earth's upper shell) consists of nearly a dozen large slabs called *plates*. These plates move relative to each other and interact at their boundaries, where most *faults* lie. The majority of earthquakes are the result of movements along faults. The San Andreas fault, which extends northwest from Southern California for about 1000 km, results from the abutment of two major plates. The San Francisco earthquakes of 1906 and 1989 were both caused by movement along this fault.

The frequency of shallow earthquakes as a function of their magnitude obeys the Gutenberg-Richter law [161], which states that the logarithm of the frequency of shocks decreases linearly with magnitude. That is,

[38] The epicenter is the point on the surface of Earth that is directly above the source of the earthquake, called the *focus*.

$$\log_{10} N(M > m) = a - bm,$$

where the left-hand side is the base 10 logarithm of the number of earthquakes whose magnitude M is larger than m. Data collected [30] from the Web site of the Southern California Earthquake Data Center[39] show that for a total number of 335,076 earthquakes recorded during the period 1984–2000, the exponent b is equal to 0.95 for magnitudes larger than 2.

Most of the various models that have been proposed to account for the Gutenberg-Richter law are based on the idea of self-organized criticality. One popular model has been suggested by Olami, Feder, and Christensen [273].[40] Consider a two-dimensional square array of blocks in which each block is connected to its four nearest neighbors by springs of elastic constant K. Additionally, each block is connected to a single rigid driving plate by another set of springs of elastic constant K_L as well as connected frictionally to a fixed rigid plate. The blocks are driven by the relative movement of the two rigid plates. When the force on one of the blocks is larger than some threshold value F_{th}, the block slips to a zero-force position. The slip of one block redefines the forces on its nearest neighbors, and this may result in further slips, starting an avalanche of slips.

If $\Delta_{i,j}$ is the displacement of block (i,j) from its zero-force position, the total force exerted by the springs on block (i,j) is

$$F_{i,j} = K(4\Delta_{i,j} - \Delta_{i-1,j} - \Delta_{i+1,j} - \Delta_{i,j-1} - \Delta_{i,j+1}) + K_L\Delta_{i,j}.$$

If v is the velocity of the moving plate relative to the fixed plate, the total force on a given block increases at a rate proportional to $K_L v$ until the force on one block reaches its threshold value and triggers an earthquake. This mechanism is referred to as a *stick-slip frictional instability*.[41] The redistribution of forces after a local slip of block (i,j) is such that

$$F_{i,j} \to 0, \quad F_{i\pm1,j} \to F_{i\pm1,j} + \frac{K}{4K + K_L} F_{i,j}, \quad F_{i,j\pm1} \to F_{i,j\pm1} + \frac{K}{4K + K_L} F_{i,j}$$

The parameter $\alpha = K/(4K + K_L)$ varies between 0 and 0.25.

The system evolves in time according to the following rules.

1. Consider an $L \times L$ array of blocks satisfying a zero-force condition on the boundary.
2. Initialize forces on blocks to a random value between 0 and F_{th}.
3. If, for a pair (i,j), $F_{i,j} > F_{\text{th}}$, then redistribute the force $F_{i,j}$ on the neighboring blocks as indicated above. An earthquake is taking place.

---

[39] http://www.scecdc.scec.org/.

[40] A purely deterministic one-dimensional version of this model had been studied by Carlson and Langer [78]. Both models are variants of the Burridge-Knopoff model [74].

[41] The earthquake is the "slip," and the "stick" is the period between seismic events during which elastic strain accumulates.

4. Repeat step 3 until the earthquake is fully evolved.
5. Locate the block on which the exerted force is maximum. If F_{max} is the value of this force, increment the force exerted on each block by $F_{th} - F_{max}$ and repeat step 3.

The numerical determination of the probability distribution of the size—*i.e.*, the total number of relaxations—of the earthquakes, which is proportional to the energy released during an earthquake, is found to exhibit self-organized criticality for a wide range of values of the parameter α.[42] The exponent values depend upon α. Lattice sizes used in the computer simulations were rather small: from 15 to 50.

As stressed by the authors, it is important to note that in this model, in contrast with the original Bak-Tang-Wiesenfeld model, only a fraction of the transported quantity is dissipated in each relaxation event: for $K_L > 0$, the redistribution of forces is *nonconservative*. This fraction is controlled by the parameter α.

Soon after the publication of the paper of Olami, Feder, and Christensen, Klein and Rundle [190], noticing that the behavior, as a function of the lattice size, of the exponential cutoff observed by the authors could not be correct, found suspect the observed critical behavior. A few years later, de Carvalho and Prado [80], revisiting the Olami-Feder-Christensen [273], found, in contrast with its authors and a subsequent study by Middleton and Tang [242], that it can be critical only in the conservative case, *i.e.*, for $\alpha = 0.25$. Christensen, Hamon, Jensen, and Lise [88], repeating the simulations of de Carvalho and Prado with the same parameter values, claimed that numerical results did not really exclude that the Olami-Feder-Christensen model remained critical also in the nonconservative case. In their reply, de Carvalho and Prado [81] agreed that on numerical evidence a true critical behavior for the nonconservative case could not be ruled out, but they thought that it was unlikely.

Recently, extensive numerical simulations [207, 208] have shown that the model displays scaling behavior on rather large square lattices (512×512). The probability density function of earthquake sizes obeys the Gutenberg-Richter law with a universal exponent $\tau = 1.8$[43] that does not depend upon the parameter α. It has been found, however, that the probability distribution of earthquakes initiated close to the boundary of the system does not obey

---

[42] The energy E released during the earthquake is given in terms of the magnitude m by the relation

$$\log_{10} E = c + dm,$$

where c and d are positive constants. From the Gutenberg-Richter law, the probability for an earthquake to have an energy larger than E behaves as E^{-B}, where $B = b/d$. In the modern scientific literature, an earthquake size is measured by its seismic moment M_0, which is defined as GAu, where G is the shear modulus, A the rupture area, and u the mean slip.

[43] This exponent τ is equal to $1 + B$, where B is the exponent characterizing the power-law behavior of the cumulative distribution function (see Footnote 42).

finite-size scaling.[44] On the other hand, numerical investigations of the Olami-Feder-Christensen model on a random graph [209] show that, for an average vertex degree equal to 4, there is no criticality in the system, as the cutoff in the probability distribution of earthquake sizes does not scale with system size. On the contrary, if each vertex has the same degree, for a weaker disorder, the system does exhibit criticality and obey finite-size scaling with universal critical exponents that are different from those of the lattice model.

Self-organized criticality is not the only mechanism that may explain the scale-free Gutenberg-Richter law. The Newman-Sneppen model [259, 318], described as a model of mass extinction on page 347, may be used as an earthquake model replacing the word 'species' by 'point of contact' in a subterranean fault. Choosing an exponential distribution for $f_{st}(x)$, the exponent of the size distribution is found equal to $1 + B = 1.84 \pm 0.03$. In order to investigate possible connections with a spatially organized system, Newman and Sneppen [259] implemented their model on a lattice and, at each time step, they eliminated not only those agents whose stress thresholds fall below the stress level but also their neighbors. As for noninteracting systems, they found an avalanche power-law distribution with an exponent close to 2.

To conclude this discussion on earthquake models, let us just mention another completely different model [83]. Its essential ingredient is the fractal nature of the solid-solid contact surfaces involved in the stick-slip mechanism. The authors relate the total contact area between the two surfaces to be proportional to the elastic strain energy grown during the sticking period as the solid-solid friction force due to interactions between asperities increases. This energy is released as one surface slips over the other. Assuming that the fractal dimension d_f of the surfaces in contact is less than 2, the exponent B (see Footnote 42) of the Gutenberg-Richter law is found to be approximately equal to $1/d_f$.

Example 68. Rainfall. Conventional methods of rain measurement are based on the idea of collecting rain in a container and measuring the amount of water after a certain time. Recently, high-resolution data have been collected by the Max-Planck Institute for Meteorology (Hamburg, Germany). Data collected at 250 meters above sea level with a one-minute resolution have been analyzed [281]. The authors define a *rain event* as a sequence of successive nonzero rain rates (*i.e.*, two rain events are separated by at least one minute). The size of a rain event is the height, in millimeters, of the rain collected in a water column during the rain event. The *drought duration* is the time, in minutes, between two consecutive rain events. They found that the number $N(M)$ of rain events of size M per year and the number $N(D)$ of drought periods of duration D per year have the power-law behaviors

$$N(M) \sim M^{-1.36} \quad \text{and} \quad N(D) \sim D^{-1.42},$$

---

[44] This feature was already apparent in the numerical results of Olami, Feder, and Christensen, and it is the reason why Klein and Rundle [190] cast doubt on the critical behavior.

which extend over at least three orders of magnitude. These power-law behaviors are, according to the authors, consistent with a self-organized critical
process. As for a sandpile to which sand grains are constantly added, the
atmosphere is driven by a slow and constant energy input from the Sun, leading to water evaporation. The energy is thus stored in the form of clouds
and suddenly released in bursts when the vapor condenses to water drops.
The power-law distribution $N(M) \sim M^{-1.36}$ is the analog of the power-law
distribution of the avalanche size of the sandpile model.

Exercises

Exercise 8.1 *(i) Show that, among all continuous probability density functions f defined
on $\mathbb{R}$ subject to the conditions*

$$\int_{\mathbb{R}} x f(x)\, dx = m, \qquad \int_{\mathbb{R}} (x-m)^2 f(x)\, dx = \sigma^2,$$

the normal distribution maximizes the entropy defined as

$$S(f) = -k \int_{\mathbb{R}} f(x) \log f(x)\, dx,$$

where k is a constant depending upon the entropy unit.

*(ii) What kind of auxiliary condition(s) should be imposed so that an a priori interesting
probability density function f would maximize the entropy? Consider the case $f(x) =
a/\pi(a^2 + x^2)$, where $x \in \mathbb{R}$.*

Exercise 8.2 *The exponential probability density function is defined, for $x \geq 0$, by*

$$f_{\exp}(x) = \lambda e^{-\lambda x} \qquad (\lambda > 0).$$

*(i) Find the probability density function of the sum of n independent identically dis-
tributed exponential random variables.*

(ii) Show that

$$\int_0^{n+x\sqrt{n}} \frac{u^{n-1}}{(n-1)!} e^{-u}\, du = \frac{1}{\sqrt{\pi}} \int_{-\infty}^{x} \exp\left(-\tfrac{1}{2}u^2\right) du.$$

*(iii) Illustrate the approach to the Gaussian distribution of the scaled sum of n inde-
pendent identically distributed exponential random variables for two values of n, one
somewhat small and the other one sufficiently large for the two distributions to be
reasonably close in an interval of a few units around the origin.*

Exercise 8.3 *Consider a sequence of independent identically distributed Bernoulli random variables (X_j) such that, for all j,*

$$P(X_j = -1) = \tfrac{1}{2} \quad and \quad P(X_j = 1) = \tfrac{1}{2}.$$

Use Theorem 22 to show that, if $0 < a < 1$,[45]

$$\lim_{n \to \infty} (P(S_n > an))^{1/n} = \frac{1}{\sqrt{(1+a)^{1+a}(1-a)^{1-a}}}.$$

Exercise 8.4 *If $(x_j)_{1 \leq j \leq n}$ is a sequence of observed values of a random variable X, in order to limit risk, one is often interested in finding the probability distributions of minimum and maximum values.*

(i) Consider a sequence of n independent identically distributed random variables $(X_j)_{1 \leq j \leq n}$; knowing the common cumulative distribution function F_X of these random variables, determine the cumulative distribution functions F_{Y_n} and F_{Z_n} of the random variables

$$Y_n = \min(X_1, X_2, \ldots, X_n) \quad \text{and} \quad Z_n = \max(X_1, X_2, \ldots, X_n).$$

(ii) Assuming that the n independent random variables have a common Cauchy probability density function centered at the origin and unit parameter, find the limit

$$\lim_{n \to \infty} P\left(Z_n > \frac{nz}{\pi}\right).$$

Exercise 8.5 *In statistical physics, the Boltzmann-Gibbs distribution applies to large systems in equilibrium with an infinite reservoir of energy. In this case, the system is said to be in thermal equilibrium, which means that its temperature is constant and its average energy is fixed. Claiming that in a closed economic system the total amount of money is conserved, Drăgulescu and Yakovenko [106] argue that, in a large system of N economic agents, money plays the role of energy and the probability that an agent has an amount of money between m and $m + dm$ is $p(m)dm$ with $p(m)$ of the form $Ce^{-m/T}$, where $C = 1/T$ is a normalizing factor and T an effective "temperature."*

(i) Assuming that m can take any nonnegative value, derive this expression of $p(m)$. What is the meaning of the "temperature" T?

(ii) What is the form of $p(m)$ if m values are restricted to the interval $[m_1, m_2]$? Discuss, in this case, the sign of T.

Exercise 8.6 *Show that a power law with an exponential cutoff distribution of the form $x^{-\alpha}(1-\varepsilon)^{-x}$, where ε is a small positive number, can be mistaken for a pure power-law distribution.*

Exercise 8.7 *(i) In order to show that the ARCH model can describe time series with sequences of data points looking like outliers, generate a time series from the ARCH process defined by (8.2) with independent identically distributed Gaussian random variables W_t with zero mean and unit variance, i.e., for all $t > 0$, $W_t = N(0,1)$.*

(ii) Determine the different statistics (mean, variance, skewness, kurtosis excess) for the generated time series X_t.

Exercise 8.8 *(i) Use the method described on page 334 to obtain an approximate graph of the cumulative distribution function from a random sample drawn from a Student's t-distribution for $\mu = 2$.*

---

[45] This example is taken from G. R. Grimmett and D. R. Stirzaker [158], p. 187, where a few more examples can be found.

(ii) Consider a rather large random sample to determine numerically the exponent characterizing the power-law behavior of the tail of the Student's t-distribution.

Exercise 8.9 *Study numerically the positive tail of the probability distribution of a random sequence generated from the ARCH process of Exercise 8.7.*

Exercise 8.10 *The ARCH process of order τ is defined by the relations*

$$X_t = \sigma_t W_t, \quad \text{with} \quad \sigma_t^2 = a_0 + a_1 X_{t-1}^2 + a_2 X_{t-2}^2 + \cdots + a_\tau X_{t-\tau}^2,$$

where $a_0 > 0$ and $a_j \geq 0$ for $j = 1, 2, \ldots, \tau$. In some applications, it appears that the order τ takes a high value, and one therefore has to estimate many parameters. It would be convenient to have a simpler model involving less parameters. The GARCH model, proposed by Bollerslev [66], involves only three parameters. It is defined by the relations

$$X_t = \sigma_t W_t, \quad \text{with} \quad \sigma_t^2 = a_0 + a_1 X_{t-1}^2 + b_1 \sigma_{t-1}^2, \tag{8.3}$$

where the W_t are independent identically distributed Gaussian random variables with zero mean and unit variance, a_0, a_1, and b_1 are positive, and $a_1 + b_1 < 1$.

(i) Show that the GARCH model may describe time series with data points looking like outliers.

(ii) Study numerically the positive tail of the probability distribution of a random sequence generated from a GARCH process.

Solutions

Solution 8.1 *(i) The probability function f that maximizes the entropy*

$$S(f) = -k \int_{\mathbb{R}} f(x) \log f(x) \, dx$$

and satisfies the conditions

$$\int_{\mathbb{R}} f(x) \, dx = 1, \quad \int_{\mathbb{R}} x f(x) \, dx = m, \quad \int_{\mathbb{R}} (x - m)^2 f(x) \, dx = \sigma^2,$$

is the solution to the equation

$$\frac{\delta}{\delta f} \left(k \int_{\mathbb{R}} f(x) \log f(x) \, dx, + \lambda_1 \int_{\mathbb{R}} f(x) \, dx \right.$$

$$\left. + \lambda_2 \int_{\mathbb{R}} x f(x) \, dx + \lambda_3 \int_{\mathbb{R}} (x - m)^2 f(x) \, dx \right) = 0,$$

where λ_1, λ_2, and λ_3 are Lagrange multipliers to be determined from the conditions above and $\delta/\delta f$ denotes the functional derivative with respect to the function f. Taking the derivative yields

$$k(\log f(x) + 1) + \lambda_1 + \lambda_2 x + \lambda_3 (x - m)^2 = 0.$$

This shows that the function f should be proportional to $e^{-\lambda_2 x - \lambda_3 (x-m)^2}$, and expressing that it has to satisfy the three conditions above, yields

$$f(x) = \frac{1}{\sqrt{2\pi}\sigma} \exp\left(-\frac{(x-m)^2}{2\sigma^2}\right).$$

f is the probability density function of the normal distribution. Its entropy is equal to

$$S_{\mathrm{normal}} = \frac{k}{2}(1 + \log 2\pi) + k\log\sigma = k(1.41894\ldots + \log\sigma).$$

To verify that the entropy of the normal distribution is maximum, it suffices to consider any other probability density with the same mean value m and standard deviation σ and check that its entropy is smaller. The entropy of the symmetric exponential distribution

$$\frac{1}{\sqrt{2}\sigma} \exp\left(-\frac{\sqrt{2}}{\sigma}|x-m|\right)$$

is

$$S_{\mathrm{exponential}} = k\left(1 + \frac{1}{2}\log 2 + \log\sigma\right) = k(1.34657\ldots + \log\sigma) < S_{\mathrm{normal}}.$$

Since the entropy measures the lack of information, the central limit theorem shows that averaging a large number of independent identically distributed random variables leads to a loss of information about the probability distribution of the random variables.

(ii) The probability density function f that maximizes the entropy

$$S(f) = -k\int_{\mathbb{R}} f(x)\log f(x)\, dx$$

subject to the n auxiliary conditions

$$\int_{\mathbb{R}} F_j(x)f(x)\, dx = c_j \quad (j = 1, 2, \ldots, n),$$

with $F_1(x) = 1$ (normalization of f), should verify

$$k(\log f(x) + 1) + \lambda_1 + \lambda_2 F_2(x) + \cdots + \lambda_n F_n(x) = 0,$$

where $\lambda_1, \lambda_2, \ldots, \lambda_n$ are Lagrange multipliers.

If we want f to be the Cauchy probability density function, the equation above becomes

$$k(\log(a/\pi) + 1) + \lambda_1 - k\log(a^2 + x^2) + \lambda_2 F_2(x) + \cdots + \lambda_n F_n(x) = 0.$$

Since all the moments of the Cauchy distribution are infinite, we cannot expect to satisfy this equation expanding $\log(a^2 + x^2)$ and writing down an infinite number of conditions for all moments of even order. The only possibility is to choose λ_1 to cancel all constant terms and $F_2(x)$ proportional to $\log(a^2 + x^2)$, which would be, as pointed out by Montroll and Schlesinger [248], a rather unlikely a priori choice.

Solution 8.2 *(i) There are at least two different methods to obtain this result.*
First method. Let us first find the probability density function of the sum of two independent identically distributed exponential random variables. It is given by

$$f_{2\exp}(s) = \int_0^s \lambda^2 e^{-\lambda x} e^{-\lambda(s-x)}\, dx$$
$$= \lambda^2 s e^{-\lambda s}.$$

For the sum of three exponential random variables, we find

$$f_{3\exp}(s) = \int_0^s \lambda^3 x e^{-\lambda x} e^{-\lambda(s-x)}\, dx$$
$$= \frac{\lambda^3 s^2}{2} e^{-\lambda s}.$$

This last result suggests that the probability density function of the sum of n independent identically distributed exponential random variables could be the Gamma distribution of index n defined by

$$f_{\mathrm{Gamma}}(s) = \frac{\lambda^n s^{n-1}}{(n-1)!} e^{-\lambda s}.$$

This is easily verified by induction since

$$\int_0^s \frac{\lambda^{n+1} x^{n-1}}{(n-1)!} e^{-\lambda x} e^{-\lambda(s-x)}\, dx = \frac{\lambda^{n+1} s^n}{n!} e^{-\lambda s}.$$

Second method. The characteristic function of an exponential random variable is given by

$$\varphi_{\exp}(t) = \int_0^\infty \lambda e^{-\lambda x} e^{itx}\, dx$$
$$= \left(1 - \frac{it}{\lambda}\right)^{-1}.$$

On the other hand, the characteristic function of a Gamma random variable of index n is

$$\varphi_{\mathrm{Gamma}}(t) = \int_0^\infty \frac{\lambda^n x^{n-1}}{(n-1)!} e^{-\lambda x} e^{itx}\, dx$$
$$= \left(1 - \frac{it}{\lambda}\right)^{-n},$$

which proves that the sum of n independent identically distributed exponential random variables is a Gamma random variable of index n.

(ii) For $\lambda = 1$, the mean m_1 and the standard deviation σ of the corresponding exponential random variable are both equal to 1. Hence, from the central limit theorem, it follows that

$$\int_0^{n+x\sqrt{n}} \frac{u^{n-1}}{(n-1)!} e^{-u}\, du = \frac{1}{\sqrt{\pi}} \int_{-\infty}^x \exp\left(-\tfrac{1}{2}u^2\right)\, du.$$

(iii) Figure 8.14 shows the approach to the Gaussian distribution of the scaled sum of 10 and 50 independent identically distributed exponential random variables.

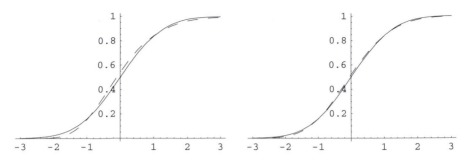

Fig. 8.14. *Convergence to the normal distribution of the scaled sum of 10 (left figure) and 50 (right figure) exponential random variables with unit parameter value. Continuous curves represent the Gaussian distribution and dashed curves distributions of scaled sums of independent exponential random variables for $\lambda = 1$.*

Solution 8.3 *The mean value of the Bernoulli random variables X_j is zero, and their moment-generating function M is given by $M(t) = \cosh t$. This function is finite in any finite neighborhood of the origin. From Theorem 22 it follows that, for $0 < a < 1$,*

$$\lim_{n \to \infty} \left(P(S_n > na) \right)^{1/n} = \inf_{t>0} \left(e^{-at} \cosh t \right).$$

As a function of e^t, $e^{-at} \cosh t$ has a minimum for $e^t = \sqrt{(1+a)/(1-a)}$, so

$$\inf_{t>0} \left(e^{-at} \cosh t \right) = \frac{1}{\sqrt{(1+a)^{1+a}(1-a)^{1-a}}},$$

which proves the required result.

Solution 8.4 *(i) By definition*

$$
\begin{aligned}
F_{Y_n}(y) &= P\big(\min(X_1, X_2, \ldots, X_n) \le y \big) \\
&= 1 - P\big(\min(X_1, X_2, \ldots, X_n) > y \big) \\
&= 1 - P\big(X_1 > y, X_2 > y, \cdots, X_n > y \big) \\
&- 1 - P(X_1 > y)P(X_2 > y) \cdots P(X_n > y) \\
&= 1 - \big(1 - F_X(y) \big)^n.
\end{aligned}
$$

Similarly,

$$
\begin{aligned}
F_{Z_n}(z) &= P\big(\max(X_1, X_2, \ldots, X_n) \le z \big) \\
&= P\big(X_1 \le z, X_2 \le z, \cdots, X_n \le z \big) \\
&= P(X_1 \le z)P(X_2 \le z) \cdots P(X_n \le z) \\
&= \big(F_X(z) \big)^n.
\end{aligned}
$$

(ii) The cumulative distribution function $F_X(x)$ of a Cauchy random variable X centered at the origin and unit parameter is given by

$$F_X(x) = \int_{-\infty}^x \frac{1}{\pi} \frac{du}{1+u^2} = \frac{1}{2} + \frac{\arctan(x)}{\pi}.$$

Hence,[46]

$$\lim_{n \to \infty} P\left(Z_n > \frac{nz}{\pi}\right) = 1 - \lim_{n \to \infty} \left(\frac{1}{2} + \frac{1}{\pi} \arctan\left(\frac{nz}{\pi}\right)\right)^n$$

$$= 1 - \lim_{n \to \infty} \left(1 - \frac{1}{\pi} \arctan\left(\frac{\pi}{nz}\right)\right)^n$$

$$= 1 - \lim_{n \to \infty} \left(1 - \frac{1}{zn} + o(n^{-1})\right)^n$$

$$= 1 - e^{-1/z}.$$

Remark 26. The function $z \mapsto e^{-1/z}$ is the cumulative distribution function of a positive random variable. As a consequence of the result above, the scaled random variable $\pi Z_n/n$ converges in distribution to a positive random variable Z such that $F_Z(z) = e^{-1/z}$. For $z = 1/\log(2)$, $e^{-1/z} = \frac{1}{2}$; therefore, the probability that, in a sequence of n realizations of a symmetric Cauchy random variable with unit parameter, the maximum of the n observed values has a probability equal to $\frac{1}{2}$ to be greater than $n/(\pi \log 2) \approx 0.4592241\,n$.

Solution 8.5 *(i) To derive the expression of $p(m)$, we proceed as in Exercise 8.1. The entropy associated with a probability density function p is*

$$S(p) = -\int_0^\infty p(m) \log p(m)\, dm.$$

Using Lagrange multipliers to maximize this entropy with respect to the function p under the constraints

$$\int_0^\infty p(m)\, dm = 1 \quad \text{and} \quad \int_0^\infty mp(m)\, dm = \frac{M}{N},$$

which express that $p(m)$ has to be normalized and that the average amount of money per agent, equal to the total amount of money M divided by the number of agents N, is fixed, we have to maximize

$$-\int_0^\infty p(m) \log p(m)\, dm + \lambda_1 \int_0^\infty p(m)\, dm + \lambda_2 \int_0^\infty mp(m)\, dm.$$

Writing that the functional derivative with respect to p of the expression above is zero yields

$$-\log p(m) - 1 + \lambda_1 + \lambda_2 m = 0.$$

This shows that p is an exponential distribution, which can be written

$$p(m) = \frac{1}{T} \exp\left(-\frac{m}{T}\right) \qquad (T > 0).$$

The meaning of the parameter T is found expressing that the average amount of money per agent is fixed. This gives

$$\int_0^\infty mp(m)\, dm = T = \frac{M}{N}.$$

---

[46] Using the relations $\arctan(x) + \arctan(1/x) = \pi/2$ and $\arctan(x) = x + o(x)$ as $x \to 0$.

That is, when the system is in equilibrium, the parameter T, which plays the role of the "temperature" in statistical physics, is the average amount of money per agent. It is always positive.

Remark 27. Maximizing the entropy means that the disorder associated with the distribution of an amount of money M between N agents under the weak constraint that fixes the average amount per agent is maximized. Taking into account more restrictive constraints such as a fixed standard deviation would give a Gaussian distribution (as in Exercise 8.1) corresponding to a much more egalitarian society. Note that, for the exponential distribution, the most probable amount of money is equal to zero!

(ii) In the derivation of the probability distribution of money above, it was assumed that the amount of money could vary from 0 to infinity. If $m \in [m_1, m_2]$, the maximization of the entropy under constraints leads again to an exponential distribution, but now we have to satisfy the conditions

$$\int_{m_1}^{m_2} Ce^{-m/T}\,dm = 1 \quad \text{and} \quad \int_{m_1}^{m_2} Cme^{-m/T}\,dm = \langle m \rangle = \frac{M}{N}.$$

Hence,

$$CT\left(e^{-m_1/T} - e^{-m_2/T}\right) = 1,$$

$$CT\left(e^{-m_1/T}(m_1 + T) - e^{-m_2/T}(m_2 + T)\right) = \langle m \rangle.$$

The first equation determines the normalizing factor C. The probability density function is then given by

$$p(m) = \frac{e^{-m/T}}{T(e^{-m_1/T} - e^{-m_2/T})}.$$

Replacing C by its expression in the second equation yields

$$\frac{e^{m_2/T}(m_1 + T) - e^{m_1/T}(m_2 + T)}{e^{m_2/T} - e^{m_1/T}} = \langle m \rangle.$$

Let

$$m_+ = \frac{m_2 + m_1}{2} \quad \text{and} \quad m_- = \frac{m_2 - m_1}{2}$$

or

$$m_1 = m_+ - m_- \quad \text{and} \quad m_2 = m_+ + m_-.$$

In terms of m_+ and m_-, the relation between T and $\langle m \rangle$ becomes

$$\coth \frac{m_-}{T} - \frac{T}{m_-} = -\frac{m_+ + \langle m \rangle}{m_-}.$$

The odd function $L : x \mapsto \coth x - 1/x$, familiar in statistical physics, is the Langevin function. Its graph is represented in Figure 8.15. The relation above shows that the sign of T is the same as the sign of $-(m_+ + \langle m \rangle)$. If, for example, $m_1 = -5$, $m_2 = 5$, and $\langle m \rangle = 4$, then $-(m_+ + \langle m \rangle) = -4$, and solving for T, we find $T = -1$. The graph of the corresponding probability density function is shown in Figure 8.16.

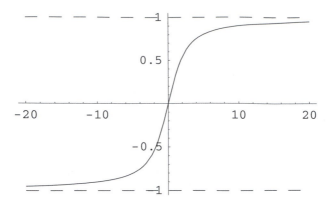

Fig. 8.15. *Graph of the Langevin function.*

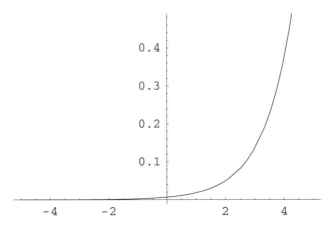

Fig. 8.16. *Probability density function of the distribution of money when the amount of money possessed by one individual is restricted to the interval* $[m_1, m_2]$.

Solution 8.6 *In order to show that a power law with an exponential cutoff distribution can be mistaken for a pure power-law distribution, consider the probability density function*

$$f(x) = Cx^{-1.5}(1 - 0.0005)^x \qquad (x \geq 1),$$

where C is a normalizing factor approximately equal to 0.520. Figure 8.17 shows a log-log plot of the sequence $(f(10\,k))$ for $k = 1, 2, \ldots, 100$. A least-squares fit of the apparent power-law behavior of this sequence gives an exponent equal to -1.6. While the random variable distributed according to f has a mean value equal to 40.2, the random variable defined for $x \geq 1$ with a power-law probability density proportional to $x^{-1.6}$ has no mean value.

Solution 8.7 *(i) Consider the ARCH model defined by the relations*

$$X_t = \sigma_t W_t, \quad \text{with} \quad \sigma_t^2 = a_0 + a_1 X_{t-1}^2 \quad (a_0 > 0,\ 0 < a_1 < 1),$$

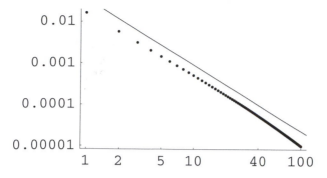

Fig. 8.17. *Log-log plot of the sequence $\big(f(10\,k)\big)$ for $k = 1, 2, \ldots, 100$, which exhibits an apparent power-law behavior, with an exponent equal to -1.6, over about two orders of magnitude. The slightly shifted straight line represents the least-squares fit.*

where, for all positive integers t, $W_t = N(0,1)$, and choose, for example, $a_0 = 0.6$, $a_1 = 0.5$, and $\sigma_1 = 1$. Figure 8.18 clearly shows the existence of outliers present in a generated sequence of 1000 data points.

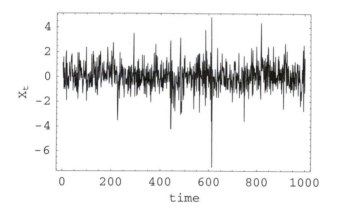

Fig. 8.18. *ARCH model: (X_t) time series for $t = 1, 2, \ldots, 1000$.*

Another interesting way to show the existence of outliers in the generated time series (X_t) is the scatter plot of X_t versus X_{t-1} represented in Figure 8.19.
By comparison, the scatter plot of the Gaussian time series is shown in Figure 8.20.
(ii) The different statistics for the time series W_t and X_t are given in the following table. To obtain a better accuracy, we generated a sequence of $50,000$ data points.

	mean	variance	skewness	kurtosis excess
W_t	-0.000854	0.997	-0.0213	0.0240
X_t	-0.00242	1.20	-0.0847	3.86

The numerical values of the statistics of the time series W_t are to be compared to the theoretical values, respectively equal to 0, 1, 0, and 0. We have also represented the

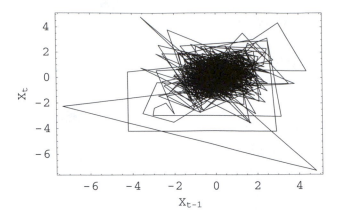

Fig. 8.19. *Scatter plot of X_t versus X_{t-1}.*

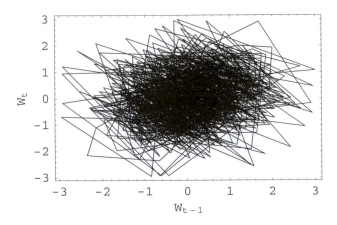

Fig. 8.20. *Scatter plot of W_t versus W_{t-1}.*

histograms of both time series in Figure 8.21. The heights of the bars being scaled so that their areas sum to unity, the histograms represent approximate probability density functions.

Solution 8.8 *(i) The probability density of the Student's t-distribution for $\mu = 2$ is given by*

$$f_t(x) = \frac{1}{(2 + x^2)^{\frac{3}{2}}}.$$

Following the method described on page 334, from a random sample of 2000 points, we obtain the plot represented in Figure 8.22.

(ii) In order to study numerically the power-law behavior of the positive tail of the probability distribution, we have generated a random sample of size $n = 100,000$ drawn from a Student's t-distribution for $\mu = 2$ and considered the last 2000 points of the ordered random sequence. Using the notation of page 334, a log-log plot of the sequence of points $(\xi_k, 1 - \eta_k)$ for $k > 98,000$ is represented in Figure 8.23. From a least-squares

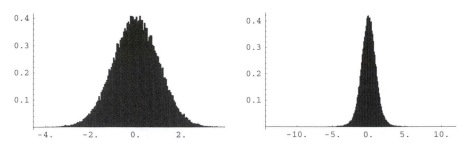

Fig. 8.21. *Histograms of time series (W_t) (left) and (X_t) (right). The heights of the bars are scaled so that their areas sum to unity.*

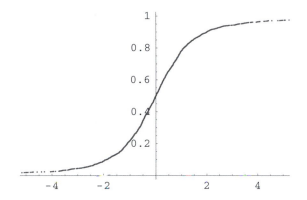

Fig. 8.22. *Approximate cumulative distribution function of the Student's t-distribution for $\mu = 2$.*

fit, the exponent characterizing the power-law behavior of the cumulative distribution function is found to be equal to -1.99, in very good agreement with the asymptotic behavior of the probability density function $f_t(x) \sim x^{-3}$ for large x. Here the tail corresponds to x-values greater than 4.8 ($\log 4.8 = 1.57$).

Solution 8.9 *If we consider the ARCH model of Exercise 8.7, from the sequence of 50,000 points generated to compute the four statistics (see solution of Exercise 8.7), and using the method described on page 334, from the log-log plot of the last 2000 points of the ordered sequence, represented in Figure 8.24, it can be shown that the tail of the distribution exhibits a power-law behavior. A least-squares fit gives $\alpha = -3.62$ for the exponent characterizing the power-law behavior for large positive x values of the cumulative distribution function (see Figure 8.24). The last 100 points have been discarded.*

Solution 8.10 *Choosing $a_0 = 0.5$, $a_1 = 0.5$, $b_1 = 0.3$, and $\sigma_1 = 1$, the time series and the scatter plot represented, respectively, in Figures 8.25 and 8.26 show the existence of outliers in a sequence of 1000 data points generated by the GARCH process defined by (8.3).*
(ii) Different statistics for a time series X_t of 100,000 data points generated by the GARCH process defined above are given in the following table.

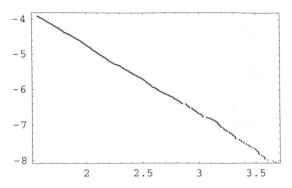

Fig. 8.23. *Power-law behavior of the cumulative distribution function of the Student's t-distribution for $\mu = 2$ determined from the last 2000 points of an ordered random sequence of size $n = 100,000$. A least-squares fit gives an exponent equal to -1.99, in very good agreement with the theoretical value -2.*

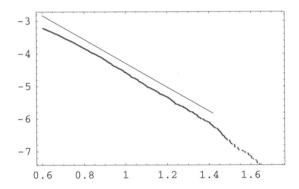

Fig. 8.24. *Power-law behavior of the positive tail of the probability distribution of the X_t sequence of $50,000$ points generated by the ARCH model studied in Exercise 8.7. The straight line representing the least-squares fit has been slightly translated for clarity. The exponent is equal to -3.62.*

	mean	variance	skewness	kurtosis excess
X_t	0.000933	2.45	0.142	10.5

Using the method described on page 334, from the log-log plot of the last 5000 points of the ordered sequence represented in Figure 8.27, show that the tail of the distribution exhibits a power-law behavior. A least-squares fit gives $\alpha = 3.62$ for the exponent characterizing the power-law behavior for large positive x values of the cumulative distribution function.

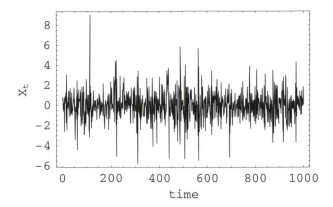

Fig. 8.25. *GARCH model: (X_t) time series for $t = 1, 2, \ldots, 1000$.*

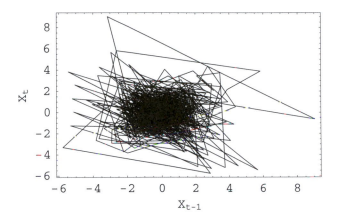

Fig. 8.26. *GARCH model: Scatter plot of X_t versus X_{t-1}.*

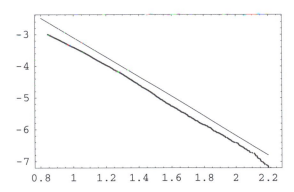

Fig. 8.27. *Power-law behavior of the positive tail of the probability distribution of the X_t sequence of $100,000$ points generated by the GARCH process. The straight line representing the least-squares fit has been slightly translated for clarity. The exponent is equal to -3.08.*

Notations

The essential mathematical notations used in the text are grouped below.

$\square$	indicates the end of a proof.
$\mathbb{R}$	set of real numbers.
$\mathbb{R}_+$	set of positive numbers.
$\mathbb{C}$	set of complex numbers.
$\mathbb{Z}$	set of all integers.
$\mathbb{Z}_L$	set of integers modulo L.
$\mathbb{N}$	set of positive integers.
$\mathbb{N}_0$	set of nonnegative integers.
$\mathbb{Q}$	set of rational numbers.
$\emptyset$	empty set.
$x \in \mathcal{X}$	x is an element of the set $\mathcal{X}$.
$\mathcal{S} \subset \mathcal{X}$	$\mathcal{S}$ is a subset of $\mathcal{X}$.
$A \cup B$	set of elements either in A or B.
$A \cap B$	set of elements that are in A and B.
A_c	complement of set A.
$A \setminus B$	set of all elements in A that are not in B.
$A \triangle B$	set of elements either in A or B but not in A and B.
$|A|$	number of elements of set A.
$\{x \in \mathcal{M} \mid P(x)\}$	all elements of $\mathcal{M}$ that have the property $P(x)$.
$a \sim b$	a is equivalent to b.
$a \Rightarrow b$	a implies b.
$a \Leftrightarrow b$	a implies b and conversely.
$\lfloor x \rfloor$	largest integer less than or equal to x.
$\lceil x \rceil$	smallest integer greater than or equal to x.
$\overline{z}$	complex conjugate of z.
$m(A)$	Lebesgue measure of the set A.
$f : X \to Y$	f is a mapping from the set X into the set Y.

$f : x \mapsto f(x)$	mapping f takes the point x to the point $f(x)$.
$f^{-1}(x)$	set of all preimages of x.
$f \circ g$	composite of mappings f and g, g being applied first.
$\mathbf{x} \in \mathbb{R}^n$	$\mathbf{x}$ is an element of the n-dimensional vector space $\mathbb{R}^n$.
$\mathbf{x} = (x_1, x_2, \ldots, x_n)$	$x_1, x_2, \ldots, x_n$ are the components of vector $\mathbf{x}$.
$\|\mathbf{x}\|$	norm of vector $\mathbf{x}$.
$\dim(\mathcal{S})$	dimension of (manifold) $\mathcal{S}$.
$[a, b]$	closed interval; i.e., $\{x \in \mathbb{R} \mid a \leq x \leq b\}$.
$]a, b[$	open interval; i.e., $\{x \in \mathbb{R} \mid a < x < b\}$.
$[a, b[$	$\{x \in \mathbb{R} \mid a \leq x < b\}$.
$]a, b]$	$\{x \in \mathbb{R} \mid a < x \leq b\}$.
$d(x, y)$	distance between points x and y in a metric space.
$N(\mathbf{x})$	neighborhood of $\mathbf{x}$.
$\dot{\mathbf{x}}$	time derivative of vector $\mathbf{x}$.
$D\mathbf{X}(\mathbf{x})$	derivative of vector field $\mathbf{X}$ at $\mathbf{x}$.
$D_{\mathbf{x}}\mathbf{X}(\mathbf{x}, \boldsymbol{\mu})$	derivative with respect to $\mathbf{x}$ of vector field $\mathbf{X}$ at $(\mathbf{x}, \boldsymbol{\mu})$.
$\operatorname{tr} \mathbf{A}$	trace of square matrix $\mathbf{A}$.
$\det \mathbf{A}$	determinant of square matrix $\mathbf{A}$.
$\operatorname{diag}[\lambda_1, \ldots, \lambda_n]$	diagonal matrix whose diagonal elements are $\lambda_1, \ldots, \lambda_n$.
$\operatorname{diag}[B_1, \ldots, B_n]$	block-diagonal matrix whose diagonal blocks are $B_1, \ldots, B_n$.
$[\mathbf{a}_1, \mathbf{a}_2, \ldots, \mathbf{a}_n]$	$n \times n$ matrix whose element a_{ij} is the j-component of the n-dimensional vector $\mathbf{a}_i$.
$\operatorname{spec}(\mathbf{A})$	spectrum of linear operator $\mathbf{A}$.
$f(x) \overset{x \to 0}{=} O\big(g(x)\big)$	there exist two positive constants A and a such that $|f(x)| \leq A|g(x)|$ for $|x| < a$.
$f(x) \overset{x \to 0}{=} o\big(g(x)\big)$	for any $\varepsilon > 0$, there exists $\delta > 0$ such that $|f(x)| \leq \varepsilon|g(x)|$ for $|x| < \delta$.
$f(x) \sim g(x)$	$f(x)$ and $g(x)$ have the same asymptotic behavior.
$f(x + 0)$	$\lim_{\varepsilon \to 0}$, where $\varepsilon > 0$.
$f(x - 0)$	$\lim_{\varepsilon \to 0}$, where $\varepsilon < 0$.
$a \approx b$	a is approximately equal to b.
$a \lesssim b$	a is less than or approximately equal to b.
$a \gtrsim b$	a is greater than or approximately equal to b.
$G(N, M)$	graph of order N and size M.
$A(G)$	adjacency matrix of graph G.
$V(G)$	set of vertices of graph G.
$E(G)$	set of edges of graph G.
$N(x)$	neighborhood of vertex x of a graph.
$d(x)$	degree of the vertex x of a graph.
$d_{\text{in}}(x)$	in-degree of the vertex x of a digraph.
$d_{\text{out}}(x)$	out-degree of the vertex x of a digraph.

L_N	characteristic path length of network N.
C_N	clustering coefficient of network N.
D_N	diameter of network N.
$\binom{n}{k}$	binomial number.
$P(X = x)$	probability that the random variable X is equal to x.
F_X	cumulative distribution function of the random variable X defined by $F_X(x) = P(X \leq x)$.
f_X	probability density function of the absolutely continuous random variable X.
$X_n \xrightarrow{\mathrm{d}} X$	sequence of random variables (X_n) converges in distribution to X.
$\widetilde{m}$	median of a distribution defined by $F(\widetilde{m}) = \frac{1}{2}$.
$\langle X \rangle$	average value of random variable X.
$m_r(X)$	moment of order r of a random variable X; i.e., $m_r(X) = \langle X^r \rangle$.
$\sigma^2(X)$	variance of a random variable X; i.e., $\sigma^2(X) = m_2(X) - m_1^2(X)$.
φ_X	characteristic function of the random variable X.
$N(m, \sigma^2)$	normal random variable on mean m and variance σ^2.
$\widehat{f}$	Fourier transform of function f.
$L_{\alpha,\beta}$	probability density function of stable Lévy distribution.
f_{t}	probability density of a Student's t-distribution.
$W = \{W_t \mid t \geq 0\}$	stochastic process.

References

1. AKIN E. and DAVIS M., *Bulgarian Solitaire*, American Mathematical Monthly **92** 237–250 (1985)
2. ALBERT R., JEONG H., and BARABÁSI A.-L., *Diameter of the World Wide Web*, Nature **401** 130–131 (1999)
3. ALBERT R. and BARABÁSI A.-L., *Topology of Evolving Networks: Local Events and Universality*, Physical Review Letters **85** 5234–5237 (2000)
4. ALBERT R., JEONG H., and BARABÁSI A.-L., *Error and Attack Tolerance of Complex Networks*, Nature **406** 378–382 (2000)
5. ALBERT R. and BARABÁSI A.-L., *Statistical Mechanics of Complex Networks*, Review of Modern Physics **74** 47–97 (2002)
6. ALLEE W. C., *Animal Aggregations: A Study in General Sociology* (Chicago: The University of Chicago Press 1931). Reprinted from the original edition by AMS Press, New York (1978)
7. ANOSOV D. V. and ARNOLD V. I. (editors), *Dynamical Systems I: Ordinary Differential Equations and Smooth Dynamical Systems* (Heidelberg: Springer-Verlag 1988)
8. ARGOLLO DE MENEZES M., MOUKARZEL C. F., and PENNA T. J. P., *First-Order Transition in Small-World Networks*, Europhysics Letters **50** 5–8 (2000)
9. ARNOL'D V. I., *Ordinary Differential Equations* (Berlin: Springer-Verlag 1992)
10. ARNOL'D V. I., *Catastrophe Theory* (Berlin: Springer-Verlag 1992)
11. ARROWSWMITH D. K. and PLACE C. M., *An Introduction to Dynamical Systems* (Cambridge, UK: Cambridge University Press 1990)
12. ASSAF D., IV, and GADBOIS S., *Definition of Chaos*, American Mathematical Monthly **99** 865 (1992)
13. AXELROD R., *Effective Choice in the Prisoner's Dilemma*, Journal of Conflict Resolution **24** 3–25 (1980)
14. AXELROD R., *More Effective Choice in the Prisoner's Dilemma*, Journal of Conflict Resolution **24** 379–403 (1980)
15. AXELROD R. and HAMILTON W. D., *The Evolution of Cooperation*, Science **211** 1390–1396 (1981)
16. AXELROD R., *The Dissemination of Culture: A Model with Local Convergence and Global Polarization*, Journal of Conflict Resolution **41** 203–226 (1997), Reprinted in Axelrod [17] pp 148–177

17. AXELROD R., *The Complexity of Cooperation: Agent-Based Models of Competition and Collaboration* (Princeton, NJ: Princeton University Press 1997)

18. AXTELL R., AXELROD R., EPSTEIN J. M., and COHEN M. D., *Aligning Simulation Models: A Case Study and Results*, Computational and Mathematical Organization Theory **1** 123–141 (1996). See an adapted version in Axelrod [17], pp. 183–205.

19. AYALA F. J., *Experimental Invalidation of the Principle of Competitive Exclusion*, Nature **224** 1076–1079 (1969)

20. AYALA F. J., GILPIN M. E., and EHRENFELD J. G., *Competition Between Species: Theoretical Models and Experimental Tests*, Theoretical Population Biology **4** 331–356 (1973)

21. BACHELIER L., *Théorie de la spéculation* (Paris: Gauthier-Villars 1900). English translation by A. James Boness, in *The Random Character of Stock Market Prices*, pp. 17–78, Paul H. Cootner (editor) (Cambridge, MA: The MIT Press 1964)

22. BADGER W. W., *An Entropy-Utility Model for the Size Distribution of Income*, in *Mathematical Models as a Tool for the Social Sciences*, pp. 87–120, B. J. West (editor) (New York: Gordon and Breach 1980)

23. BAHR D. B. and BEKOFF M., *Predicting Flock Vigilance from Simple Passerine Interactions: Modelling with Cellular Automata*, Animal Behavior **58** 831–839 (1999)

24. BAK P., TANG C., and WIESENFELD K., *Self-Organized Criticality: An Explanation of $1/f$ noise*, Physical Review Letters **59** 381–384 (1987)

25. BAK P., TANG C., and WIESENFELD K., *Self-Organized Criticality* Physical Review A **38** 364–374 (1988)

26. BAK P., CHEN K., and TANG C., *A Forest-Fire Model and Some Thoughts on Turbulence*, Physics Letters A **147** 297–300 (1990)

27. BAK P. and CHEN K., *Self-Organized Criticality*, Scientific American **264** 46–53 (1991)

28. BAK P. and SNEPPEN K., *Punctuated Equilibrium and Criticality in a Simple Model of Evolution*, Physical Review Letters **24** 4083–4086 (1993)

29. BAK P., *How Nature Works: The Science of Self-Organized Criticality* (New York: Springer-Verlag 1996)

30. BAK P., CHRISTENSEN K., DANON L., and SCANLON T., *Unified Scaling Law for Earthquakes*, preprint:cond-mat/0112342

31. BANKS J., BROOKS J., CAIRNS G., DAVIS G., and STACEY P., *On Devaney's Definition of Chaos*, American Mathematical Monthly **99** 332–334 (1992)

32. BARABÁSI A.-L. and ALBERT R., *Emergence of Scaling in Random Networks*, Science **286** 509–512 (1999)

33. BARABÁSI A.-L., ALBERT R., and JEONG H., *Mean-Field Theory for Scale-Free Random Networks*, Physica A **272** 173–187 (1999)

34. BARABÁSI A.-L., ALBERT R., and JEONG H., *Scale-Free Characteristics of Random Networks: the Topology of the World Wide Web*, Physica A **281** 69–77 (2000)

35. BARABÁSI A.-L., JEONG H., NÉDA Z., RAVASZ E., SCHUBERT A., and VICSEK T., *Evolution of the Social Network of Scientific Collaborations*, Physica A **311** 590–614 (2002)

36. BARRAT A., *Comment on "Small-World Networks: Evidence for a Crossover Picture"*, preprint:cond-mat/9903323

37. BARTHÉLÉMY M. and AMARAL L. A. N., *Erratum: Small-World Networks: Evidence for a Crossover Picture [Phys. Rev. Lett. 82, 3180 (1999)]*, Physical Review Letters **82** 5180 (1999)

38. BEDDINGTON J. R., FREE C. A., and LAWTON J. H., *Dynamic Complexity in Predator-Prey models Framed in Difference Equations*, Nature **255** 58–60 (1975)

39. BEGON M., HARPER J. L., and TOWNSEND C. R., *Ecology: Individuals, Populations and Communities* (Oxford: Blackwell 1986)

40. BENDIXSON I., *Sur les courbes définies par des équations différentielles*, Acta Mathematica **24** 1–88 (1901)

41. BERRYMAN M. J., ALLISON A., and ABBOTT D., *Classifying Ancient Texts by Inter-Word Spacing*, preprint:physics/0206080

42. BIDAUX R., BOCCARA N., and CHATÉ H., *Order of the Dimension Versus Space Dimension in a Family of Cellular Automata*, Physical Review A **39** 3094–3105 (1989)

43. BLANCHARD F., *β-expansions and Symbolic Dynamic*, Theoretical Computer Science **65** 131–141 (1989)

44. BOCCARA N., *On the Microscopic Formulation of Landau Theory of Phase Transitions*, Solid State Communications **11** 39–40 (1972)

45. BOCCARA N., *Symétries brisées* (Paris: Hermann 1976)

46. BOCCARA N., *Functional Analysis: An Introduction for Physicists* (Boston: Academic Press 1990)

47. BOCCARA N., NASSER J., and ROGER M., *Annihilation of Defects During the Evolution of Some One-Dimensional Class-3 Deterministic Cellular Automata*, Europhysics Letters **13** 489–494 (1990)

48. BOCCARA N., NASSER J., and ROGER M., *Particlelike Structures and Their Interactions in Spatiotemporal Patterns Generated by One-Dimensional Deterministic Cellular Automaton Rules*, Physical Review A **44** 866–875 (1991)

49. BOCCARA N. and CHEONG K., *Automata Network SIR Models for the Spread of Infectious Diseases in Populations of Moving Individuals*, Journal of Physics A: Mathematical and General **25** 2447–2461 (1992)

50. BOCCARA N. and ROGER M., *Period-Doubling Route to Chaos for a Global Variable of a Probabilistic Automata Network*, Journal of Physics A: Mathematical and General **25** L1009–L1014 (1992)

51. BOCCARA N., *Transformations of One-Dimensional Cellular Automaton Rules by Translation-Invariant Local Surjective Mappings*, Physica D **68** 416–426 (1993)

52. BOCCARA N. and CHEONG K., *Critical Behaviour of a Probabilistic Automata Network SIS Model for the Spread of an Infectious Disease in a Population of Moving Individuals*, Journal of Physics A: Mathematical and General **26** 3707–3717 (1993)

53. BOCCARA N., NASSER J., and ROGER M., *Critical Behavior of a Probabilistic Local and Nonlocal Site-Exchange Cellular Automaton*, International Journal of Modern Physics C **5** 537–545 (1994)

54. BOCCARA N., CHEONG K., and ORAM M., *Probabilistic Automata Epidemic Model with Births and Deaths Exhibiting Cyclic Behaviour*, Journal of Physics A: Mathematical and General **27** 1585–1597 (1994)

55. BOCCARA N., ROBLIN O., and ROGER M., *An Automata Network Predator-Prey Model with Pursuit and Evasion*, Physical Review E **50** 4531–4541 (1994)

56. BOCCARA N., *Probabilités* (Paris: Ellipses 1995)

57. BOCCARA N., *Intégration* (Paris: Ellipses 1995)

58. BOCCARA N., FUKŚ H., and GEURTEN S., *A New Class of Automata Networks*, Physica D **103** 145–154 (1997)

59. BOCCARA N. and FUKŚ H., *Cellular Automaton Rules Conserving the Number of Active Sites*, Journal of Physics A: Mathematical and General **31** 6007–6018 (1998)

60. BOCCARA N. and FUKŚ H., *Critical Behaviour of a Cellular Automaton Highway Traffic Model*, Journal of Physics A: Mathematical and General **33** 3407–3415 (2000)

61. BOCCARA N., *On the Existence of a Variational Principle for Deterministic Cellular Automaton Models of Highway Traffic Flow*, International Journal of Modern Physics C **12** 1–16 (2001)

62. BOCCARA N. and FUKŚ H., *Number-Conserving Cellular Automaton Rules*, Fundamenta Informaticae **52** 1–13 (2002),

63. BOER J. DE, DERRIDA B., FLYVBERG H., JACKSON A. D., and WETTIG T., *Simple Model of Self-Organized Biological Evolution*, Physical Review Letters **73** 906–909 (1994)

64. BOINSKI S. and GARBER P. A. (editors), *On the Move: How and Why Animals Travel in Groups*, (Chicago: The University of Chicago Press 2000)

65. BOLLOBÁS B., *Modern Graph Theory* (New York: Springer-Verlag 1998)

66. BOLLERSLEV T., *Generalized Autoregressive Conditional Heteroskedasticity*, Journal of Econometrics **31** 307–327 (1986)

67. BONABEAU E., TOUBIANA L., and FLAHAUT A., *Evidence for Global Mixing in Real Influenza Epidemics*, Journal of Physics A: Mathematical and General **31** L361–L365 (1998)

68. BOUCHAUD J.-P. and POTTERS M., *Theory of Financial Risks: From Statistical Physics to Management Risks* (Cambridge, UK: Cambridge University Press 2000)

69. BOUCHAUD J.-P., *An Introduction to Statistical Finance*, Physica A **313** 238–251 (2002)

70. BOYD R. and LORBERBAUM J. P., *No Pure Strategy is Evolutionarily Stable in the Repeated Prisoner's Dilemma Game*, Nature **327** 58–59 (1987)

71. BRANDT J., *Cycles of Partitions*, Proceedings of the American Mathematical Society **85** 483–487 (1982)

72. BROADBENT S. R. and HAMMERSLEY J. M., *Percolation Processes: I. Crystal and Mazes*, Proceedings of the Cambridge Philosophical Society **53** 629–641 (1957)

73. BULMER M. G., *A Statistical Analysis of the 10-Year Cycle in Canada*, Journal of Animal Ecology **43** 701–718 (1974)

74. BURRIDGE R. and KNOPOFF L., *Models of Theoretical Seismicity*, Bulletin of the Seismological Society of America **57** 341–371 (1967)

75. BURSTEDDE C., KLAUCK K., SCHADSCHNEIDER A., and ZITTARTZ J., *Simulation of Pedestrian Dynamics Using a Two-Dimensional Cellular Automaton*, Physica A **295** 507–525 (2001)

76. CANTOR C., *De la puissance des ensembles parfaits de points*, Acta Mathematica **4** 381–392 (1884)

77. CARDY J. L. and GRASSBERGER P., *Epidemic Models and Percolation*, Journal of Physics A: Mathematical and General **18** L267–L271 (1985)

78. CARLSON J. M. and LANGER J. S., *Mechanical Model of an Earthquake Fault*, Physical Review A **40** 6470–6484 (1989)

79. CARR J., *Applications of Centre Manifold Theory* (New York: Springer-Verlag 1981)

80. CARVALHO J. X. DE and PRADO C. P. C., *Self-Organized Criticality in the Olami-Feder-Christensen Model*, Physical Review Letters **84** 4006–4009 (2000)

81. CARVALHO J. X. DE and PRADO C. P. C., *de Carvalho and Prado Reply*, Physical Review Letters **87** 039802 (1p) (2001)

82. CASTRIGIANO D. P. L. and HAYES S. A., *Catastrophe Theory* (Reading, MA: Addison-Wesley 1993)

83. CHAKRABARTI B. K. and STINCHCOMBE R. B., *Stick-Slip Statistics for Two Fractal Surfaces: A Model for Earthquakes*, preprint:cond-mat/9902164

84. CHATÉ H. and MANNEVILLE P., *Collective Behaviors in Spatially Extended Systems with Local Interactions and Synchronous Updating*, Progress of Theoretical Physics **87** 1–60 (1992)

85. CHAU H. F., YAN K. K., WAN K. Y., and SIU L. W., *Classifying Rational Densities Using Two One-Dimensional Cellular Automata*, Physical Review E **57** 1367–1369 (1998)

86. CHAU H. F., SIU L. W., and YAN K. K., *One-Dimensional n-ary Density Classification Using Two Cellular Automaton Rules*, International Journal of Modern Physics C **10** 883–889 (1999)

87. CHOWDHURY D., SANTEN L., and SCHADSCHNEIDER A., *Statistical Physics of Vehicular Traffic and Some Related Systems*, Physics Reports **329** 199–329 (2000)

88. CHRISTENSEN K., HAMON D., JENSEN H. J., and LISE S., *Comment on "Self-Organized Criticality in the Olami-Feder-Christensen Model"*, Physical Review Letters **87** 039801 (1p) (2001)

89. CLARK C. W., *Mathematical Bioeconomics: The Optimal Management of Renewable Resources* (New York: John Wiley & Sons 1990)

90. CLUTTON-BROCK T. H., O'RIAIN M. J., BROTHERTON P. N. M., GAYNOR D., KANSKY R., GRIFFIN A. S., and MANSER M., *Selfish Sentinels in Cooperative Mammals*, Science **284** 1640–1644 (1999)

91. COLLET P. and ECKMANN J.-P., *Iterated Maps on the Interval as Dynamical Systems* (Boston: Birkhäuser 1980)

92. COOTNER P. H., *Comments on the Variation of Certain Speculative Prices* in *The Random Character of Stock Market Prices*, pp. 333–337, Paul H. Cootner (editor) (Cambridge, MA: The MIT Press 1964)

93. COSTA FILHO R. N., ALMEIDA M. P., ANDRADE JR J. S., and MOREIRA J. E., *Scaling Behavior in a Proportional Voting Process*, Physical Review E **60** 1067–1068 (1999)

94. COSTA FILHO R. N., ALMEIDA M. P., ANDRADE JR J. S., and MOREIRA J. E., *Brazilian Elections: Voting for a Scaling Democracy*, preprint: cond-mat/0211212

95. COULLET P. and TRESSER C., *Itérations d'endomorphismes et groupe de renormalisation*, Journal de Physique Colloque **39** C5–C25 (1978)

96. CRESSMAN R., *The Stability Concept of Evolutionary Game Theory: A Dynamic Approach* (Berlin: Springer-Verlag 1992)

97. CUSHING J. M., *Integrodifferential Equations and Delay Models in Population Dynamics* (Berlin: Springer-Verlag 1977)

98. CZIRÓK A., STANLEY H. E., and VICSEK T., *Spontaneously Ordered Motion of Self-Propelled Particles*, Journal of Physics A: Mathematical and General **30** 1375–1385 (1997)

99. D'ANCONA U., *The Struggle for Existence* (Leiden: Brill 1954)

100. DAVIDSEN J., EBEL H., and BORNHOLDT S., *Emergence of a Small World from Local Interactions: Modeling Acquaintance Networks*, Physical Review Letters **88** 128701 (4p) (2002)

101. DERRIDA B., GERVOIS A., and POMEAU Y., *Universal Metric Properties of Bifurcations and Endomorphisms*, Journal of Physics: Mathematical and General A **12** 269–296 (1979)

102. DEVANEY R. L., *Chaotic Dynamical Systems* (Redwood City, CA: Addison-Wesley 1989)

103. DOMANY E. and KINZEL W., *Equivalence of Cellular Automata to Ising Models and Directed Percolation*, Physical Review Letters **53** 311–314 (1984)

104. DOROGOVTSEV S. N., MENDES J. F. F., and SAMUKHIN A. N., *Structure of Growing Networks with Preferential Linking*, Physical Review Letters **85** 4633–4636 (2000)

105. DOROGOVTSEV S. N. and MENDES J. F. F., *Language as an Evolving Word Web*, Proceedings of the Royal Society B **268** 2603–2606 (2001)

106. DRĂGULESCU A. A. and YAKOVENKO V. M., *Statistical Mechanics of Money*, The European Physical Journal B **17** 723–729 (2000)

107. DRĂGULESCU A. A. and YAKOVENKO V. M., *Evidence for the Exponential Distribution of Income in the USA*, The European Physical Journal B **20** 585–589 (2001)

108. DRĂGULESCU A. A. and YAKOVENKO V. M., *Exponential and Power-Law Probability Distributions of Wealth and Income in the United Kingdom and the United States*, Physica A **299** 213–221 (2001)

109. DROSSEL B. and SCHWABL F., *Self-organized Critical Forest-Fire Model*, Physical Review Letters **69** 1629–1632 (1992)

110. DUBOIS M. A., FAVIER C., and SABATIER P., *Mathematical Modelling of the Rift Valley Fever in Senegal*, in *Simulation and Industry 2001*, pp. 530–531, Norbert Giambiasi and Claudia Frydman (editors) (Ghent: SCS Europe Bvba, 2001)

111. DULAC M. H., *Sur les cycles limites*, Bulletin de la Société Mathématique de France **51** 45–188 (1923)

112. DURAND B., FORMENTI E., and RÓKA Z., *Number-Conserving Cellular Automata: From Decidability to Dynamics*, preprint nlin.CG/0102035

113. EBEL H., MIELSCH M. -I., and BORNHOLDT S., *Scale-Free Topology of e-mail Networks*, Physical Review E **66** 035103(R) (4p) (2002)

114. EDEN M., *A Two-Dimensional Growth Process*, in *The Proceedings of the 4th Berkeley Symposium on Mathematical Statistics and Probability*, Volume 3, pp. 223–239, Jerzy Neyman (editor) (Berkeley: University of California Press 1961)

115. ELDREDGE N. and GOULD S. J., *Punctuated Equilibria: An Alternative to Phyletic Gradualism*, in *Models in Paleobiology*, T. J. M. Schopf (editor), pp. 82–115 (San Francisco: Freeman and Cooper 1972)

116. ENGLE R. F., *Autoregressive Conditional Heteroskedasticity with Estimates of the Variance of the United Kingdom Inflation*, Econometrica **50** 987–1008 (1982)

117. EPSTEIN J. M. and AXTELL R., *Growing Artificial Societies: Social Science from the Bottom Up* (Cambridge, MA: The MIT Press 1996)

118. ETIENNE G., *Tableaux de Young et Solitaire Bulgare*, Journal of Combinatorial Theory A **58** 181–197 (1991)

119. FAMA E. F., *Mandelbrot and the Stable Paretian Hypothesis* in *The random character of stock market prices*, pp 297–306, Paul H. Cootner (editor), (Cambridge, MA: The MIT Press 1964)

120. FARMER J. D., OTT E., and YORK J. A., *The Dimension of Chaotic Attractors*, Physica D **7** 153–180 (1983)

121. FEIGENBAUM M. J., *Quantitative Universality for a Class of Nonlinear Transformations*, Journal of Statistical Physics **19** 25–52 (1978)

122. FEIGENBAUM M. J., *Universal Behavior in Nonlinear Systems*, Los Alamos Science **1** 4–27 (1980)

123. FELLER W., *On the Logistic Law of Growth and Its Empirical Verification in Biology*, Acta Biotheoretica **5** 51–65 (1940)

124. FERMAT, P. DE, *Varia Opera Mathematica D. Petri de Fermat, Senatoris Tolosani* (Tolosæ: Johannem Pech 1679)

125. FERRER I CANCHO R. and SOLÉ R. V., *Two Regimes in the Frequency of Words and the Origins of Complex Lexicons*, Journal of Quantitative Linguistics **8** 165–173 (2001)

126. FERRER I CANCHO R. and SOLÉ R. V., *The Small-World of Human Language*, Proceedings of the Royal Society of London B **268** 2261–2266 (2001)

127. FINERTY J. P., *The Population Ecology of Cycles in Small Mammals*, (New Haven: Yale University Press 1980)

128. FLYVBERG H., SNEPPEN K., and BAK P., *Mean Field Theory for a Simple Model of Evolution*, Physical Review Letters **71** 4087–4090 (1993)

129. FRANSES P. H., *Time Series Models for Business and Economic Forecasting* (Cambridge, UK: Cambridge University Press 1998)

130. FRETTE V., CHRISTENSEN K., MALTHE-SØRENSSEN A., FEDER J., JØSSANG T., and MEAKIN P., *Avalanche Dynamics in a Pile of Rice*, Nature **379** 49–52 (1996)

131. FRISH H. L. and HAMMERSLEY J. M., *Percolation Processes and Related Topics*, Journal of the Society for Industrial and Applied Mathematics **11** 894–918 (1963)

132. FUJIWARA Y., SOUMA W., AOYAMA H., KAIZOJI T., and AOKI M., *Growth and Fluctuations of Personal Income*, preprint:cond-mat/0208398

133. FUKŚ H., *Solution of the Density Classification Problem with Two Cellular Automaton Rules*, Physical Review E **55** R2081–R2084 (1997)

134. FUKUI M. and ISHIBASHI Y., *Traffic Flow in 1D Cellular Automaton Model Including Cars Moving with High Speed*, Journal of the Physical Society of Japan **65** 1868–1870 (1996)

135. FUKUI M. and ISHIBASHI Y., *Self-Organized Phase Transitions in Cellular Automaton Models for Pedestrians*, Journal of the Physical Society of Japan **68** 2861–2863 (1999)

136. FUKUI M. and ISHIBASHI Y., *Jamming Transition in Cellular Automaton Models for Pedestrians on Passageway*, Journal of the Physical Society of Japan **68** 3738–3739 (1999)

137. GARDNER M., *The Fantastic Combinations of John Conway's New Solitaire Game "Life"*, Scientific American **223** 120–123 (1970)

138. GARDNER M., *Mathematical Games*, Scientific American **249** 12–21 (1983)
139. GAVALAS G. E., *Nonlinear Differential Equations of Chemically Reacting Systems* (New York: Springer-Verlag 1968)
140. GAYLORD R. J. and D'ANDRIA L. J., *Simulating Society: A Mathematica® Toolkit for Modeling Socioeconomic Behavior* (New York: Springer-Verlag 1998)
141. GENNES P.-G. DE, *La percolation: un concept unificateur*, La Recherche **7** 919–926 (1976)
142. GIBRAT R., *Les inégalités économiques* (Paris: Sirey 1931)
143. GILPIN M. E. and JUSTICE K. E., *Reinterpretation of the Invalidation of the Principle of Competitive Exclusion*, Nature **236** 273–274 and 299–301 (1972)
144. GILPIN M. E., *Do Hares Eat Lynx?*, American Naturalist **107** 727–730 (1973)
145. GNEDENKO B. V. and KOLMOGOROV A. N., *Limit Distributions for Sums of Independent Random Variables*, translated from Russian by K. L. Chung (Cambridge, MA: Addisson-Wesley 1954)
146. GOLES E., *Sand Piles, Combinatorial Games and Cellular Automata*, in *Instabilities and Nonequilibrium Structures*, pp. 101–121, E. Tirapegui and W. Zeller (editors) (Dordrecht: Kluwer Academic Publishers 1991)
147. GOMPERTZ B., *On the Nature of the Function Expressing Human Mortality* Philosophical Transactions **115** 513–585 (1825)
148. GOPIKRISHNAN P., MEYER M., AMARAL L. A. N., and STANLEY H. E., *Inverse Cubic Law for the Distribution of Stock Price Variations*, The European Physical Journal B **3** 139–140 (1998)
149. GOPIKRISHNAN P., PLEROU V., AMARAL L. A. N, MEYER M., and STANLEY H. E., *Scaling of the Distribution of Fluctuations of Financial Market Indices*, Physical Review E **60** 5305–5316 (1999)
150. GORDON D. M., *Dynamics of Task Switching in Harvester Ants*, Animal Behavior **38** 194–204 (1989)
151. GORDON D. M., *The Organization of Work in Social Insect Colonies*, Nature **380** 121–124 (1996)
152. GORDON D. M., *Ants at Work: How an Insect Society is organized* (New York: The Free Press 1999)
153. GRASSBERGER P., *On the Critical Behavior of the General Epidemic Process and Dynamical Percolation*, Mathematical Biosciences **63** 157–172 (1983)
154. GRASSBERGER P., *Chaos and Diffusion in Deterministic Cellular Automata*, Physica D **10** 52–58 (1984)
155. GRASSBERGER P. and KANTZ H., *On a Forest-Fire Model with Supposed Self-Organized Criticality*, Journal of Statistical Physics **63** 685–700 (1991)
156. GRASSBERGER P., *On a Self-Organized Critical Forest-Fire Model*, Journal of Physics A: Mathematical and General **26** 2081–2089 (1993)
157. GRASSBERGER P., *Critical Behavior of the Drossel-Schwabl Forest-Fire Model*, New Journal of Physics **4** 17 (15p) (2002), preprint :cond-mat/0202022
158. GRIMMETT G. R. and STIRZAKER D. R., *Probability and Random Processes* (Oxford: Clarendon Press 1992)
159. GRIMMETT G., *Percolation* (Berlin: Springer-Verlag 1999)
160. GUCKENHEIMER J. and HOLMES P., *Nonlinear Oscillations, Dynamical Systems, and Bifurcations of Vector Fields* (New York: Springer-Verlag 1983)
161. GUTENBERG B. and RICHTER C. F., *Seismicity of the Earth*, Geological Society of America Special Papers, Number 34 (1941)
162. GUTOWITZ H. A., VICTOR J. D., and KNIGHT B. W., *Local Structure Theory for Cellular Automata*, Physica D **28** 18–48 (1987)

163. HALE J. K., *Ordinary Differential Equations* (New York: Wiley-Interscience 1969)

164. HALE J. K. and KOÇAK H., *Dynamics and Bifurcations* (Heidelberg: Springer-Verlag 1991)

165. HARRISON G. W., *Comparing Predator-Prey Models to Luckinbill's Experiment with Didinium and Paramecium*, Ecology **76** 357–374 (1995)

166. HASSARD B. D., KAZARINOFF N. D., and WAN Y.-H, *Theory and Applications of Hopf Bifurcation* (Cambridge, UK: Cambridge University Press 1981)

167. HASSEL M. P., *Density-Dependence in Single-Species Populations*, Journal of Animal Ecology **44** 283–295 (1975)

168. HAUSDORFF F., *Dimension und äußeres Maß*, Mathematische Annalen **79** 157–179 (1918)

169. HÉNON M., *A Two-dimensional Mapping with a Strange Attractor*, Communications in Mathematical Physics **50** 69–77 (1976)

170. HETHCOTE H. W. and YORKE J. A., *Gonorrhea Transmission Dynamics and Control* (Heidelberg: Springer-Verlag 1984)

171. HIRSCH M. W. and SMALE S., *Differential Equations, Dynamical Systems and Linear Algebra* (New York: Academic Press 1974)

172. HIRSCH M. W., PUGH C. C., and SHUB M., *Invariant Manifolds* (Heidelberg: Springer-Verlag 1977)

173. HOLLING C. S., *The Components of Predation as Revealed by a Study of Small-Mammal Predation of the European Pine Sawfly*, Canadian Entomologist **91** 293–320 (1959)

174. HOLLING C. S., *The Functional Response of Predators to Prey Density and Its Role in Mimicry and Population*, Memoirs of the Entomological Society of Canada **45** 1–60 (1965)

175. HUANG Z. F. and SOLOMON S., *Power, Lévy, Exponential and Gaussian-like Regimes in Autocatalytic Financial Systems*, The European Physical Journal B **20** 601–607 (2001)

176. HUTCHINSON G. E., *Circular Cause Systems in Ecology*, Annals of the New York Academy of Sciences **50** 221–246 (1948)

177. HUTCHINSON G. E., *An Introduction to Population Ecology*, (New Haven: Yale University Press 1978)

178. IGUSA K., *Solution of the Bulgarian Solitaire Conjecture*, Mathematics Magazine **58** 259–271 (1985)

179. JANSON S., ŁUCZAK T., and RUCIŃSKI A., *Random Graphs* (New York: Academic Press 2000)

180. JIANHUA W., *The Theory of Games* (Oxford: Clarendon Press 1988)

181. KATOK A. and HASSELBLATT B., *Introduction to the Modern Theory of Dynamical Systems* (Cambridge, UK: Cambridge University Press 1975)

182. KERMACK W. O. and MCKENDRICK A. G., *A contribution to the Mathematical Theory of Epidemics*, Proceedings of the Royal Society A **115** 700–721 (1927)

183. KERNER B. S. and REHBORN H., *Experimental Features and Characteristics of Traffic Jams*, Physical Review E **53** R1297–R1300 (1996)

184. KERNER B. S. and textscRehborn H., *Experimental Properties of Complexity in Traffic Flows*, Physical Review E **53** R4275–R4278 (1996)

185. KINGSLAND S. E., *The Refractory Model: The Logistic curve and the History of Population Ecology*, The Quaterly Review of Biology **57** 29–52 (1982)

186. KINGSLAND S. E., *Modeling Nature: Episodes in the History of Population Ecology* (Chicago: University of Chicago Press 1995)

187. KINZEL W., *Directed Percolation*, in *Percolation, Structures and Processes*, Annals of the Israel Physical Society, **5**, pp. 425–445 (1983)

188. KIRCHNER A. and SCHADSCHNEIDER A., *Simulation of Evacuation Processes Using a Bionics-Inspired Cellular Automaton Model for Pedestrian Dynamics*, Physica A **312** 260–276 (2002)

189. KIRCHNER J. W. and WEIL A., *No fractals in Fossil Extinction Statistics*, Nature **395** 337–338 (1998)

190. KLEIN W. and RUNDLE J., *Comment on "Self-Organized Criticality in a Continuous, Nonconservative Cellular Automaton Modeling Earthquakes"*, Physical Review Letters **71** 1288–1288 (1993)

191. KOLB M., BOTET R., and JULLIEN R., *Scaling of Kinetically Growing Clusters*, Physical Review Letters **51** 1123–1126 (1983)

192. KOLMOGOROV A. N., *Grundbegriffe der Wahrscheinlichkeitsrechnung* (Berlin: Verlag von Julius Springer 1933), English translation edited by Nathan Morrison: *Foundations of the Theory of Probability* (New York: Chelsea Publishing Company 1950)

193. KOLMOGOROV A. N., *Sulla teoria di Volterra della lotta per l'esistenza*, Giornale dell'Istituto Italiano degli Attuari **7** 74–80 (1936)

194. KOPONEN I., *Analytic Approach to the Problem of Convergence of Truncated Lévy Flights Towards the Gaussian Stochastic Process*, Physical Review E **52** 1197–1199 (1995)

195. KRAPIVSKY P. L., REDNER S., and LEYVRAZ F., *Connectivity of Growing Random Networks*, Physical Review Letters **85** 4629–4632 (2000)

196. KRAPIVSKY P. L., RODGERS G. J., and REDNER S., *Degree Distributions of Growing Networks*, Physical Review Letters **86** 5401–5404 (2001)

197. KRAPIVSKY P. L. and REDNER S., *Organization of Growing Random Networks*, Physical Review E **63** 066123 (14p) (2001)

198. LAGARIAS J. C., *The $3x+1$ Problem and Its Generalizations*, American Mathematical Monthly **92** 3–23 (1985)

199. LAND M. and BELEW R. K., *No Perfect Two-State Cellular Automata for Density Classification*, Physical Review Letters **74** 5148–5150 (1995)

200. LANG S., *Undergraduate Analysis* (Heidelberg: Springer-Verlag 1983)

201. LESLIE P. H., *Some Further Notes on the Use of Matrices in Population Mathematics*, Biometrika **35** 213-245 (1948)

202. LÉVY P., *Théorie de l'addition des variables aléatoires*, 2ème édition (Paris: Gauthier-Villars 1954)

203. LÉVY P., *Processus stochastiques et mouvement brownien* (Paris: Gauthier-Villars 2ème édition 1965)

204. LI T.-Y. and YORK J. A., *Period Three Implies Chaos*, American Mathematical Monthly **82** 985–992 (1975)

205. LI T.-Y., MISIUREWICZ M., PIANIGIANI G., and YORK J. A., *Odd Chaos*, Physics Letters **87 A** 271–273 (1982)

206. LILJEROS F., EDLING C. R., AMARAL L. A. N., STANLEY H. E., and ÅBERG Y., *The Web of Human Sexual Contacts*, Nature **411** 907–908 (2001)

207. LISE S. and PACZUSKI M., *Self-Organized Criticality and Universality in a Nonconservative Earthquake Model*, Physical Review E **63** 036111 (5p) (2001)

208. LISE S. and PACZUSKI M., *Scaling in a Nonconservative Earthquake Model of Self-Organized Criticality*, Physical Review E **64** 046111 (5p) (2001)

209. LISE S. and PACZUSKI M., *A Nonconservative Earthquake Model of Self-Organized Criticality on a Random Graph*, preprint:cond-mat/0204491

210. LOÈVE M., *Probability Theory* (Berlin: Springer-Verlag 1977)
211. LOMBARDO M. P., *Mutual Restraint in Tree Swallows: A Test of the Tit for Tat Model of Reciprocity*, Science **227** 1363–1365 (1985)
212. LORENZ E. N., *Deterministic Nonperiodic Flow*, Journal of Atmospheric Sciences **20** 130–141 (1963)
213. LORENZ E. N., *The Essence of Chaos*, (Seattle: University of Washington Press 1993)
214. LOTKA A. J., *Elements of Physical Biology* (Baltimore: Williams and Wilkins 1925). Reprinted under the new title: *Elements of Mathematical Biology* (New York: Dover 1956)
215. LUCKINBILL L. S., *Coexistence in Laboratory Populations of Paramecium aurelia and its Predator Didinium nasutum*, Ecology **54** 1320–1327 (1973)
216. LUDWIG D., ARONSON D. G., and WEINBERG H. F., *Spatial Patterning of the Spruce Budworm*, Journal of Mathematical Biology **8** 217–258 (1979)
217. LUDWIG D., JONES D. D., and HOLLING C. S., *Qualitative Analysis of Insect Outbreak Systems: The Spruce Budworm and Forest*, Journal of Animal Ecology **47** 315–332 (1978)
218. LUX T and MARCHESI M., *Scaling and Criticality in a Stochastic Multi-Agent Model of a Financial Market*, Nature **397** 498–500 (1999)
219. MACKAY G. and JAN N., *Forest Fires as Critical Phenomena*, Journal of Physics A: Mathematical and General **17** L757–L760 (1984)
220. MALESCIO G., DOKHOLYAN N. V., BULDYREV S. V., and STANLEY H. E., *Hierarchical Organization of Cities and Nations*, preprint:cond-mat/0005178
221. MALTHUS T. R., *An Essay on the Principle of Population, as it Affects the Future Improvement of Society, with Remarks on the Speculations of Mr. Goodwin, M. Condorcet, and Other Writers* (London: Johnson 1798)
222. MANDELBROT B. B., *The Variation of Certain Speculative Prices*, Journal of Business **35** 394–419 (1963) reprinted in *The Random Character of Stock Market Prices*, pp. 307–332, Paul H. Cootner (editor) (Cambridge, MA: The MIT Press 1964)
223. MANDELBROT B. B., *Information, Theory and Psycholinguistics: A Theory of Words Frequencies* in *Readings in Mathematical Social Science*, P. Lazarfeld and N. Henry (editors) (Cambridge, MA: The MIT Press, 1966)
224. MANDELBROT B. B., *Fractals: Form, Chance, and Dimension* (San Francisco: Freeman 1977)
225. MANDELBROT B. B., *The Fractal Geometry of Nature* (San Francisco: Freeman 1982)
226. MANTEGNA R. N. and STANLEY H. E., *Stochastic Process with Ultraslow Convergence to a Gaussian: The Truncated Lévy Flight*, Physical Review Letters **73** 2946–2949 (1994)
227. MANTEGNA R. N. and STANLEY H. E., *An Introduction to Econophysics: Correlations and Complexity in Finance* (Cambridge, MA: Cambridge University Press 1999)
228. MARSILI M., MASLOV S., and ZHANG Y.-C., *Comment on "Role of Intermittency in Urban Development: A Model of Large-Scale City Formation,"* Physical Review Letters **80** 4830 (1998)
229. MAY R. M., *Stability and Complexity in Model Ecosystems* (Princeton: Princeton University Press 1974)
230. MAY R. M., *Simple Mathematical Models with Very Complicated Dynamics*, Nature **261** 459–467 (1976)

231. MAY R. M. and OSTER G. F., *Bifurcations and Dynamic Complexity in Simple Ecological Models*, American Naturalist **110** 573–599 (1976)
232. MAY R. M., *More Evolution of Cooperation*, Nature **327** 15–17 (1987)
233. MAYNARD SMITH J., *Mathematical Ideas in Biology* (Cambridge, UK: Cambridge University Press 1968)
234. MAYNARD SMITH J., *Models in Ecology* (Cambridge, UK: Cambridge University Press 1974)
235. MAYNARD SMITH J., *Evolution and the Theory of Games* (Cambridge, UK: Cambridge University Press 1982)
236. MAYNARD SMITH J., *The Theory of Evolution* (Cambridge, UK: Cambridge University Press 1993)
237. MEAKIN P., *Diffusion-Controlled Cluster Formation in Two, Three, and Four Dimensions*, Physical Review A **27** 604–607 (1983)
238. MEAKIN P., *Diffusion-Controlled Cluster Formation in 2–6 Dimensional Spaces*, Physical Review A **27** 1495–1507 (1983)
239. MEAKIN P., *Formation of Fractal Clusters and Networks of Irreversible Diffusion-Limited Aggregation*, Physical Review Letters **51** 1119–1122 (1983)
240. MERMIN N. D. and WAGNER H., *Absence of Ferromagnetism or Antiferromagnetism in One- and Two-Dimensional Isotropic Heisenberg Models*, Physical Review Letters **17** 1133–1136 (1966)
241. METROPOLIS M., STEIN M. L., and STEIN P. R., *On Finite Limit Sets for Transformations of the Unit Interval*, Journal of Combinatorial Theory, **15** 25–44 (1973)
242. MIDDLETON A. A. and TANG C., *Self-Criticality in Nonconserved Systems*, Physical Review Letters **74** 742–745 (1995)
243. MILGRAM S., *The Small-World Problem*, Psychology Today **1** 61–67 (1967)
244. MILINSKI M., *Tit for Tat in Sticklebacks and the Evolution of Cooperation*, Nature **325** 433-435 (1987)
245. MIYAZIMA S., LEE Y., NAGAMINE T., and MIYAJIMA H., *Power-Law Distribution of Family Names in Japanese Societies*, Physica A **278** 282–288 (2000)
246. MONTEMURRO M. A., *Beyond the Zipf-Mandelbrot Law in Quantitative Linguistics*, preprint:cond-mat/0104066
247. MONTEMURRO M. A. and ZANETTE D., *Entropic Analysis of the Role of Words in Literary Texts*, Advances in Complex Systems **5** 7–17 (2002)
248. MONTROLL E. W. and SHLESINGER M. F., *Maximum Entropy Formalism, Fractals, Scaling Phenomena, and $1/f$ Noise: A Tale of Tails*, Journal of Statistical Physics **32** 209–230 (1983)
249. MOREIRA A., *Universality and Decidability of Number-Conserving Cellular Automata*, Theoretical Computer Science, **292** 711–721 (2003)
250. MOSSNER W. K., DROSSEL B., and SCHWABL F., *Computer Simulations of the Forest-Fire Model*, Physica A **190** 205–217 (1992)
251. MUELLER L. D. and AYALA F. J., *Dynamics of Single-Species population Growth: Stability or Chaos*, Ecology **62** 1148–1154 (1981)
252. MURAMATSU M., IRIE T., and NAGATANI T., *Jamming Transition in Pedestrian Counter Flow*, Physica A **267** 487–498 (1999)
253. MURAMATSU M. and NAGATANI T., *Jamming Transition in Two-Dimensional Pedestrian Traffic*, Physica A **275** 281–291 (2000)
254. MURAMATSU M. and NAGATANI T., *Jamming Transition of Pedestrian Traffic at a Crossing with Open Boundaries*, Physica A **286** 377–390 (2000)

255. MURRAY J. D., *Mathematical Biology* (Heidelberg: Springer-Verlag 1989)
256. MYRBERG P. J., *Sur l'itération des polynômes quadratiques*, Journal de Mathématiques Pures et Appliquées **41** 339–351 (1962)
257. NAGEL K. and SCHRECKENBERG M., *A Cellular Automaton Model for Freeway Traffic*, Journal de Physique I **2** 2221–2229 (1992)
258. NEUMANN J. VON and MORGENSTERN O., *Theory of Games and Economic Behavior*, 3d edition (Princeton, NJ: Princeton University Press 1953)
259. NEWMAN M. E. J. and SNEPPEN K., *Avalanches, Scaling, and Coherent Noise*, Physical Review E **54** 6226–6231 (1996)
260. NEWMAN M. E. J., *A Model of Mass Extinction*, Journal of Theoretical Biology **189** 235–252 (1997)
261. NEWMAN M. E. J. and WATTS D. J., *Scaling and Percolation in the Small-World Network Model*, Physical Review E **60** 7332–7342 (1999)
262. NEWMAN M. E. J. and WATTS D. J., *Renormalization Group Analysis of the Small-World Model*, Physics Letters A **263** 341-346 (1999)
263. NEWMAN M. E. J., *The Structure of Scientific Collaboration Networks*, Proceedings of the National Academy of Sciences of the United States of America **98** 404-409 (2001)
264. NEWMAN M. E. J., *Scientific Collaboration Networks. I. Network Construction and Fundamental Results* Physical Review E **64** 016131 (8p) (2001)
265. NEWMAN M. E. J., *Scientific Collaboration Networks. II. Shortest Paths, Weighted Networks, and Centrality*, Physical Review E **64** 016132 (7p) (2001)
266. NEWMAN M. E. J. and ZIFF R. M., *Fast Monte Carlo Algorithm for Site and Bond Percolation*, Physical Review E **64** 016706 (16p) (2001)
267. NEWMAN M. E. J., STROGATZ S. H., and WATTS D. J., *Random Graphs with Arbitrary Degree Distributions and Their Applications*, Physical Review E **64** 026118 (17p) (2001)
268. NICHOLSON A. J. and BAILEY V. A., *The Balance of Animal Populations. Part I*, Proceedings of the Zoological Society of London **3** 551–598 (1935)
269. NIESSEN W. VON and BLUMEN A., *Dynamics of Forest Fires as a Directed Percolation*, Journal of Physics A: Mathematical and General **19** L289–L293 (1986)
270. NISHIDATE K., BABA M., and GAYLORD R. J., *Cellular Automaton Model for Random Walkers*, Physical Review Letters **77** 1675–1678 (1997)
271. NOWAK M. A. and MAY R. M., *Evolutionary Games and Spatial Chaos*, Nature **359** 826–829 (1992)
272. OKUBO A., *Diffusion and Ecological Problems: Mathematical Models* (Heidelberg: Springer-Verlag 1980)
273. OLAMI Z., FEDER H. J. S., and CHRISTENSEN K., *Self-Organized Criticality in a Continuous, Nonconservative Cellular Automaton Modeling Earthquakes*, Physical Review Letters **68** 1244–1247 (1992)
274. OONO Y., *Period $\neq 2^n$ Implies Chaos*, Progress in Theoretical Physics (Japan) **59** 1028–1030 (1978)
275. ORTUÑO M, CARPENA P., BERNAOLA-GALVÁN P., MUÑOZ E., and SOMOZA A. M., *Keyword Detection in Natural Languages and DNA*, Europhysics Letters **57** 759–764 (2002)
276. PALIS J. and MELO W. DE, *Geometric Theory of Dynamical Systems, An Introduction* (New York: Springer-Verlag 1982)
277. PARETO V., *Cours d'Économie Politique Professé à l'Université de Lausanne* (Lausanne: F. Rouge 1896–1897)

278. PARRISH J. K. and HAMMER W. M. (editors), *Animal Groups in Three Dimensions* (Cambridge, UK: Cambridge University Press 1997)
279. PASTOR-SATORRAS R. and VESPIGNANI A., *Corrections to Scaling in the Forest-Fire Model*, Physical Review E **61** 4854–4859 (2000)
280. PEARL R. and REED L. J., *On the Rate of Growth of the Population of the United States since 1790 and Its Mathematical Representation*, Proceedings of the National Academy of Sciences of the United States of America **21** 275–288 (1920)
281. PETERS O., HERTLEIN C., and CHRISTENSEN K., *A Complexity View of Rainfall*, Physical Review Letters **88** 018701(4p) (2002)
282. PHILIPPI T. E., CARPENTER M. P., CASE T. J., and GILPIN M. E., *Drosophila Population Dynamics: Chaos and Extinction*, Ecology **68** 154–159 (1987)
283. PIELOU E. C., *Mathematical Ecology* (New York: John Wiley 1977)
284. PLEROU V., GOPIKRISHNAN P., AMARAL L. A. N., MEYER M., and STANLEY H. G., *Scaling of the Distribution of Price Fluctuations of Individual Companies*, Physical Review E **60** 6519–6529 (1999)
285. PODOBNIK B., IVANOV P. C., LEE Y., and STANLEY H. E., *Scale-Invariant Truncated Lévy Process*, Europhysics Letters **52** 491–497 (2000)
286. PODOBNIK B., IVANOV P. C., LEE Y., CHESSA A., and STANLEY H. E., *Systems with Correlations in the Variance: Generating Power-Law Tails in Probability Distributions*, Europhysics Letters **50** 711–717 (2000)
287. POINCARÉ H., *Mémoire sur les courbes définies par une équation différentielle*, Journal de Mathématiques, 3ème série **7** 375–422 (1881), and **8** 251–296 (1882). Reprinted in *Œuvres de Henri Poincaré*, tome I (Paris: Gauthier-Villars 1951)
288. POINCARÉ H., *Sur l'équilibre d'une masse fluide animée d'un mouvement de rotation*, Acta Mathematica **7** 259–380 (1885). Reprinted in *Œuvres de Henri Poincaré*, tome VII (Paris: Gauthier-Villars 1952)
289. POINCARÉ H., *Sur le problème des trois corps et les équations de la dynamique*, Acta Mathematica **13** 1–270 (1890). Reprinted in *Œuvres de Henri Poincaré*, tome VII (Paris: Gauthier-Villars 1952)
290. POINCARÉ H., *Science et Méthode*, (Paris: Flammarion 1909). English translation by G. B. HALSTED, in *The Foundations of Science: Science and Hypothesis, The Value of Science, Science and Method* (Lancaster, PA: The Science Press 1946)
291. POL B. VAN DER, *Forced Oscillations in a Circuit with Nonlinear Resistance*, London, Edinburgh and Dublin Philosophical Magazine **3** 65–80 (1927)
292. POMEAU Y. and MANNEVILLE P., *Intermittent Transition to Turbulence in Dissipative Dynamical Systems*, Communications in Mathematical Physics **74** 189–197 (1980)
293. RAPOPORT R. and CHAMMAH A. M. with the collaboration of ORWANT C. J., *Prisoner's Dilemma: A Study in Conflict and Cooperation* (Ann Arbor: The University of Michigan Press 1965)
294. RAUP D. M. and SEPKOSKI J. J., *Periodicity of Extinctions in the Geologic Past*, Proceedings of the National Academy of Sciences of the United States of America, **81** 801–805 (1984)
295. REDNER S., *How Popular Is Your Paper? An Empirical Study of the Citation Distribution*, The European Physical Journal B **4** 131–134 (1998)
296. REDNER S., *Aggregation Kinetics of Popularity*, Physica A **306** 402–411 (2002)
297. RESNICK M., *Turtles, Termites, and Traffic Jams, Explorations in Massively Parallel Microworlds* (Cambridge, MA: The MIT Press 1997)

298. RICHTER C. F., *An Instrumental Earthquake Magnitude Scale*, Bulletin of the Seismological Society of America **25** 1–32 (1935)

299. ROBINSON C., *Dynamical Systems: Stability, Symbolic Dynamics and Chaos* (Boca Raton: CRC Press 1995)

300. RUELLE D. and TAKENS F., *On the Nature of Turbulence*, Communications in Mathematical Physics **20** 167–192 (1971)

301. RUELLE D., *Strange Attractors*, The Mathematical Intelligencer **2** 126–137 (1980)

302. RUELLE D., *Chaotic Evolution and Strange Attractors* (Cambridge, UK: Cambridge University Press 1989)

303. RUELLE D., *Deterministic Chaos: The Science and the Fiction*, Proceedings of the Royal Society A **427** 241–248 (1990)

304. RUSSELL D. A., HANSON J. D., and OTT E., *Dimension of Strange Attractors*, Physical Review Letters **45** 1175–1178 (1980)

305. SACHS L., *Applied Statistics: A Handbook of Techniques*, translated from German by Zenon Reynarowych (New York: Springer-Verlag 1984)

306. SCHELLING T., *Models of Segregation*, American Economic Review, Papers and Proceedings **59** (2) 488–493 (1969)

307. SCHELLING T., *Micromotives and Macrobehavior* (New York: Norton 1978)

308. SCHENK K., DROSSEL B., CLAR S., and SCHWABL F., *Finite-Size Effects in the Self-Organized Critical Forest-Fire Model*, The European Physical Journal B **15** 177–185 (2000)

309. SCHENK K., DROSSEL B., and SCHWABL F., *Self-Organized Critical Forest-Fire Model on Large Scales*, Physical Review E **65** 026135 (8p) (2002)

310. SCUDO F. M., *Vito Volterra and Theoretical Ecology*, Theoretical Population Biology **2** 1–23 (1971)

311. SEGEL L. A. and JACKSON J. L., *Dissipative Structures: An Explanation and an Ecological Example*, Journal of Theoretical Biology **37** 545–559 (1972)

312. SETON E. T., *Life Histories of Northern Mammals: An Account of the Mammals of Manitoba* (New York: Arno Press 1974)

313. SHIELDS P. C., *The Ergodic Theory of Discrete Sample Paths*, Graduate Studies in Mathematics Series, volume 13, (Prividence, RI: American Mathematical Society 1996

314. SKELLAM J. G., *Random Dispersal in Theoretical Populations*, Biometrika **38** 196–218 (1951)

315. SMALE S., *Differentiable Dynamical Systems*, Bulletin of the American Mathematical Society **73** 747–817 (1967)

316. SMITH F. E., *Population Dynamics of Daphnia magna and a New Model for Population Growth*, Ecology **44** 651–663 (1963)

317. SNEPPEN K., BAK P., FLYVBJERG H., and JENSEN M. H., *Evolution as a Self-Organized Critical Phenomenon*, Proceedings of the National Academy of Sciences of the United States of America, **92** 5209–5213 (1995)

318. SNEPPEN K. and NEWMAN M. E. J., *Coherent Noise, Scale Invariance and Intermittency in Large Systems*, Physica D **110** 209–223 (1997)

319. SOLÉ R. V. and MANRUBIA S. C., *Extinction and Self-Organized Criticality in a Model of Large-Scale Evolution*, Physical Review E **54** R42–R45 (1996)

320. SOLÉ R. V., MANRUBIA S. C., BENTON M., and BAK P., *Self-Similarity of Extinction Statistics in the Fossil Record*, Nature **388** 765–767 (1997)

321. SORNETTE D., JOHANSEN A., and BOUCHAUD J.-P., *Stock Market Crashes, Precursors and Replicas*, Journal de Physique I **6** 167–175, (1996)

322. SOUMA W., *Physics of Personal Income*, `preprint:cond-mat/02022388`
323. SPARROW C., *The Lorenz Equations: Bifurcations, Chaos, and Strange Attractors* (New York: Springer-Verlag 1982)
324. SPENCER J., *Balancing Vectors in the Max Norm*, Combinatorica **6** 55-65 (1986)
325. STANLEY S. M., *Macroevolution, Pattern and Process* (San Francisco: W. H. Freeman (1979)
326. ŠTEFAN P., *A Theorem of Šarkovskii on the Existence of Periodic Orbits of Continuous Endomorphisms of the Real Line*, Communications in Mathematical Physics **54** 237-248 (1977)
327. "STUDENT", *On the Probable Error of the Mean*, Biometrika **6** 1-25 (1908)
328. THOM R., *Stabilité structurelle et morphogenèse* (Paris: Interéditions 1972). English translation: *Structural Stability and Morphogenesis* (Reading, MA: Addison-Wesley 1972)
329. THOM R., *Prédire n'est pas expliquer* (Paris: Flammarion 1993)
330. THOMPSON J. N., *Interaction and Coevolution* (New York: John Wiley 1982)
331. TONER J. and TU Y., *Long-Range Order in a Two-Dimensional Dynamical XY Model: How Birds Fly Together*, Physical Review Letters **75** 4326-4329 (1995)
332. TRESSER C., COULLET P., and ARNEODO A., *Topological Horseshoe and Numerically Observed Chaotic Behaviour in the Hénon Mapping*, Journal of Physics A: Mathematical and General **13** L123-L127 (1980)
333. TSALLIS C. and ALBUQUERQUE M. P. DE, *Are Citations of Scientific Papers a Case of Nonextensivity*, The European Physical Journal B **13** 777-780 (2000)
334. TURING A. M., *The Chemical Basis of Morphogenesis*, Philosophical Transaction of the Royal Society B **237** 37-72 (1952)
335. VÁZQUEZ A., *Knowing a Network by Walking on it: Emergence of Scaling*, Europhysics Letters **54** 430-435 (2001)
336. VÁZQUEZ A., *Statistics of Citation Networks*, `preprint:cond-mat/0105031`
337. VELLEKOOP M. and BERGLUND R., *On Intervals, Transitivity = Chaos*, American Mathematical Monthly **101** 353-355 (1994)
338. VERHULST F., *Notice sur la loi que la population suit dans son accroissement*, Correspondances Mathématiques et Physiques **10** 113-121 (1838)
339. VICSEK T., CZIRK A., BEN-JACOB E., COHEN I., and SHOCHET O., *Novel Type of Phase Transition in a System of Self-Driven Particles*, Physical Review Letters **75** 1226-1229 (1995)
340. VOLTERRA V., *Variazioni e fluttuazioni del numero d'individui in specie animali conviventi*, Rendiconti dell'Accademia dei Lincei, **6** (2) 31-113 (1926). An abridged English version has been published in *Fluctuations in the Abundance of a Species Considered Mathematically*, Nature **118** 558-560 (1926)
341. WALTMAN P., *Deterministic Threshold Models in the Theory of Epidemics*, (Heidelberg: Springer-Verlag 1974)
342. WATTS D. J. and STROGATZ S. H., *Collective Dynamics of 'Small-World' networks*, Nature **393** 440-442 (1998)
343. WEINREICH G., *Solids: Elementary Theory for Advanced Students* (New York: John Wiley 1965)
344. WHITTAKER J. V., *An Analytical Description of Some Simple Cases of Chaotic Behavior*, American Mathematical Monthly **98** 489-504 (1991)

345. WILKINSON D. and WILLEMSEN J. F., *Invasion Percolation: A New Form of Percolation Theory*, Journal of Physics A: Mathematical and General **16** 3365–3376 (1983)
346. WILKS S. S., *Mathematical Statistics* (New York: John Wiley 1962)
347. WITTEN T. A. and SANDER L. M., *Diffusion-Limited Aggregation,, a Kinetic Critical Phenomenon*, Physical Review Letters **47** 1400–1403 (1981)
348. WOLFRAM S., *Statistical Physics of Cellular Automata*, Reviews of Modern Physics, **55** 601–644 (1983)
349. WOLFRAM S., *Cellular Automata as Models of Complexity*, Nature **311** 419–424 (1984)
350. WU F. Y., *The Potts Model*, Reviews of Modern Physics **54** 235–268 (1982)
351. ZANETTE D. H. and MANRUBIA S. C., *Role of Intermittency in Urban Development: A Model of Large-Scale City Formation*, Physical Review Letters **79** 523–526 (1997)
352. ZANETTE D. H. and MANRUBIA S. C., *Zanette and Manrubia Reply*, Physical Review Letters **80** 4831 (1998)
353. ZANETTE D. H. and MANRUBIA S. C., *Vertical Transmission of Culture and the Distribution of Family Names*, Physica A **295** 1–8 (2001)
354. ZEEMAN C., *Catastrophe Theory: Selected Papers 1972–1977* (Reading, MA: Addison-Wesley 1977)
355. ZIPF G. K., *Human Behavior and the Principle of Least Effort: An Introduction to Human Ecology* (Reading, MA: Addison Wesley 1949), reprinted by (New York: Hafner 1965)
356. ZIPF G. K., *The Psycho-biology of Language: An Introduction to Dynamic Philology* (Cambridge, MA: The MIT Press 1965)

Index

α-limit point, 64
α-limit set, 64
β-expansions, 221
ω-limit point, 64
ω-limit set, 64
ε-C^1-close, 70, 117
ε-C^1-perturbation, 70, 117
ε-neighborhood, 70, 117
n-block, 209
n-input rule, 191

absolutely continuous probability
 measure, 318
accumulation point, 162
active site, 192
age distribution, 12
agent-based modeling, 187, 243
 moving agents, 242
Allee effect, 136
animal groups, 2
ants, 1–2
apparent power-law behavior, 311, 324,
 341, 361
ARCH model, 337–339, 353, 354, 360,
 363
Arnol'd, V. I., 62
artificial society, 243
atlas, 43
attachment rate, 296
attracting set, 161
attractive focus, 54
attractive node, 54
automata network, 187
avalanche, 342

duration, 342
size, 342
Axelrod model of culture dissemination,
 245–248
Axelrod, R., 243, 245, 246, 253, 255,
 256, 272
Ayala, F. J., 103, 177
Ayala-Gilpin-Ehrenfeld model, 103

Bachelier, L., 335
backward orbit, 108
baker's map, 165
basin of attraction, 64
Beddington-Free-Lawton model, 115
Bendixson criterion, 85
Bethe lattice, 261
bidirectional pedestrian traffic
 model, 205
bifurcation
 diagram, 73
 flip, 123
 fold, 54
 Hopf, 27, 83–85, 125–127, 310
 period-doubling, 123–125, 143
 pitchfork, 73, 74, 79–80, 122–123
 point, 27, 29
 saddle-node, 73–77, 120, 139, 140
 symmetry breaking, 74
 tangent, 54, 73
 transcritical, 74, 77–79, 121–122, 268,
  301, 308
bifurcations
 of maps, 120–127
 of vector fields, 71–85

sequence of period-doubling, 127–135, 143
binomial distribution, 322
bird flocks, 2
block probability distribution, 210
boundary of a set, 162
boundary point, 162
budworm outbreaks, 7–9, 99
Bulgarian solitaire, 15
burst, 155
Burstedde-Klauck-Schadschneider-Zittartz pedestrian traffic model, 206

C^0
  conjugate, 56, 112
  distance, 70
  function, 42
  norm, 69
  topology, 69, 70
C^1
  class, 42
  distance, 70
C^k
  class, 42
  diffeomorphism, 42, 43
  distance, 70
  function, 42, 43
  vector field, 43
C-interval, 215
Cantor diagonal process, 163
Cantor function, 163
Cantor set, 165, 166, 169, 218
capacity, 165
carrying capacity, 6, 8, 9, 13, 21–25, 28
cascade, 108
catastrophe, 85–91
Cauchy distribution, 326, 352, 355
Cauchy sequence, 110
cellular automaton, 33
  evolution operator, 191
  generalized rule 184, 198, 201
  global rule, 191
  limit set, 193, 194, 199, 200, 202, 203, 209, 213, 234, 258, 263
  rule 184, 192–194, 198, 202, 205, 208
  rule 18, 208, 263
  rule 232, 264
  totalistic rule, 224, 234, 235, 272

center, 54
center manifold, 59
central control, 2
central limit theorem, 320–323, 330, 337, 356
characteristic path length, 281, 285, 299, 302
characteristic polynomial, 54
chart, 43
citation network, 290, 291, 294, 297
Clark, C. W., 98
closed orbit, 44
clustering coefficient, 281, 299, 302
cobweb, 108
codimension, 118
collaboration network of movie actors, 276, 279, 282, 283, 287, 288, 292, 294
Collatz
  conjecture, 15
  original problem, 15
collective stability, 256
community matrix, 19
competition, 66, 67, 99, 102
competitive exclusion principle, 67
complete metric space, 110
conditional heteroskedasticity, 337
configuration, 191
conforming, 261
conjugate
  flows, 49
  maps, 108
contracting map, 110
cooperative effect, 188
critical behavior, 188, 225
critical car density, 265
critical density, 203
critical exponents, 188, 199, 226
critical patch size, 94
critical point, 133, 188
critical probability, 221
critical space dimension
  lower, 189, 223
  upper, 33, 189, 200, 223
critical state, 188
cross-section, 118
cumulative distribution function, 219, 220, 303, 311, 317
  approximate, 334, 362

cusp catastrophe, 85
cylinder set, 215

damped pendulum, 71
de Gennes, P.-G., 221
decentralized systems, 4
defect, 194
defective tiles, 201–203
degree of cultural similarity, 246, 247
degree of mixing, 228–239
degree of separation, 276
delayed logistic model, 126
demographic distribution, 339
dense orbit, 152, 156, 161, 169
dense periodic points, 176, 180
dense subset, 147
density classification problem, 259
derivative, 42
Devaney, R. L., 127, 153
Didinium, 71, 83
diffeomorphism, 42, 43
diffusion, 91
diffusion equation, 91
diffusion-induced instability, 95
diffusion-limited aggregation, 227
dimensionless, 7–9
directed percolation, 223
    probability, 224
discrete diffusion equation, 230
discrete one-population model, 113,
    124, 128, 161, 169
disease-free state, 138, 142, 240, 301,
    308, 309
distance, 14, 69
    between two evolution operators, 218,
    219
    between two linear operators, 51
Domany-Kinzel cellular automaton, 224
drought duration, 351
Dulac criterion, 85
dynamical system, 9–14

earthquake
    stick-slip frictional instability, 349
earthquakes, 348–351
    Burridge-Knopoff model, 349
    epicenter, 348
    focus, 348
    Gutenberg-Richter law, 348

lithosphere, 348
    magnitude number, 348
    Olami-Feder-Christensen model, 349
    plate tectonics, 348
    Richter scale, 348
    seismic moment, 350
econophysics, 335
Eden model, 260, 267
elementary cellular automaton, 192
e-mail networks, 292
emergent behavior, 2–4
endemic state, 138, 142, 240, 301, 308,
    309
epidemic model
    Boccara-Cheong SIR, 236–239
    Boccara-Cheong SIS, 239–242
    Grassberger, 224
    Hethcote-York, 46
equilibrium point, 44
equivalence relation, 50
Erdös number, 276
ergodic map, 156, 158
ergodic theory, 156
error and attack tolerance, 294
Euler Γ function, 327
event, 317
evolution law, 9, 14
evolution operator, 213–216
evolutionarily stable strategy, 255
existence and unicity of solutions, 10,
    43
exponential cutoff, 353, 360
exponential distribution, 313, 358
exponential of an operator, 51
extinction, 101, 140

family names, 340
Feigenbaum number, 130, 172, 235
financial markets, 335–339
finite vertex life span, 328
first return map, 118
fish schools, 2
fixed point, 20, 23, 35, 36, 44, 108
flocking behavior, 207
floor field, 206
    static, 206
flow, 43, 44, 49, 50, 56, 66, 72, 107–109,
    118
forest fires, 226–227

critical probability, 227
direction of travel, 226
intensity, 226
rate of spread, 226
forward orbit, 108
Fourier series, 94
fractal, 165, 227
free empty cells, 201, 202
free-moving phase, 198, 199, 201
fruit flies, 99
Fukui-Ishibashi pedestrian traffic
    model, 204
Fukui-Ishibashi traffic flow model, 198
fundamental diagram, 260
fundamental solution, 92

game of life, 3
game theory, 248
    minimax theorem, 252
    mixed strategies, 251
    optimal mixed strategies, 254
    optimal strategies, 251, 252
    payoff matrix, 249
    saddle point, 251
    solution of the game, 251, 252
    strategy, 250
    value of the game, 250, 252
    zero-sum game, 249
Gamma distribution, 356
GARCH model, 354, 363
Gauss distribution, 320, 327
general epidemic process, 224–226
generalized eigenvector, 53
generalized homogeneous function, 92,
    189
generating function, 300, 306
Gibrat index, 312
global evolution rule, 193, 213, 217
Gompertz, B., 138, 176
Gordon, D. M., 1, 2
graph, 3, 187, 276
    adjacency matrix, 278
    adjacent vertex, 276
    arc, 276
    bipartite graph, 279
    chain, 277
    characteristic path length, 281
    chemical distance between two
        vertices, 277

clustering coefficient, 281
complete, 277
component, 278
connected, 278
cycle, 278
degree of a vertex, 277
diameter, 281
directed edge, 276
distance between two vertices, 277
empty, 277
giant component, 276, 280, 281, 302,
in-degree of a vertex, 277, 288, 289,
    291, 295–297
isolated vertex, 277
isomorphism, 277
link, 276
neighborhood of a vertex, 276
order, 277
out-degree of a vertex, 277, 288, 289,
    291, 295–297
path, 277
phase transition, 280
random, 279
    binomial, 280
    characteristic path length, 281
    clustering coefficient, 281
    diameter, 281
    uniform, 280
regular, 277
size, 277
transitive linking, 298
tree, 278
vertex, 276
graphical analysis, 108
Grassberger, P., 224, 225, 259, 343–345
Green's function, 92
growing network model
    Barabási-Albert, 292–294
    Dorogovtsev-Mendes-Samukhin, 295
    extended Barabási-Albert, 294
    Krapivsky-Rodgers-Redner, 295–297
    Vázquez, 297–298
growth rate, 66

Hénon map, 170
Hénon strange attractor, 172
habitat size, 36
Hamming distance, 14, 231
Harrison, G. W., 25

Hartman-Grobman theorem, 56
harvesting, 98, 137, 140
Hassel one-population model, 113
Hausdorff dimension, 164, 172, 227
Hausdorff outer measure, 164
highway car traffic, 196–204
histogram, 148
Holling, C. S., 25
homeomorphism, 42, 50
host-parasitoid
    Beddington-Free-Lawton model, 115
    Nicholson-Bailey model, 114
host-parasitoid model, 114
Hutchinson, G. E., 6, 12, 13, 22, 28
Hutchinson time-delay model, 12, 28
hyperbolic point, 56, 112
hysteresis, 87
hysteresis loop, 87

implicit function theorem, 72, 75–78,
    120–123
infective individual, 45, 137, 187, 236,
    239
infinite-range interactions, 137
interior of a set, 162
interior point, 162
Internet Movie Database, 276, 282
invariant measure, 148, 156
invariant probability density, 156,
    176–178, 181
invariant probability measure, 157
invariant set, 64
invasion percolation, 261, 271
iterated prisoner's dilemma, 253, 256

Jacobian, 56, 72, 112, 120
jammed phase, 198, 199

Kermack-McKendrick threshold
    theorem, 238, 309
Kingsland, S. E., 18
Kolmogorov, A. N., 67, 317
kurtosis excess, 322

Lévy distribution, 327, 328
Lévy flight, 328
laminar phase, 154
laminar traffic flow, 197
Langevin function, 359

large deviations, 323
law of large numbers, 333
Leslie, P. H., 25
Leslie's model, 48
Li, T.-Y., 154, 159
Li-York
    condition, 160
    theorem, 159
limit cycle, 22, 25, 27, 35, 64, 309
limit point, 162
limit set, 209
linear
    differential equation, 56
    function, 56, 112
    part, 56, 112
    recurrence equation, 112
    system, 56
    transformation, 42
linearization, 51
local evolution rule, 33, 187, 191, 192,
    216, 243, 246, 247, 254
local jam, 201
local structure theory, 209
local transition rule, 187
logistic model, 66
    fluctuating environments, 27
    time-continuous, 6–9, 12–14
    time-delay, 12, 28
lognormal distribution, 312, 324
long-range move, 228, 237, 239, 244
look-up table, 192
Lorentz distribution, 326
Lorenz, E. N., 145, 146, 174
Lorenz equations, 174
Lorenz strange attractor, 174
Lotka, J., 17
Lotka-Volterra
    competition model, 66
    modified model, 23
    time-continuous model, 17, 98
    time-discrete model, 33
Luckinbill's experiment, 26
Ludwig-Jones-Holling spruce budworm
    model, 7, 88, 95, 99
Lyapunov exponent, 158–159
    logistic map, 158
Lyapunov function, 98
    strong, 60
    weak, 60, 100

Lyapunov theorem, 60
lynx-hare cycle, 21

majority rule, 264
Malthus, T. R., 22
Mandelbrot, B. B., 165, 291, 335
Manhattan distance, 221
manifold, 11, 43
Manneville, P., 154
market, 3
market index
    Hang Seng, 337
    NIKKEI, 337
    S&P 500, 337
mass extinctions, 347–348
May, R. M., 24, 113, 146, 256, 257, 272
Maynard Smith, J., 4, 103, 126, 255
mean-field approximation, 35, 208–209,
    237–240, 301, 307
metric, 51, 69
Milgram, S., 275, 276
model, 4–14
moment-generating function, 323
motion representation, 197
movie actors, 3
multiple random walkers, 13, 229, 231,
    242, 260, 267
    amiable, 242
    timorous, 242 243, 247, 254
Muramatsu-Irie-Nagatani pedestrian
    traffic model, 206
Murray, J. D., 95
muskrat, 93
Myrberg, P. J., 128

Nagel-Schreckenberg traffic flow model,
    196
neighborhood, 189
    Moore, 33
    von Neumann, 33
Newman, M. E. J., 282, 286, 287, 347,
    348, 351
Newton's method, 111
Nicholson-Bailey model, 114
nilpotent operator, 53
nonlinear oscillator, 11
norm, 69
norm of a linear operator, 51
normal distribution, 92, 320, 352

null cline, 66, 99, 102
number-conserving cellular automaton,
    194, 260, 265

off-lattice models, 207
open cluster, 222
open set, 70
orbit, 43, 108
order parameter, 189

Paramecium, 71, 83
Pareto
    exponent, 311–313, 328
    law, 311
particle-hopping model, 196
particlelike structures, 194
pedestrian traffic, 204–207
pendulum
    damped, 57, 61
    simple, 10, 60
percolation, 223–226
    bond, 223
    probability, 225
    site, 223
percolation 221–224
    bond 221
    probability 223
    site 221
    threshold, 222
perfect set, 166
perfect tile, 199, 201–203
period of a closed orbit, 44
period of a periodic point, 108
period-doubling bifurcation, 129
period-doubling route to chaos, 234
periodic orbit, 108, 147, 149–151, 159,
    160, 176
periodic point, 108
Perron-Frobenius equation, 157
perturbed center, 65
perturbed harmonic oscillator, 58, 71,
    99
phase flow, 43
phase portrait, 20, 43
phase space, 9–15, 42
Pielou, E. C., 24
Poincaré-Bendixson theorem, 64, 99
Poincaré, H., 41, 145
Poincaré map, 118

Poincaré recurrence theorem, 152
Pomeau, Y., 154, 172
population growth, 5
positively invariant set, 65
predation, 17
predator's functional response, 24, 25
predator's numerical response, 24
preferential attachment, 290, 293
prisoner's dilemma, 252
probabilistic cellular automaton, 196
probabilistic cellular automaton rule,
    192
probability density, 318
    Cauchy, 326, 330, 352, 355, 357
    exponential, 313, 358
    Gamma, 356
    Gauss, 320, 327
    Lévy, 327, 328
    lognormal, 312, 324
    Lorentzian, 326
    normal, 320, 352
    Student, 330, 362
    truncated Lévy, 328, 330
probability distribution
    Bernoulli, 228
    binomial, 322
    Poisson, 280, 299, 300, 302, 306, 307,
        321
probability measure, 156, 317
probability space, 317

radii of a rule, 33, 191
rain event, 351
rainfall, 351–352
random dispersal, 91
    critical patch size, 94
    induced instability, 95–97
    one-population models, 92
random network, 279–284, 292, 294
    epidemic model on a, 301
random variable, 317
    absolutely continuous, 318
    average value, 318
    Bernoulli, 228, 323, 352
    binomial, 322
    Cauchy, 326, 357
    characteristic function, 319
    convergence almost sure, 338
    convergence in distribution, 319

convergence in probability, 333
    discrete, 318
    exponential, 352
    Gaussian, 320–325
    kurtosis, 322
    kurtosis excess, 322, 338, 353, 361,
        363
    Lévy, 325–328, 335
    lognormal, 324
    Lorentzian, 326
    maximum of a sequence, 353
    mean value, 318, 353, 361, 363
    median, 322
    minimum of a sequence, 353
    mode, 322
    moment-generating function, 323
    moment of order r, 318
    normal, 325
    sample of observed values, 332
    skewness, 327, 353, 361, 363
    standard deviation, 319
    truncated Lévy, 328
    variance, 318, 353, 361, 363
random walk
    Cauchy, 328
random walk and diffusion, 91
refuge, 94, 102
removed individual, 45, 137, 236
repulsive focus, 54
repulsive node, 54
Resnick, M, 4
return, 335
    normalized, 336
rich get richer phenomenon, 293
Rift Valley Fever, 242
rotating hoop, 81
route to chaos
    intermittency, 154
    period-doubling, 154
Ruelle, D., 156, 167
rule code number, 192

saddle, 54, 58
saddle node, 54
sample space, 317
Šarkovskii theorem, 150, 151, 159, 176
scaled variables, 7
scaling relation, 189, 200, 285
scatter plot, 361

scientific collaboration networks, 282, 283, 289, 294
second-order phase transition, 188, 198, 199, 308
self-consistency conditions, 210
self-organization, 2–3
self-organized criticality, 341–352
self-propelled particles, 207
self-similar function, 191
self-similar set, 164
sensitive dependence on initial conditions, 153, 161, 176, 180
sentinel, 2
Seton, E. T., 21
sexual contact network, 291
sexually transmitted disease, 46, 138
short-range move, 228, 237, 239, 241, 244
simulation, 4
singular measure, 219, 318
sink, 57
SIR epidemic model
    Boccara-Cheong, 137
    Kermack-McKendrick, 45, 137
SIS epidemic model
    Boccara-Cheong, 138
    Hethcote-York, 47
site-exchange cellular automata, 228–243
Skellam, J. G., 93
skewness, 322
small-world model
    Davidsen-Ebel-Bornholdt, 298–299
    highly connected extra vertex, 287
    Newman-Watts, 286–287
    shortcut, 286
    Watts-Strogatz, 283–286
    Watts-Strogatz modified, 285
Smith model, 98, 100
smooth function, 42
social networks, 3
source, 58
space
    metric, 69
    normed vector, 69
space average, 156, 158
space of elementary events, 317
spatial segregation model, 243–245
spectrum, 120

stability
    asymptotic, 23, 44, 110
    Lyapunov, 44, 110
    structural, 24, 69, 118
stable manifold, 56, 59
stable manifold theorem, 59
stable Paretian hypothesis, 335
stable probability distribution, 326
StarLogo, 4
start-stop waves, 197
state
    at a site, 191
    of a cellular automaton, 191
state space, 9
statistics
    arithmetic mean, 332
    kurtosis excess, 332
    median, 332
    skewness, 332
    standard deviation, 332
stock exchange, 3
stock market, 3
strange attractor, 167, 169, 172, 174
strange repeller, 167
street gang control, 87–91
Strogatz, S. H., 283, 285, 286, 292, 294
Student's test, 333
    confidence interval, 334
subgraph, 276
superstable, 133
surjective mapping, 259
susceptible individual, 45, 47, 137, 187, 236, 239
suspicious tit for tat, 261
symmetrical difference, 219
symmetry-breaking field, 199

T-point cycle, 108
tent map, 109
    asymmetric, 176, 178
    binary, 149
    symmetric, 176
tentative move, 229
termites, 4
theta model, 177
three-spined sticklebacks, 256
threshold phenomenon, 45
time-discrete analogue, 32
tit for tat, 255, 256

tit for two tats, 261
topological space, 70
topological transitivity, 152
topologically equivalent flows, 50, 82
topology, 70
traffic, 3
traffic flow models, 212–213
trajectory, 43
trapping region, 161
tree swallow, 256
truncated Lévy distribution, 328, 330
truncation operator, 209
Turing effect, 95
two-dimensional linear flows, 54
two-lane car traffic flow model, 259, 265
two-species competition model, 99

U-sequences, 134
unfolding, 82
unimodal map, 131, 133
universality, 131–135, 172, 227, 235,
     294, 337, 339
universality class, 188, 207, 225
unstable, 44

unstable manifold, 56, 59

van der Pol oscillator, 61
variational principle, 201, 202
vector field, 42
velocity configuration, 197
velocity probability distribution, 213
velocity rule, 197
Verhulst, P. F., 6
volatility, 336
Volterra, V., 17
voting, 340

Watts, D. J., 283, 285, 266, 292, 294
Web pages, 288
Whittaker, J. V., 149
word network, 316
World Wide Web, 3, 288, 289, 294, 295,
     298

York, J. A., 154, 159

Zipf law, 314

Graduate Texts in Contemporary Physics

B.M. Smirnov: **Clusters and Small Particles: In Gases and Plasmas**

B.M. Smirnov: **Physics of Atoms and Ions**

M. Stone: **The Physics of Quantum Fields**

F.T. Vasko and A.V. Kuznetsov: **Electronic States and Optical Transitions in Semiconductor Heterostructures**

A.M. Zagoskin: **Quantum Theory of Many-Body Systems: Techniques and Applications**